VOLUME FIFTEEN

Radioactivity in the Environment

AIRBORNE RADIOACTIVE CONTAMINATION IN INHABITED AREAS

RADIOACTIVITY IN THE ENVIRONMENT

A companion series to the Journal of Environmental Radioactivity

Series Editor
M.S. Baxter
Ampfield House
Clachan Seil
Argyll, Scotland, UK

Volume 1: Plutonium in the Environment
(A. Kudo, Editor)

Volume 2: Interactions of Microorganisms with Radionuclides
(F.R. Livens and M. Keith-Roach, Editors)

Volume 3: Radioactive Fallout after Nuclear Explosions and Accidents
(Yu.A. Izrael, Author)

Volume 4: Modelling Radioactivity in the Environment
(E.M. Scott, Editor)

Volume 5: Sedimentary Processes: Quantification Using Radionuclides
(J. Carroll and I. Lerche, Authors)

Volume 6: Marine Radioactivity
(H.D. Livingston, Editor)

Volume 7: The Natural Radiation Environment VII
(J.P. McLaughlin, S.E. Simopoulos and F. Steinhäusler, Editors)

Volume 8: Radionuclides in the Environment
(P.P. Povinec and J.A. Sanchez-Cabeza, Editors)

Volume 9: Deep Geological Disposal of Radioactive Waste
(R. Alexander and L. McKinley, Editors)

Volume 10: Radioactivity in the Terrestrial Environment
(G. Shaw, Editor)

Volume 11: Analysis of Environmental Radionuclides
(P.P. Povinec, Editor)

Volume 12: Radioactive Aerosols
(C. Papastefanou, Author)

Volume 13: U–Th Series Nuclides in Aquatic Systems
(S. Krishnaswami and J. Kirk Cochran, Editors)

Volume 14: Remediation of Contaminated Environments
(G. Voigt and S. Fesenko, Editors)

Volume 15: Airborne Radioactive Contamination in Inhabited Areas
(K.G. Andersson, Editor)

VOLUME FIFTEEN

Radioactivity in the Environment

AIRBORNE RADIOACTIVE CONTAMINATION IN INHABITED AREAS

Editor

Kasper G. Andersson

Risoe National Laboratory for Sustainable Energy,
Technical University of Denmark, Roskilde, Denmark

ELSEVIER

Amsterdam • Boston • Heidelberg • London • New York • Oxford
Paris • San Diego • San Francisco • Singapore • Sydney • Tokyo

Elsevier
Radarweg 29, PO Box 211, 1000 AE Amsterdam, The Netherlands
Linacre House, Jordan Hill, Oxford OX2 8DP, UK

First edition 2009

British Library Cataloguing in Publication Data
A catalogue record for this book is available from the British Library

Library of Congress Cataloging-in-Publication Data
A catalog record for this book is available from the Library of Congress

ISBN: 978-0-08-044989-0
ISSN: 1569-4860

Printed and bound in Great Britain

09 10 11 12 13 10 9 8 7 6 5 4 3 2 1

Contents

Contributors

Kasper G. Andersson
Risoe National Laboratory for Sustainable Energy, Technical University of Denmark, DK-4000 Roskilde, Denmark

Miriam A. Byrne
School of Physics, National University of Ireland, Galway, Ireland

Thomas W. Charnock
Health Protection Agency, Radiation Protection Division, Chilton, Didcot, Oxon OX11 0RQ, UK

Ellen-Marie Forsberg
Work Research Institute, St Olavs plass, N-0130 Oslo, Norway

J. Arthur Jones
Health Protection Agency, Radiation Protection Division, Chilton, Didcot, Oxon OX11 0RQ, UK

Raimo Mustonen
STUK – Radiation and Nuclear Safety Authority, 00881 Helsinki, Finland

Ken W. Nicholson
Nicholson Environmental, The Cottage, Victoria Road, Quenington, Gloucestershire GL7 5BW, UK

Deborah H. Oughton
Norwegian University of Life Sciences, N-1432 Aas, Norway

Peter Zombori
KFKI Atomic Energy Research Institute, H-1525 Budapest, Hungary

Introduction

This publication is aimed at providing an overview of the different aspects that determine the consequences in inhabited areas of airborne radioactive contamination from a radiological or nuclear emergency and the applicability of countermeasure strategies to reduce these. It is the first-ever extensive and holistic account to have been published on this subject, considering in detail, for example, physical, chemical, radiological, social, ethical, economic and environmental processes and factors. It is structured in 10 chapters that are written by leading experts on the different topics of importance. Without their contributions this project would not have been possible.

It needs to be emphasised that the term 'inhabited areas' should be understood in a broad sense as areas where humans are living, working or in other ways spending time. Likewise, the term 'airborne radioactive contamination' is in this publication taken to mean radioactive matter originating from an atmospheric release in connection with an emergency situation. Exposure to naturally occurring radionuclides in the inhabited environment (e.g. radon concentrations in buildings) is only mentioned to provide context to examples of exposure to radionuclides in emergency situations. Similarly, this publication does not deal with problems like the consequences of routine releases of radionuclides due to coal-burning or discharges from processing in various industries. Nuclear bomb explosions are also omitted since environmental contamination would here be expected to pose only a small threat to a population compared to the massive effect of the shock wave, instantaneously caused radiation injuries and burns directly or indirectly caused by the heat of the explosion (Kucan, 2004).

For many decades, investigations of the behaviour and implications of airborne radioactive contamination in the environment almost exclusively focused on agricultural areas and food production. This was due to the erroneous assumption that the consequences of credible contaminating emergencies would be restricted to rural areas with very low population density. However, due to the Chernobyl accident in 1986, several hundreds of thousands of persons were removed from their homes (Hubert et al., 1996), demonstrating a great need for knowledge and instruments that could be applied to minimise the manifold adverse consequences of contamination in inhabited areas.

The world of today is not only facing the risk of accidents (e.g. in nuclear or radiological facilities or during transportation of sources). As described in Chapter 1, also other types of emergencies, notably involving radiological terrorism, pose a threat. Acts of terror would be most likely to take place in a city, where the greatest number of people could become directly affected. A recent report from the US Commission on the Prevention of Weapons of Mass Destruction Proliferation and Terrorism (Graham et al., 2008) considers it most likely that, unless the world community acts decisively and with great urgency, a large attack using a weapon of mass destruction – quite possibly involving nuclear materials – will occur in a major city somewhere in the world within the next 4 years. Due to the significantly lesser difficulties in acquiring the components and knowledge needed to construct a potent radiological dispersal device, a radiological terror attack might well be even more likely to occur.

Considering the various processes that govern the outcome of different types of emergencies, it is clear that decision support systems that are traditionally equipped with computing methodologies and data libraries that are relevant to certain types of nuclear accidents would, for many other types of emergency scenarios involving the spread of radioactive material, be inadequate to provide reasonably reliable consequence prognoses and allow optimisation of intervention to take place. In recent years an effort has been initiated to address this shortcoming, and new and important findings are described in this volume.

Its second chapter describes the atmospheric dispersion processes of airborne contaminants following an emergency and gives an account of the mechanisms that determine the contaminant deposition under different circumstances on the various outdoor surfaces in an inhabited area (e.g. open grassed spaces, road surfaces, outer walls and roofs of buildings and vegetation). Also resuspension of contaminants from outdoor surfaces may in some cases lead to significant exposure, for example through inhalation, and the mechanisms of resuspension in the inhabited complex are described in detail.

The third chapter deals with the fate of the airborne contamination inside dwellings. Filtration and deposition in dwellings are examples of important factors that can be strongly influenced by the physicochemical characteristics of the contaminants. These processes, together with the air exchange rate, largely determine the equilibrium relationship between indoor and outdoor contaminant air concentrations. Also indoor contaminant resuspension and clearance, and a wide range of factors influencing air concentrations and deposition on surfaces inside dwellings are discussed.

The fourth chapter discusses mechanisms through which humans may become contaminated, either through inhalation, through deposition of airborne contaminants on the human body, or by contact transfer, having touched a contaminated surface. The mechanisms of contamination and clearance, of radionuclides in the respiratory tract as well as on human skin, hair and clothing, are described in detail, outlining dependencies of various factors. Whereas the ICRP lung model is well established through many years of use, the knowledge of the processes and parameters governing the contamination of outer human body surfaces has in recent years been greatly expanded.

In the fifth chapter, the processes of migration and fixation of contaminants on the different types of outdoor surfaces in an inhabited area are explained. The physicochemical characteristics of the contaminant determine its affinity to different materials. For instance, radiocaesium can in cationic form be fixed very strongly to certain minerals that are present in many urban construction materials as well as in soils. Without action, this type of immobilisation can maintain a high level of external exposure in the area. Large and insoluble particles, as would be produced in some perceivable malicious radionuclide dispersion scenarios, would much more easily be removed from impermeable surfaces, both through natural processes and by force.

Knowledge of such processes as deposition, resuspension, clearance and migration on/in environmental surfaces and humans is combined in Chapter 6 with parameters describing dose responses for calculation of external dose contributions to humans located in different geometrical configurations in an inhabited area. Largely depending on the type of release scenario, a host of other dose contributions may also be important in connection with an airborne contamination in an inhabited complex, and the features of these are outlined with a series of examples. Current trends in dose modelling for inhabited areas are discussed in relation to commonly used and newly developed consequence estimation models for use in emergency decision support.

Due to case-specific characteristics and processes, which will in general to a great extent govern the exposure situation after an emergency leading to an airborne release of radioactive matter, measurements are paramount for interpreting adverse consequences and optimally designing and implementing measures to counter them. The question is very much how, where and when to measure, to apply available resources in the best possible way and enable the most reliable consequence estimates. Chapter 7 gives an introduction to the requirements and implications of developing tailored measurement strategies for the various tasks in different time phases following different types of emergencies affecting inhabited areas.

On the basis of the newest practical experience, Chapter 8 gives an account of countermeasures that may be useful under different circumstances in reducing doses to humans from airborne contamination in an inhabited area. Different types of countermeasures are considered, for example for reduction of doses received from different pathways, in different time phases, and in relation to different types of emergencies. Also management processes for the wastes arising from countermeasure implementation are discussed. An introduction is given to a new European decision-aiding methodology involving systematic and holistic descriptions of the highly diverse important features of each dose-reductive countermeasure option.

Acknowledging the ICRP principles of justification and optimisation, a wide range of non-radiological perspectives of an emergency need careful consideration in selecting the right countermeasure strategy for a specific situation. Indeed, some potentially highly attractive countermeasures for airborne contamination scenarios affecting inhabited areas do not at all bring about a reduction in dose. The complex issues of including social and ethical criteria in emergency response are discussed in the ninth chapter. This comprises engagement of representatives of concerned parties (stakeholders) in a dialogue to secure ethically and rationally defendable solutions, and the use of a specially developed methodology is suggested to transparently and systematically present criteria of importance to the decision process.

In the tenth chapter, the overall purposes and implications of countermeasure strategies are outlined, and illustrated with practical examples demonstrating the effect of different steps of an integrated dose-reduction strategy for a given area. Also examples of serious pitfalls in the implementation phase are highlighted. To support in the decision-making process, various types of tools may be valuable instruments in reaching an optimised solution. More explicitly, the use of decision-support handbooks, computerised consequence assessment systems and multi-criteria analysis tools is discussed.

The authors would like to acknowledge valuable remarks received from Professor M.S. Baxter throughout the drafting process of this publication.

REFERENCES

Graham, B., J. Talent, G. Allison, R. Cleveland, S. Rademaker, T. Roemer, W. Sherman, H. Sokolski, and R. Verma. (2008). *World at Risk – The Report of the Commission on the Prevention of WMD Proliferation and Terrorism*. Vintage Books, A Division of Random House, Inc., New York, ISBN 978-0-307-47326-4.

Hubert, P., L. Annisomova, G. Antsipov, V. Ramsaev, and V. Sobotovitch (Eds) (1996). *Strategies of Decontamination*. Final Report APAS COSU 1991-1995: ECP4 Project. European Commission, Report EUR 16530 EN, ISBN 92-827-5195-3.

Kucan, J. O. (2004). Hiroshima and Nagasaki: Review of the consequences, implications in the post 9/11 world, http://www.journalofburnsandwounds.com/volume03/volume03_article05.pdf

Kasper G. Andersson
Editor

CHAPTER 1

Potential Sources of Contamination in Inhabited Areas

Raimo Mustonen*

Contents

1. Introduction

There are several types of events that could result in dispersion of radioactive substances in an inhabited environment. These include both intentional and unintentional events. Releases of radioactive substances could range from major events involving a nuclear facility or a nuclear weapon to small events such as a transportation accident. The extent of the contamination and impact on the environment and people depend greatly on the specific event and the radionuclides involved. However, many aspects of assessing and remediating the consequent situation and of protecting people will be the same or similar regardless of the spatial scale and specific radionuclides involved. Of course, the time needed to prepare

*Corresponding author. Tel.: +358 9 7598 8492; Fax: +358 9 7598 8498
E-mail address: raimo.mustonen@stuk.fi

STUK – Radiation and Nuclear Safety Authority, 00881 Helsinki, Finland

Radioactivity in the Environment, Volume 15
ISSN 1569-4860, DOI 10.1016/S1569-4860(09)00401-X

these actions depends on the event. This chapter deals with potential sources of radioactive contamination of urban or inhabited areas. It is aimed at providing an overview of different types of conceivable scenarios that could lead to contamination on different scales and the radionuclides that would be of primary concern. Examples are given of experience from the past.

2. Background

Worldwide there are hundreds of medical, industrial and academic applications using radioactive sources of significant strength. These applications encompass *inter alia* sterilisation of foodstuffs and pharmaceutical products, industrial and medical radiography, teletherapy, exploring for oil, research, etc. Seven reactor-produced radionuclides are of particular concern due to their radiotoxicity, their widespread use and their sufficiently long half-lives: ^{241}Am (432 years), ^{252}Cf (2.6 years), ^{137}Cs (30 years), ^{60}Co (5.3 years), ^{192}Ir (74 days), ^{238}Pu (88 years) and ^{90}Sr (29 years) (Ferguson et al., 2003). These radionuclides are used in many applications with different activities as shown in Figure 1 (IAEA, 2001).

There are many factors decreasing the overall security of a radioactive source during its life cycle. There can be weaknesses in organisations, in regulations and procedures, in the proper working of regulatory bodies and in regulatory enforcement. Often a lack of knowledge and awareness is the reason for getting the source lost. The follow-up of registries can be a

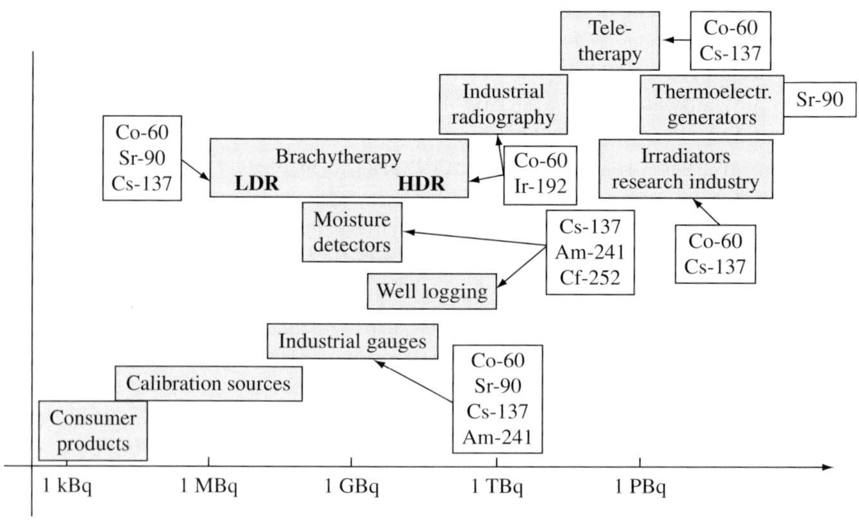

Figure 1 Commercially available radionuclides and their activities in different applications in industry, medicine and research.

problem, especially in the case of bankruptcy or change of ownership. In addition, security, for example during transport, can be deficient. Furthermore, there can be inhibitions to legal disposal because of the high costs involved; or worse, legal disposal can be non-existent. It is estimated that in the USA there are between 500,000 and 2,000,000 sources which are no longer needed and up to 375 sources are yearly reported to be orphaned. The figures for the European Union are 30,000 disused sources and up to 70 sources per year reported to be orphaned. In the former Soviet Union it is estimated that there are thousands of orphan sources of high-risk category (Ferguson et al., 2003).

The past six decades have shown that various accidents related to the use of radioactive and nuclear materials must be taken into consideration although today the likelihood of major accidents is small and releases of radioactive substances into the environment are minimised with effective safety and security systems. As a consequence of the terrorist attacks on the US World Trade Center and the Pentagon Headquarters, the subway systems in Spain, Japan and the UK, embassies, nightclubs and hotels, political leaders and authorities have become more aware of the need to re-assess existing threats and our preparedness for them. There are several lessons learned from the recent attacks and other events where radioactive or nuclear materials have been involved. The first of these is the intent of terrorists to stage multiple events simultaneously, and this must be taken into account in emergency planning today. The second factor, which is very difficult to predict, is the concept of suicide scenarios. A third lesson is that terrorists deliberately choose improbable or unexpected events and we can no longer rely on historical factors such as the probability of failure rates of various components to predict the likelihood of an event. The fourth lesson is the realisation of a terrorist event combining multiple hazardous agents. Thus, planning for a radiological incident alone is an outmoded concept and authorities need to be able to recognise and respond to a situation where there is a combined chemical, biological and radiological hazard.

3. Accidents at Nuclear Installations

In 2006, there were 442 commercial nuclear power reactors operating in 31 countries, 11 reprocessing plants in 9 countries, 284 nuclear research reactors in 56 countries and 220 nuclear-powered ships and submarines (World Nuclear Association, 2008). Most of them are situated quite close to inhabited areas and accidents occurring with them might have severe consequences for the local population. There are different types of nuclear reactors. Most are used for power generation, but some can also produce plutonium for weapons and fuel. Two components are common to all reactors: control rods and a coolant. Control rods determine the rate of fission by regulating the number of neutrons. These rods consist of neutron-absorbing elements such as boron. The coolant removes the heat generated by fission reactions. Water is the most common coolant, but pressurised water, helium gas and liquid sodium have also been used. Slow-neutron reactors operate on the principle that ^{235}U undergoes fission more readily with thermal or slow neutrons. Therefore, these reactors require a moderator to slow down neutrons from high speeds upon emerging from fission reactions. The most common moderators are graphite (carbon), light water (H_2O) and heavy water (D_2O). Since slow reactors are highly efficient in producing fission of ^{235}U, slow-neutron reactors operate with slightly enriched uranium. Light-water reactors are classified as either pressurised-water reactors (PWR) or boiling-water reactors (BWR), depending on whether the coolant water is kept under pressure or not.

Fuel amounts in reactors vary with their size. Small research reactors may contain only a few kilograms of uranium, plutonium or thorium, whereas reactors used in submarines and surface vessels can contain a few hundreds of kilograms, and power plant reactors up to more than 100 tons of uranium. The composition of nuclear fuel varies depending on the reactor type and application. Natural uranium, containing about 99.3% ^{238}U and about 0.7% ^{235}U, is used in some power reactors to produce plutonium. Low-enriched uranium (enriched to increase the concentration of ^{235}U up to some 20%) is used in most commercial power reactors to produce electricity and in some research reactors. Highly enriched uranium (^{235}U from 20% up to more than 90%) is used in many research reactors, naval propulsion reactors, nuclear weapons and reactors producing tritium and plutonium. Mixed plutonium–uranium oxide (MOX) fuel is used in some research and experimental reactors and also in some power reactors.

The consequences of severe accidents at nuclear installations relate to the amount of fuel in addition to the type of accident. Hundreds of accidents and incidents have occurred with small research reactors and nuclear-powered ships and submarines. Some of them have resulted in loss of lives and human exposure to radiation at different levels. Accidents in nuclear submarines and vessels may lead to serious consequences in inhabited areas only if they happen in harbours. A damaged reactor may result in dispersion of radioactive materials within an area of a few tens of square kilometres calling for protective actions and later on also some clean-up actions. Small research reactors are normally close to or inside inhabited areas and their severe accidents may also contaminate areas of a few tens of square kilometres; hence, protective and clean-up actions might be needed.

A few severe accidents have happened in nuclear power reactors, the best known being at the Chernobyl plant in Ukraine in 1986, at the Three Mile Island plant in Pennsylvania in the USA in 1979 and at Windscale in Cumbria in Northern England in 1957. The reactor of the Windscale accident was using natural uranium as the fuel, graphite as the moderator and air as the cooling medium. The accident occurred during annealing of the graphite. Annealing is an operation which is periodically necessary to liberate the energy built up in the graphite as a result of exposure to neutrons. The operation leads to a significant liberation of energy. The event initiating the accident was an erroneous temperature indication. As a result of an underestimation of the temperatures, the nuclear power was raised until two fuel elements, which already had damaged cans, ignited. The fire spread over 20% of the reactor core. Fission product release was not monitored, because of an instrumentation failure. The fire was finally detected visually. After 12 h of great effort, the fire was brought under control and finally extinguished. A significant fraction of the affected core inventory was released over a period of 21 h. The releases of the most important radionuclides are summarised in Table 1. The maximum individual dose received was estimated at 9 mSv.

The Three Mile Island accident began early in the morning on 28 March 1979 when the plant experienced a failure in the secondary, non-nuclear section of the plant (U.S. Nuclear Regulatory Commission, 2008). The main feeding water pumps stopped running, caused by either a mechanical or electrical failure, which prevented heat removal from the steam generators. First the turbine and then the reactor automatically shut down. Immediately after that, the pressure in the primary system, inside the nuclear portion of the plant, began to increase. In order to prevent that pressure from becoming excessive, the pilot-operated relief valve opened. The valve should have closed when the pressure decreased by a certain amount, but it did not. As a result, cooling water poured out of the open valve and caused the core of the reactor to overheat. Because adequate cooling was not available, the nuclear fuel overheated to the point at which the zirconium metal tubes, holding the nuclear fuel pellets, ruptured and the fuel pellets began to melt. A considerable part of the reactor core melted during the early stages of the accident. A dose rate of 60 Gy h^{-1} was

Table 1 Estimation of release of some radionuclides from the accidents at Windscale (1957), Three Mile Island (1979) and Chernobyl (1986) (Sohier, 2002).

Radionuclide	Released activity (10^{12} Bq)		
	Windscale	Three Mile Island	Chernobyl
^{133}Xe	14,000	370,000	6,500,000
^{131}I	740–1,200	0.5	1,800,000
^{137}Cs	46	–	85,000
^{90}Sr	0.2	–	8,000
^{239}Pu	1.6	–	30

measured in the building. The major risk was linked to presence of hydrogen in the pressure vessel and its possible explosion if contacted with oxygen. This did not happened. A leak of contaminated liquids at the beginning of the accident caused a release of rare gases (see Table 1). The most important exposure route was by direct exposure to the passing radioactive cloud. The low speed and variable direction of the wind caused very complex atmospheric dispersion. Evacuation of pregnant women and young children was imposed. This decision gave rise to the spontaneous evacuation of more than 100,000 persons. The maximum dose received per individual was estimated at 0.4 mSv. Melting of nuclear fuel did not however lead to a breach of the walls of the containment building, which would have led to a release of great quantities of radioactive materials to the environment. The accident led to no deaths or injuries to plant workers or members of the nearby community.

The Chernobyl accident started on the night of 26 April 1986 in the town of Pripyat in Ukraine (Wikipedia, 2008). It is regarded as the worst accident in the history of nuclear power. It had been decided to shut down the reactor for maintenance and the management had decided to use this occasion as an opportunity to test the ability of the reactor's turbine generator to produce sufficient electricity to power the reactor's safety systems in the event of a loss of external electric power. The test led to the reactor power jumping to around 30 GW, 10 times the normal operational output. The fuel rods began to melt and the steam pressure rapidly increased, causing a large steam explosion. Generated steam travelled vertically along the rod channels in the reactor, displacing and destroying the reactor lid, rupturing the coolant tubes and then blowing a hole in the roof. The explosions that ruptured the Chernobyl reactor vessel and the consequent fire that continued for 10 days resulted in large amounts of radioactive materials being released into the environment. The cloud from the burning reactor spread numerous types of radioactive materials, especially iodine and caesium isotopes, over much of Europe. The greatest deposits of radionuclides occurred over large areas of Belarus, the Russian Federation and Ukraine. Table 1 gives the most important emissions from the accident.

Of course, the accident resulted in huge radioactive contamination at the accident site itself and also in the neighbouring areas. A total of 600,000 'liquidators' participated in emergency and recovery operations of the Chernobyl accident (The Chernobyl Forum, 2006). Among them about 350,000 workers were initially involved in containing and cleaning up the accident in 1986–1987, and about 240,000 of them took part in major mitigation activities at the reactor and within the 30 km zone surrounding the reactor. Worst contamination took place at the power plant site and, for example, about 50,000 men were needed to clean up only the roof of the neighbouring reactor and the turbine building (Nosov, 1993). Figure 2 shows the radiation map on that roof in February 1987, ten months after

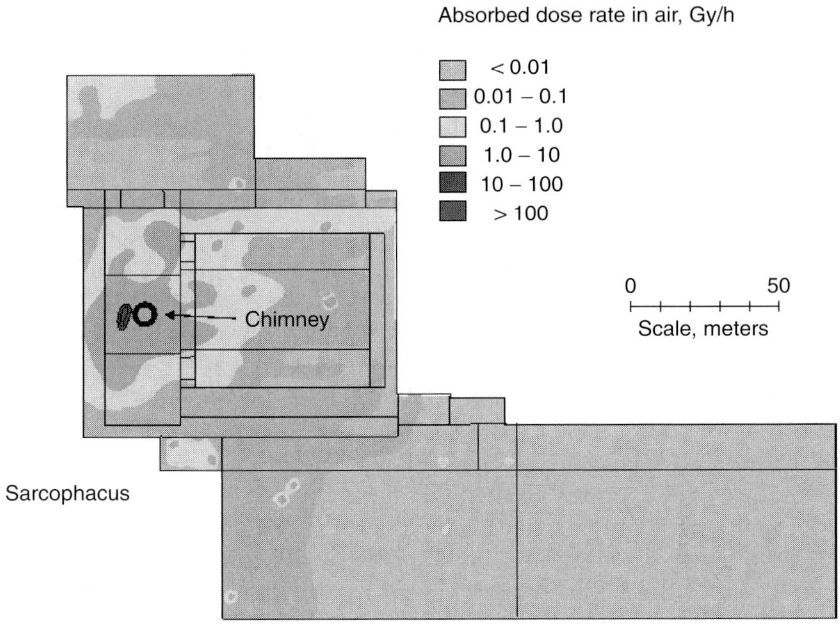

Figure 2 Measured absorbed dose rate in air on the roof of the Chernobyl reactor Unit 3 and the turbine building in January–February 1987, ten months after the accident.

the accident. It was necessary to monitor the radiation situation on the roof beforehand to be able to plan the clean-up activities in order to keep the radiation doses to the 'liquidators' at an acceptable level. Although the radiation dose rates were measured 10 months after the accident, when the short-lived radionuclides had already decayed, the highest dose rate absorbed in air was more than $100\,\mathrm{Gy\,h^{-1}}$.

More than five million people live in areas of Belarus, Russia and Ukraine that are classified as contaminated with radionuclides due to the Chernobyl accident (areas with ^{137}Cs deposition more than $37\,\mathrm{kBq\,m^{-2}}$). Amongst them, about 400,000 lived in more contaminated areas (above $555\,\mathrm{kBq\,m^{-2}}$) and 116,000 people were evacuated in the spring and summer of 1986 from the area surrounding the Chernobyl power plant to non-contaminated areas. Another 220,000 people were relocated in the subsequent years.

4. Accidents with Highly Radioactive Sources

Highly radioactive sources are used for a variety of purposes, such as medicine, research, measurement and detection in industry, and instrument calibration. Experience worldwide shows that, despite the existence of a regulatory framework, control of such high-activity sources may nevertheless be lost, even in countries with rigorous regulatory systems. A large number of incidents involving loss of control have been reported over the last 50 years. Following the terrorist attacks of the past few years, there have been heightened concerns about terrorist activity on, *inter alia*, radioactive sources, and levels of control and regulation have been raised. For example in the European Union, the member states have implemented control of high-activity sealed radioactive sources and orphan sources in their national legislation, based on the directive on high-activity sealed sources and orphan sources of the Council of the European Union (2003) (so-called HASS Directive). To prevent radiological accidents and injuries, the location of each high-activity source must be known, recorded and verified from the time the source is manufactured or imported into the community till the time it is placed in a recognised installation for its long-term storage or disposal or is exported from the community.

A high-activity radioactive source is a source having an activity of some 10^9 Bq or greater, depending on the radionuclide. This kind of source is used in the civil nuclear industry, hospitals, universities and other research institutes, industrial radiography and irradiation companies, defence organisations, offshore oil/gas service providers, etc. A number of incidents and accidents with this kind of source have resulted in excessive radiation exposure of members of the public, leading to death or serious injury. In other cases, lost sources have found their way into metal scrap destined for recycling, resulting in widespread environmental contamination requiring costly remediation (Gonzales, 1999). For example, in the USA alone, the Nuclear Regulatory Commission annually receives about 200 reports of lost, stolen or abandoned radioactive sources (Dicus, 1999).

The International Atomic Energy Agency (IAEA) has compiled a database of major accidents, which have resulted in radiation exposure sufficient to cause death or serious injury, drawing upon those accidents reported in the open literature (Lubenau and Ström, 2002; World Nuclear Association, 2008). The database contains a total of 136 accidents during the period 1945–1999, the majority of which (89 cases) involved radioactive sources. The nuclides involved were mainly ^{192}Ir, ^{60}Co and ^{137}Cs in the form of radiography and medical sources. Some of the accidents are listed below:

- In 1979, an unshielded ^{192}Ir source was accidentally left at a site in California and was picked up by a worker, resulting in severe radiation injury.
- In 1985, an abandoned teletherapy unit containing 5×10^{13} Bq of ^{137}Cs was stolen and the source capsule ruptured in Goiania in Brazil. Four severely exposed persons died, many others were seriously injured and

the consequent decontamination of buildings and land lasted for 6 months. The decontamination work had a major economic impact on the region.

- In 1994, a strong ^{137}Cs source was stolen from a radioactive waste facility in Estonia. One person died and two others were seriously injured (see Figure 3).
- In 1996, a worker in Iran dismantled the lagging of an abandoned ^{192}Ir source and put the source in his pocket. The consequent over-exposure led to a severe injury.
- In 1997, a group of border guards in the Republic of Georgia was severely exposed to radiation from several sources containing ^{137}Cs or ^{60}Co abandoned in a former military barrack.
- In 1999, some abandoned ^{60}Co teletherapy sources in Turkey were dismantled, resulting in 10 persons receiving severe radiation injuries.
- In 1999, a construction worker in Peru inadvertently picked up an uncontrolled ^{192}Ir radiography source and put it in his pocket, resulting in severe radiation burns.

There are tens of reports of cases where radioactive materials were smelted with scrap metal (The Council of the European Union, 2003).

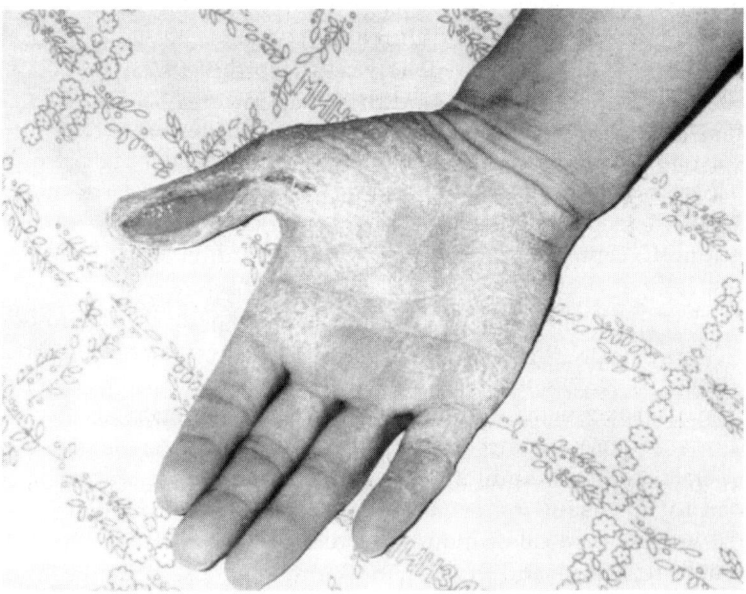

Figure 3 In 1994, three brothers stole a strong ^{137}Cs source from a radioactive waste facility in Estonia and took the source home. All family members received remarkable radiation doses and one of the brothers died as a consequence of irradiation. The photograph shows deterministic skin injuries in a hand of one of the brothers.

The source activity was not known in most of the cases, but, where reported, was generally greater than 10^9 Bq. Smelting of contaminated scrap metal may spread radioactive material into the environment, but normally environmental contamination will be minimal and the consequent health effects to people are not detectable. On the contrary, the financial consequences may be very large because of the costs resulting from decontamination, waste disposal and lost revenue during temporary shutdown of the plant.

5. Transport Accidents

About 20 million transportations of radioactive or nuclear material take place worldwide each year (World Nuclear Transport Institute, 2008). These may be of either a single package or a number of packages sent from one location to another at the same time. The transported package may contain radioactive or nuclear material for a number of applications in medicine, industry, research and nuclear power production, or wastes from these applications. Transport of nuclear materials has been carried out safely and routinely for over 40 years. Since 1971, there have been more than 20,000 shipments of spent fuel and high-level wastes (HLW; over 50,000 tonnes) over more than 30 million kilometres.

The principal assurance of safety in the transport of nuclear materials is the design of the packaging, which must withstand foreseeable accidents. The consignor bears primary responsibility for this. Many different nuclear materials are transported and the degree of potential hazard from these materials varies considerably. Different packaging standards have been developed to recognise that increased potential hazard calls for increased protection. 'Excepted' packages are for low-hazard materials, with activities below specified levels. The packages are of simple design and used mainly to transport low-activity materials used for medical diagnostic tests. If severely damaged it is assumed that most of the material might be lost from the package, but it would represent a low hazard to workers dealing with the package debris. 'Industrial' packages normally consist of steel drums or a freight container used to transport bulk materials such as wastes or mineral ores and uranium-ore concentrate. 'Type A' packages are designed to withstand minor accidents and are used for medium-activity materials such as medical or industrial radioisotopes. Packages for HLW and spent fuel are robust and very secure containers and they are known as 'Type B' packages. They also maintain shielding from gamma and neutron radiation, even under extreme conditions. Packages used for the transport of nuclear materials are designed to retain their integrity during the various conditions that may be encountered while they are being transported and to ensure that an accident will not have any major adverse consequences.

Regulatory performance tests include fire, impact, immersion, pressure, heat and cold.

Thanks to strict international and national regulations, there has never been a transport accident with Type B packages that has caused significant radiological consequences to people or the environment (Gonzales, 1999). Due to the great number of annual transportations of radioactive and nuclear materials, hundreds of smaller incidents have happened and will happen. The incidents and accidents can be classified into three categories:

- Administrative events consisting mainly of breaches of the regulations concerning the consignor's certificate or other shipping documents, or training.
- General shipment events including all the physical occurrences involving packages and conveyances. This includes traffic events, lost or damaged packages and incorrectly prepared packages.
- All the physical occurrences and contamination events involving irradiated nuclear fuel (INF) flasks, or the conveyances used to transport them.

For example in the UK, between 1958 and 2004, 806 incidents occurred involving transport of radioactive or nuclear materials (Hughes et al., 2006). Most of these occurred in transport of medical and industrial radioisotopes. In 768 cases there was no or extremely low radiological consequence and in 19 cases radiation doses exceeding 1 mSv were reported. Radiological consequences of transport accidents are normally restricted to very limited areas and the number of exposed people will remain less despite the fact that transport often takes place in inhabited areas.

6. Nuclear-Powered Satellites Entering the Atmosphere

For more than 40 years, nuclear power has been used to meet the energy requirements of spacecraft. Today there are some 50 nuclear satellites constantly orbiting above our heads, including deactivated but still radioactive reactors that have been boosted to "nuclear safe" higher dumping altitudes. During these four decades, several launch failures, failures to achieve the desired orbit and accidental re-entries through the earth's atmosphere have occurred. None of these accidents has caused measurable health effects in the human population, although some environmental contamination has occurred.

Two different techniques have been used to generate electric power for satellites: nuclear generators usually referred to as radioisotopic thermo-electric generators (RTGs), and nuclear reactors. Unshielded small nuclear reactors are normally used in military satellites orbiting in low orbits at an altitude of about 250–400 km. Because the atmospheric drag causes the orbit of even satellites with small cross-sections to diminish quickly at such low altitudes, a compact nuclear reactor is used for power generation rather than large solar panels.

The high decay heat of ^{238}Pu enables its use as an electricity source in nuclear thermoelectric generators of spacecraft, satellites, navigation beacons, etc. (Uranium Information Centre, 2008). Heat from the oxide fuel is converted into electricity through static thermoelectric elements (solid-state thermocouples), with no moving parts. Nuclear generators are safe, reliable and maintenance-free and can provide heat or electricity for decades under very harsh conditions, particularly where solar power is not feasible. However, their capacity to produce electricity is quite limited. Figure 4 gives an example of a RTG (Kulcinski, 2008).

Space nuclear reactors based on nuclear fission have distinct advantages of cost and electrical power generation relative to RTGs. Several different types of nuclear reactors have been developed for space applications producing electric power from 1 kW to about 100 kW (Bennett, 2008). The fuel is highly enriched uranium with total mass ranging from a few kilograms to a few tens of kilograms and is cooled with NaK coolant. They utilise thermoelectric or thermionic converters to produce electricity.

Several accidents have occurred with nuclear satellites and spacecrafts (The Pro-Nuclear Space Movement, 2008). In 1964, the US weather satellite carrying an RTG failed to achieve the planned orbit. It was designed to burn up and disperse its nuclear inventory in the upper atmosphere during re-entry and this happened as planned. Small amounts of radioactive material were measured by scientists in air and in soil samples. Later on, in 1968, a meteorological satellite failed in launch and two RTGs were later recovered from Santa Barbara Channel. No release of radioactive material was detected. In 1970, the Lunar Module of Apollo XIII re-entered the atmosphere and impacted in the South Pacific Ocean, where it remains today. No evidence of radioactive material was found. In January 1969, the launch of the Soviet COSMOS 305 lunar mission powered by RTGs created detectable amounts of radioactivity in the upper atmosphere. In April 1973, another COSMOS mission launch failure resulted in the return of the power source (nuclear reactor) into the Pacific Ocean, north of Japan. Small amounts of radioactivity were detected in air samples. Between 1978 and 1988 three incidents occurred, in which the COSMOS satellites failed to boost the reactor core into a nuclear-safe orbit as planned. In 1978, radioactive materials from the reactor of COSMOS 954 spread over a wide area in the northwest of Canada. In 1983, the reactor core of the COSMOS 1402 returned to the South Atlantic Ocean, leaving a radioactive trail through the atmosphere. In 1988, the redundant system of the COSMOS 1900 succeeded in boosting the reactor core into a storage orbit at about 720 km altitude, after the first attempt to boost it earlier in the same year had failed.

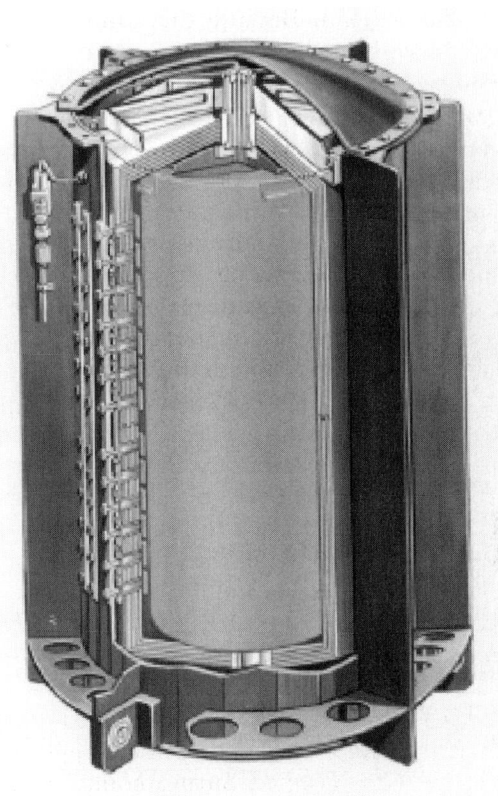

Figure 4 Cutaway of a multi-hundred watt radioisotope thermoelectric generator (MHW-RTG). Its length is 58.31 cm and the overall diameter is 39.73 cm. For Lincoln Experimental Satellites LES-8/9 the average power at the beginning of mission (BOM) was ~154 We per RTG with a mass of 39.69 kg (average). For Voyager 1/2 the average BOM power was ~158 We per RTG with a mass of 37.69 kg (average) (Kulcinski, 2008).

7. Malicious Use of Radiation and Radiological Terrorism

Malicious use of radiation and radioactive materials and terrorism using radioactive or nuclear materials are possibilities that we have to take into consideration today. As terrorism has become a daily topic of news headlines, the threat of radiological terrorism has become more apparent. Terrorists can achieve certain goals through radiological terrorism that conventional, chemical or biological terrorism cannot achieve. Although to date there have been few attempted attacks on nuclear facilities and no actual uses of radiation dispersion devices (RDDs), governments around the world and their citizens have become increasingly anxious about the threat of such attacks.

In principle there are two types of radiological terrorism: the malicious use of radiological devices and materials, and attacks on nuclear facilities, such as nuclear power plants. Both types have certain features in common, such as attempting to cause radioactive contamination and playing off public fear of radiation. The consequences of both types of attacks could also have similarities, such as slightly elevated health risks, peoples' increased fear and anxiety about their own and children's health, or massive economic costs. Already a threat to use radioactive material as a weapon, without causing any actual radiological impact, has serious consequences. There are several thinkable scenarios where radiation and radioactive or nuclear materials can be misused in order to create fear and serious health or other consequences (ICRP, 2006):

- Radiation sources may be stolen and used to deliberately and maliciously irradiate specific individuals, or people at random.
- Some conceivable scenarios for a radiological attack are based upon a covert exposure or dispersal of radioactive material, whereby the first indication that an event has occurred may be people reporting to hospitals with symptoms of radiation sickness, burns or other symptoms. Also radiation monitors used for various purposes could be the first indicators of a covert radiological attack.
- A detonation of conventional explosives to disperse an ordinary radioactive source used, for example in medicine and industry (RDD, or 'dirty bomb').
- A deliberate act to contaminate with radioactive material foodstuffs or water supplies, other particular commodities, or a specific site or environment.
- A sabotage of safety-related systems at nuclear facilities holding large amounts of radioactive materials, including nuclear fission and/or activation products and radioactive wastes.
- An extreme but possible scenario is the diversion of fissile nuclear materials and the development and use of a crude nuclear weapon, usually known as an 'improvised nuclear device'.

From the perspective of a malevolent person willing to make and use a RDD, there are three conflicting requirements regarding the radioactive source. First, the source should be 'sufficiently' radioactive to perform either direct radiological damage or at least societal damage or disruption. Second, the source should be transportable, meaning that it should not be too heavy while still providing enough shielding to the maker and carrier. Third, the source should be sufficiently dispersible to effectively contaminate the environment. For these reasons effective RDDs could contain moderate beta or gamma sources from a few to a few thousands of Ci ($\sim 10^{11}$–10^{14} Bq) and even more active alpha or beta sources could be used.

Rather than using a radiological device, terrorists might plan to attack a nuclear facility. Nuclear facilities include nuclear power plants, cooling ponds for spent nuclear fuel rods, nuclear reactors used for research or other non-electricity purposes, nuclear reprocessing facilities and nuclear waste sites. Another potential scenario for attack is during the transport of spent nuclear fuel.

The extent and severity of injuries resulting from a 'dirty bomb' attack or an attack on a nuclear facility depend on a number of factors. For a 'dirty bomb' attack, factors would include the type and amount of radioactive material and conventional explosives used, the amount of radioactive material aerosolised and the area in which the attack took place. The effects of an attack on a nuclear power plant would depend on whether radiation is 'successfully' released and the amount released.

Most predictions have assumed that to maximise the effect, terrorists would conduct a 'dirty bomb' attack in densely populated cities. Immediate deaths resulting from a 'dirty bomb' attack are more likely to be caused by the conventional bomb blast and the consequent panic than due to the radioactive material itself. Even in the case of high-intensity radioactive materials, scattering the material will dilute some of its intensity. This does not mean, however, that the radioactive material will cause little or no damage. Those people injured by the conventional explosives may also be contaminated by the radioactive material. This could lead to slower recovery or permanent internal and external damage. People, close to the area of the explosion, who are not physically wounded can still inhale microscopic particles and be directly exposed to radiation. Since a 'dirty bomb' blast may appear at first to be simply a conventional bomb attack, people covered in radioactive dust may unwittingly spread this material to hospitals and homes as they try to evacuate the area.

Although anxiety has recently increased over potential radiological terrorist attacks, history has seen very few actual attempts at radiological terrorism. Some examples of events which can be classified as radiological or nuclear terrorism are given below (Christie, 2006; Nuclear Threat Initiative, 2008; Health Protection Agency, 2008).

- In 1982, two relatively untrained amateur white persons smuggled explosive mines into the Koeberg nuclear power station in South Africa. As an employee of the station, another of them was able to easily gain access to the power plant and place the mines near the reactor heads. The explosions were designed not to breach the reactor heads, according to a published interview with the terrorist.
- In November 1995, Chechen rebels partially buried a container with a small quantity of ^{137}Cs in Moscow's Ismailovsky Park. The Chechen

leader then notified a Russian television crew, which located the container.

- In the fall of 1996, the Russian nuclear regulatory agency was warned about a planned attack on the Balakovo Nuclear Power Plant by an armed group of Chechens. Russian special forces were placed on alert and security around Russian nuclear facilities increased, but the attack never took place.
- In December 2001, two people scavenging for lead in a lighthouse on Russia's northern coast were hospitalised after being exposed to the powerful radioactive strontium source used to power the lighthouse. Radiation levels were found to be elevated along the path where the scavengers carried the leaking lead container.
- In December 2002, a criminal gang in Ecuador stole five radioactive sources from a technical company. The gang held these sources for ransom. After receiving the ransom money, however, the criminals returned only three of the sources. The other two remain missing.
- In May 2002, Georgian police conducting a routine search of a taxi found two boxes of ^{137}Cs and ^{90}Sr, and another box containing nerve gas concentrate. The taxi driver was released after police determined that he was unaware of the cargo, but two other suspects, including a man living on the Turkish border, were arrested. This event was one in a series of recoveries of radioactive sources in Georgia.
- In September 2003, a Chinese nuclear medicine expert was given a suspended death sentence after being convicted of placing radioactive ^{192}Ir pellets in a colleague's office. The expert worked at a Chinese hospital and used official papers to buy the ^{192}Ir. He then placed the pellets in his colleague's ceiling as an act of revenge. Soon after, the poisoned colleague began complaining of memory loss, fatigue, loss of appetite, headache, vomiting and bleeding gums. Before the radioactive pellets were uncovered, another 74 hospital staff members were found to have similar symptoms.
- On 1 November 2006, the former KGB officer Mr. Alexander Litvinenko suddenly fell ill in London and died 3 weeks later of acute radiation effects in hospital. Tests established that Mr. Litvinenko had a significant quantity of the radioactive isotope ^{210}Po in his body. It is not yet clear how this entered his body. Police investigations continue.

National governments are increasingly working together to establish greater safety and security for radioactive sources around the world. In 2003, more than 700 delegates from over 100 countries attended the International Conference on Security of Radioactive Sources in Vienna (IAEA, 2003). The conference was sponsored by the USA and Russia and organised by the IAEA. The conference's main findings emphasised the importance of the Basic Safety Standards and the Code of Conduct on the

Safety and Security of Radioactive Sources. The conference also called for an international effort to locate, recover and secure orphaned sources throughout the world. Several global and regional follow-up conferences have been arranged after the Vienna conference and international programmes have been launched to mitigate the threat of malicious use of radiation.

8. Special Problems Related with Radioactive Contamination of Residential Areas

Radioactive contamination of residential areas is problematic in many senses. First of all the contamination areas might vary from very large territories, if the source is a major nuclear power plant accident, to small places, if the there is a malicious use of limited amounts of radionuclides. The contamination might also contain only radionuclide or a mixture of several radionuclides depending on the source. And of course the strength of contamination may vary in large scales. Contamination of urban residential areas is of special concern because both the contamination and subsequent countermeasures and potential clean-up actions pertain directly to great number of people, their work and everyday life. Contamination of people's homes and gardens and other living environment with agents, which are not detectable with any human sense, can easily create suspense, worry and anxiety that must be taken into account in responding to the contamination. In addition, especially contamination of urban environment can be so complicated that technical means and capacity of decontamination and clean-up might be restricted and return of the living environment to the 'normal' living pattern can take months or even years. Finally there are of course also economic and financial aspects that must be taken into consideration. Normally inhabited areas contain a lot of production facilities that will suffer from radioactive contamination and also the loss in value of private property might be significant, and, of course, the costs of actions taken after contamination.

Atmospheric dispersion of radioactive pollution in an urban environment is not as well predictable as in a rural environment, especially if the source itself is inside the urban environment. The next chapter deals with dispersion of radioactive elements in urban environment and describes the special problems related to the complex dispersion environment. Big cities have very dense structure of high buildings that influence the behaviour of dispersing release of radioactivity. Urban environments also have a great variety of surfaces, on which radioactive elements can deposit and attach. Surfaces like streets, roads, building surfaces, gardens, agricultural areas and other soils, trees and bushes have different properties to catch radioactive elements and they also need different means and methods of clean-up.

Weather conditions also affect the dispersion and deposition of radionuclides (rain, humidity, wind, etc.).

Most emergency planning focuses on the emergency phase and most emergency exercises stop once emergency countermeasures have been successfully implemented and the source of the release has been brought under control. However, this can create a false impression of the major decision-making problems. In contrast, the late phase after the emergency phase is a very different decision-making problem. The later phase is not naturally time bounded. As demonstrated by the accident at Chernobyl, the longer-term consequences of an accident may last for decades. Self-styled experts will offer advice and gain credibility with some sections of the public but, on the other hand, may create confusion if their advices are not in line with the actions taken and recommended by the authorities. Companies and the self-employed may press for compensation.

If actions such as decontamination of buildings are implemented, then contaminated waste will be generated. In the absence of appropriate contingency planning, the infrastructure for the storage, treatment and disposal of waste may well not be adequate to cope with the amounts of waste generated. Recovery actions will also require workers and equipment to implement them; these may or may not be straightforward to find. If key infrastructure has been contaminated, then the consequences of delaying decisions on late-phase measures may extend far beyond the contaminated area. If people were evacuated during the emergency phase, then security of their property needs to be maintained until the advice to stay away is lifted. In short, the costs of delaying decisions are high, and the costs of getting the response 'wrong' are even higher.

The radiation protection community has traditionally taken the leading role in situations where radiation protection has had some role to play, and emergency and post-accident management of nuclear or radiological emergencies is no exception. This is not surprising given the way in which the optimisation principle, one of the main pillars of radiation protection, has been framed. A requirement 'to reduce doses to as low as reasonably achievable, social and economic factors taken into account' encourages the view that societal aspects should be integrated into radiation protection decisions. The more enlightened have long recognised that, at least for more complex issues such as post-accident management, the opposite is true – radiation protection aspects should be integrated into the broader social and political decision-making process. This was one of the main conclusions of the International Symposium on Off-Site Nuclear Emergency Management, held in Salzburg in 2003 (Kelly and Hohenberg, 2004). It is now increasingly recognised that radiation protection is only one of a large number of considerations that must be taken into account in nuclear emergency management, and in many cases it is not the most important. Today we have tools and methods to involve different stakeholders in decision-making on protective actions and countermeasures after an area has contaminated with radioactive substances. The experience gained within several exercises strongly supports the format of a facilitated workshop for tackling decision problems that concerns many different stakeholders, and the use of modern decision analysis techniques in formulating the problems and finding optimised solutions to them (Mustonen and Sinkko, 2005).

REFERENCES

Bennett, G. L. (2008). *Space nuclear power: Opening the final frontier.* Metaspace Enterprises, Emmett, ID, USA, http://www.fas.org/nuke/space/bennett0706.pdf

Christie, R. (2006). Eskom in the Democratic South Africa, *Workshop on Technopolitics in Africa, Conference Centre of Ithala Game Reserve Kwazulu Natal South Africa*, 13–16 July, 2006.

Dicus, G. J. (1999). Safety and security of radioactive sources in the United States. *IAEA Bulletin*, 41(3), 22–27.

Ferguson, C. D., T. Kazi, and J. Perera. (2003). Commercial radioactive sources: Surveying the Security Risks. Occasional Paper No. 11. Center for Nonproliferation Studies, Monterey, CA.

Gonzales, A. J. (1999). Strengthening the safety of radiation sources and the security of radioactive materials. *IAEA Bulletin*, 41(3), 2–17.

Health Protection Agency. (2008). HPA Press Statement, 24 November 2006, http://www.hpa.org.uk/hpa/news/articles/press_releases/2006/241106_litvinenko.htm

Hughes, J. S., D. Roberts, and S. J. Watson. (2006). *Review of Events Involving the Transport of Radioactive Materials in the UK, from 1958 to 2004, and their Radiological Consequences.* HPA-RPD-014, Health Protection Agency.

IAEA. (2001). *Categorization of Radiation Sources.* IAEA-TECDOC-1191, Version March 2001.

IAEA. (2003). *International Conference on Security of Radioactive Sources*, Vienna, Austria, 10–13 March.

ICRP. (2006). *Protecting People against Radiation Exposure in the Event of a Radiological Attack.* ICPR Publication 96, *Annals of the ICRP.* Elsevier, Amsterdam.

Kelly, N., and J.-K. Hohenberg (Eds). (2004). *Proceedings of the Off-Site Nuclear Emergency Management–Capabilities and Challenges, International Symposium*, 29 September to 3 October 2003, Salzburg, Austria; *Radiation Protection Dosimetry*, 109, No. 1–2.

Kulcinski, G. L. (2008). Nuclear Power Sources in Space, http://fti.neep.wisc.edu

Lubenau, J. O., and D. J. Ström. (2002). Safety and security of radiation sources in the aftermath of 11 September 2001. *Health Physics*, 83(2), 155–164.

Mustonen, R., and K. Sinkko. (2005). Methods and techniques to arrange stakeholders' decision making workshops in radiological emergencies, Radiation Protection in Transition. *Proceedings of the XIV Regular Meeting of the Nordic Society for Radiation Protection*, 27–31 August 2005, SSI Rapport 2005:15, Swedish Radiation Protection Authority, Rättvik, Sweden.

Nosov, V. (1993). Personal communication (Mr. Nosov was Head of Laboratory of Operative Radiation Control performing the radiation mapping at the Chernobyl nuclear power plant).

Nuclear Threat Initiative. (2008). Introduction to Radiological Terrorism, http://www.nti.org/h_learnmore/radtutorial/

Sohier, A. (Ed.) (2002). *A European Manual for 'Off-site Emergency Planning and Response to Nuclear Accidents'*. SCK•CEN Report R-3594. Belgian Nuclear Research Centre, Belgium.

The Chernobyl Forum: 2003–2004 (Second revised version). (2006). Chernobyl's Legacy: Health, Environmental and Socio-economic Impacts and Recommendations to the Governments of Belarus, the Russian Federation and Ukraine, http://www.iaea.org/Publications/Booklets/Chernobyl/Chernobyl.pdf

The Council of the European Union. (2003). Council Directive 2003/122/Euratom on the control of high-activity sealed radioactive sources and orphan sources. *Official Journal of the European Union* L 346/57, 31.12.2003.

The Pro-Nuclear Space Movement. (2008). http://www.nuclearspace.com/past_accidents.htm

Uranium Information Centre. (2008). Nuclear Reactors for Space, http://www.uic.com.au/nip82.htm

US Nuclear Regulatory Commission. (2008). Fact Sheet on the Accident at Three Mile Island, http://www.nrc.gov/reading-rm/doc-collections/fact-sheets/3mile-isle.html

Wikipedia. (2008). The Free Encyclopedia, http://en.wikipedia.org/wiki/Chernobyl_accident

World Nuclear Association. (2008). http://www.world-nuclear.org/info/reactors.htm

World Nuclear Transport Institute. (2008). http://www.wnti.co.uk/facts_and_figures

CHAPTER 2

The Dispersion, Deposition and Resuspension of Atmospheric Contamination in the Outdoor Urban Environment

Ken W. Nicholson*

Contents

*Corresponding author. Tel./Fax: 44-1285-750589
E-mail address: ken@nicholson-environmental.co.uk

Nicholson Environmental, The Cottage, Victoria Road, Quenington, Gloucestershire GL7 5BW, UK

Radioactivity in the Environment, Volume 15
ISSN 1569-4860, DOI 10.1016/S1569-4860(09)00402-1

1. Introduction

Atmospheric transport can be broken down into dispersion and deposition. Dispersion is the dilution process that occurs when material is transported downwind from a source and it is important in determining atmospheric concentration. In turn, atmospheric concentration is important in determining inhalation dose. Deposition is the process that results in the surface uptake of atmospheric material. It is an important consideration when estimating plume depletion and the potential for downwind transport. It is also important in determining external radiation dose. The deposition of atmospheric material can occur by its direct interaction with a surface or by its incorporation into precipitation (i.e. rainfall, snow, etc.). It is usual to refer to the former process as dry deposition (even when it is raining) and to the latter one as wet deposition. In addition, deposition in fog can occur and this comprises mechanisms apparent in both wet and dry deposition processes. Once deposited, contaminants can become airborne again due to various mechanical effects (see Figure 1). This process is called resuspension (or re-entrainment), although the term is used often to express the general process in which any contaminant might become airborne (i.e. including those materials deposited by other than atmospheric processes such as spills etc.). Usually, resuspension involves contaminants that have become associated with host particles, although the evaporation (re-volatilisation) of certain gases can occur also. In this chapter, the dispersion, deposition and resuspension of material in the urban environment are discussed, along with

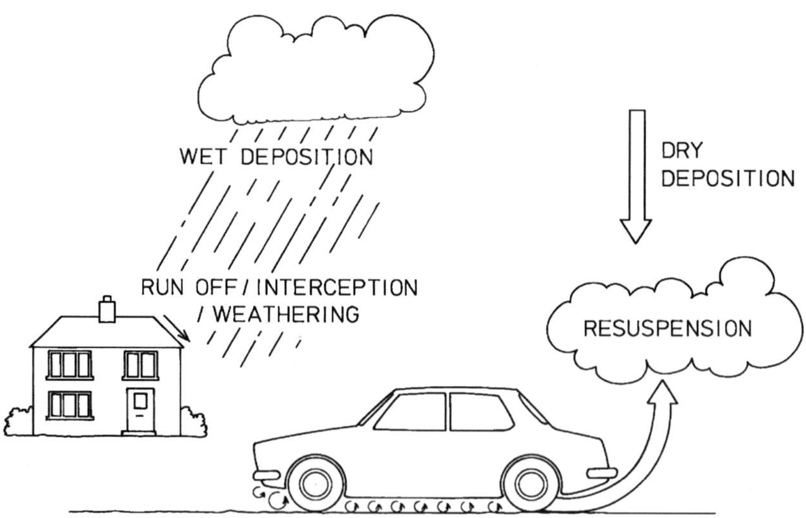

Figure 1 Atmosphere–surface exchange processes in an urban environment.

the physical factors that can affect these processes. Where appropriate, experimental data are reviewed and the requirements for further knowledge are identified.

2. Modelling of Radionuclide Dispersion

Gaussian plume and puff models are the two main types of model used for predicting the dispersion of radionuclides in the environment. The most commonly used models are probably Gaussian plume models and these are based on the statistical distribution of crosswind air concentrations downwind of a source. They involve the prediction of horizontal and vertical diffusion in the atmosphere, and they have proved very useful for determining time-averaged concentrations from point, line and area sources.

Puff modelling involves tracking finite parcels of air along trajectories downwind from a source and predicting the effects of diffusion and deposition on atmospheric concentration. Puff models necessarily require real-time data relating to meteorology across the area of study. Consequently, they tend to be much more expensive in computing time than Gaussian plume models and can be more complicated to use. On the other hand, puff models have advantages in that they can provide a more accurate prediction of the impacts of releases, especially over relatively short timescales.

The approach adopted in puff modelling can be either according to the time lapsed since the release into a given parcel of air, or on a surface grid, which requires an assessment of the amount advected into and out of that area. The former approach is termed Lagrangian, while the latter is termed Eulerian. Eulerian models are often considered better if there are complex sources or atmospheric transfer processes, although the range of models developed is extensive and includes hybrid approaches for certain applications.

Both Gaussian plume and puff models incorporate dispersion parameters that are determined by wind speed and atmospheric stability, and there have been various algorithms developed to predict these parameters. The models incorporate factors to allow for wet and dry deposition and an accurate assessment of such deposition effects is important when considering surface contamination and downwind exposure. Deposition parameters are discussed again in detail later.

There are a significant number of models available that incorporate calculations based on Gaussian plume or puff modelling. In addition, there is a suite of models available that combine elements of both types of approach. The difference between most of the Gaussian plume models lies in the characterisation of the atmospheric boundary layer and, in particular, how its structure affects thermal buoyancy (and thus vertical diffusion). Puff models are often the models of choice in decision support systems for nuclear or chemical emergency responses (e.g. RODOS, ARGOS).

However, their expense in terms of computer time and their requirement for detailed real-time meteorological data can be a problem. For the modelling of relatively short-duration releases (i.e. accident scenarios) over long distances (as in the case of Chernobyl, for example), puff modelling can hold distinct advantages over Gaussian plume modelling. However, for routine discharges, the reverse is often true.

There are a number of Gaussian plume models that have been developed commercially including AERMOD (US EPA, 2007), R91 (Clarke, 1979) and ADMS (Hall et al., 2000). The models can be adapted usually to allow for point and area sources. Their application for point sources might be appropriate in many cases involving routine and accidental releases, while assessing the impact of resuspension, for example, might be described best in terms of a multiple point or area source. The complex terrain in an urban environment has resulted in Gaussian plume models being developed to allow for street canyon effects (e.g. OSPM; Berkowicz, 2000). This development has been driven by the need to assess kerbside air concentrations of pollutants originating from road vehicles, although their application for determining the impact of traffic-generated resuspension is possible also. Every Gaussian plume model has its advantages and disadvantages in terms of simplicity of input data, source description or performance under different conditions. Consequently, there can be no universal recommendation for a specific model.

Puff models that have been developed include RIMPUFF (Thykier-Nielsen and Mikkelsen, 1991) and CALPUFF (Coulter and Eckhoff, 1998), amongst others. Some puff models have been developed to determine dispersion in the urban environment successfully and their potential for use in evaluating the impact of scenarios, such as a 'dirty bomb' explosion, has been acknowledged (Astrup et al., 2005). There have been hybrid models that have incorporated aspects of both Gaussian plume and puff modelling (e.g. DREAM; Brandt et al., 1996) and these have been useful for a number of specific applications. As in the case of Gaussian plume models, however, there can be no general recommendation of a specific model, as the model of choice will depend on the available meteorological data and the specific application for the model.

3. Physical Forms of Radionuclides in the Environment

There is a range of radionuclides that are of potential environmental interest (e.g. SCOPE50, 1993), although most are particulate in form. Of these, it is probably Pu, Sr, Cs, Am and Ru that have traditionally received most attention (see Chapter 1 for information on different types of potential sources of radioactive contamination in inhabited areas). Some of these radionuclides are classified as volatile. However, they are only present in the atmosphere as vapours if there is an accident involving a significant amount of heat (e.g. a reactor fire) and they will condense rapidly on to pre-existing particles. There are other radionuclides that can be released to the environment in gaseous form (e.g. ^{3}H, ^{14}C, ^{35}S and the noble gases). However, most of these are either unlikely to be released in large quantities or are of little radiological significance.

In addition to those radionuclides present in the atmosphere in either gaseous or particulate form, iodine is semi-volatile and can be present in both phases simultaneously. Of the various isotopes of radioiodine, ^{131}I with a half-life of 8 days can be a significant contributor to dose in the immediate aftermath of a reactor or reprocessing accident and is one of the radionuclides of major concern. ^{129}I is another isotope of iodine that can contribute to dose, although it is released in much smaller quantities than ^{131}I in most scenarios. However, the much longer half-life of ^{129}I (1.6×10^7 years) enables it to accumulate in the environment and this means that it has been recognised as a potentially important contributor to dose from routine reprocessing discharges. Predicting the environmental behaviour of iodine is further complicated since it can be present in various chemical as well as physical forms. Iodine in both reactive (I_2 and HI) and organic (e.g. CH_3I) forms has been identified in atmospheric discharges.

When describing the atmospheric deposition of a particulate species, the size distribution of the material is important (see below). Typically, most species can cover a wide size range and it is some measurement of that distribution, such as the mass median aerodynamic diameter, that is incorporated into modelling exercises.

The particle size of an atmospheric radionuclide depends on its source. For materials that are formed by condensation (e.g. isotopes of Ru and Cs in the event of a fire), their size distributions will approximate to that of the natural atmospheric aerosol, especially at significant distances from the source. This was demonstrated in the aftermath of the Chernobyl reactor accident (e.g. Jost et al., 1986; Kauppinen et al, 1986). In this respect, it is reasonable to assume that the deposition rates of such radionuclides will approximate to those of natural atmospheric materials also. Less-volatile elements, however, might be associated with airborne particles produced by mechanical effects (e.g. Pu in the case of a core explosion or in a 'dirty bomb') with much larger dimensions. It is very difficult to measure the concentration in air of a material that includes large particles (Garland and Nicholson, 1991). Moreover, it is difficult to model its behaviour, since its deposition rate will vary with distance from the source as the largest particles become preferentially deposited.

4. Dry Deposition

Dry deposition is usually expressed in terms of a dry deposition velocity (v_g) which was defined originally by Chamberlain (1953).

$$v_g\ (\mathrm{m\ s^{-1}}) = \frac{\text{Deposition flux } (\mathrm{Bq\ m^{-2}\ s^{-1}})}{\text{Atmospheric concentration at reference height, } c\ (\mathrm{Bq\ m^{-3}})} \qquad (1)$$

Dry deposition velocity was introduced in order to predict levels of deposition that could occur for a given air concentration. This has practical uses, since it is invariably air concentrations that are either measured experimentally or modelled. However, it is important to note that deposition velocity (as well as concentration) is a function of height. In addition, it can be used strictly for predicting deposition to a given type of underlying surface only. Consequently, there are inherent problems ascribing deposition velocities to large areas that might incorporate a range of surface element types and sizes.

4.1. Atmospheric dry deposition mechanisms

To understand the factors that control dry deposition and the complexities of predicting dry deposition in the urban environment, it is important to consider the structure of the atmosphere in terms of the types of boundary layer that develop (Nicholson, 1995). The planetary boundary layer is the uppermost layer which extends up to about 1–2 km, although it can subside to as low as 100 m at night. Flow in this layer is influenced by large-scale surface characteristics. Thermal effects (i.e. convection and damping) and mixing are important within this layer. The turbulent surface layer lies beneath the planetary boundary layer. This can extend up to about 100 m during the day, but like the planetary boundary layer, it subsides at night. Mixing within the turbulent surface layer is controlled by convection and turbulence originating from surface-generated drag. Finally, the lowest layer is described as the laminar (or viscous) sub-layer. This is a thin film of air within the extreme vicinity of a surface and vertical mixing within this layer is greatly restricted. In reality, turbulent eddies penetrate this layer periodically so that the description laminar is partly misleading.

In order to describe dry deposition, it has been broken down traditionally into three steps (e.g. Hosker and Lindberg, 1982):

- transport through the atmospheric turbulent boundary layer
- transport through the laminar sub-layer
- uptake at the surface

Transport through the atmospheric turbulent boundary layer is dominated by turbulent diffusion. The deposition flux ($Bq\,m^{-2}\,s^{-1}$) can be related to a diffusion coefficient (K_p) which is assumed generally to be equal for particles and gases, the atmospheric concentration gradient (dc/dz) and the height (z) above the underlying surface. Where the boundary layer is fully developed:

$$\text{Deposition flux} = -K_p \frac{dc}{dz} \tag{2}$$

The negative sign on the right-hand side of Equation (2) is included to denote that the deposition flux is in the opposite direction to the concentration gradient. When thermal buoyancy effects are negligible (i.e. during thermally neutral atmospheric conditions), K_p can be approximated to:

$$K_p = k^2 z^2 \frac{du}{dz} \tag{3}$$

where k is the von Karman constant ($= 0.41$) and u is the horizontal wind speed (du/dz is the wind speed gradient according to height). Equations (2) and (3) are often modified using ($z-d$) in place of (z), where d is the zero plane displacement. Here d is introduced to allow for an apparent displacement of the height at which the wind speed extrapolates to zero, because of the size of the surface elements. In addition, various empirical correction terms have been included (Nicholson, 1995) to allow for conditions when thermal buoyancy effects are significant and vertical mixing is either enhanced (in unstable atmospheric conditions) or restricted (during thermally stable conditions).

When considering particles, a sedimentation velocity, v_s (the terminal velocity of a particle falling in air under gravity), is added to the turbulent flux to get the total deposition so that:

$$\text{Deposition flux} = -K_p \frac{dc}{dx} + v_s c \tag{4}$$

Sedimentation velocity is highly dependent on particle size and the sedimentation rate can be significant for large particles. However, it is usually much lower than the turbulent diffusion rate for most airborne particles, except perhaps in the close proximity to certain types of source (e.g. a fire or explosion). Nicholson (1995) noted that the size at which the sedimentational component might become important for transport through the turbulent surface layer might be of the order of 80 μm aerodynamic diameter, although this is clearly dependent on atmospheric conditions (especially wind speed).

Beneath the turbulent surface layer, transport through the laminar sub-layer relies on the effects of molecular diffusion for gases. Particle transport through this layer is controlled by Brownian diffusion and sedimentation. Inertial mechanisms also can play an important role in particle deposition if the particles have sufficient momentum to cross the laminar sub-layer and come in contact with the surface (this process is termed impaction). In addition, where the surface elements protrude the atmospheric turbulent boundary layer, a particle following the streamlines might approach that element sufficiently close to come into direct contact with it (this process is called interception).

For many environmental particles, especially in the size range 0.1–1.0 μm aerodynamic diameter, transport through the laminar sub-layer can determine the efficiency of the dry deposition process. The overall effect is for a 'V'-shaped deposition velocity curve according to particle size, with diffusional effects dominating the transport of the smallest particles and sedimentation, impaction and interception controlling the deposition of the larger particles (see Figure 2). The minimum deposition velocity is typically in the range 0.1–1.0 μm aerodynamic diameter. However, there are considerable uncertainties that remain over precise values of the deposition velocity within this size range (e.g. Nicholson, 1988a).

The uptake of gases or particles at a surface relies on the affinity of that surface for the depositing species. Usually the resistance is assumed to be

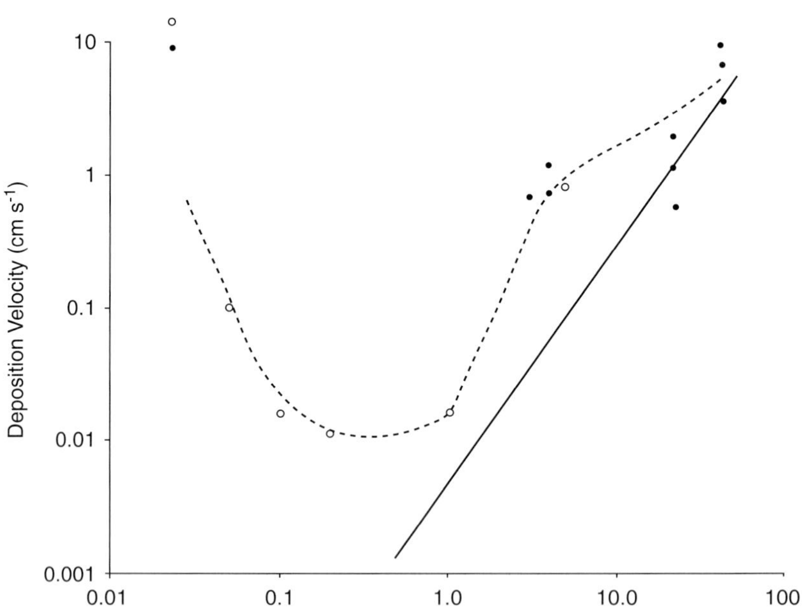

Figure 2 Representation of the dry deposition velocity measurements of particles to grass. *Note*: The points represent reported experimental data and the solid line represents the sedimentation velocity.

very low or zero for particles. In effect, particles can sometimes bounce off surfaces, although this process is poorly understood and is thought to be important only for quite large particles in windy conditions.

The efficiency of gas uptake at a surface depends strongly on the chemical nature of the gaseous material and the affinity of the surface for uptake. The stomatal activity of vegetation can be an important consideration in determining the surface uptake and deposition rate of many gases, including SO_2, CO_2 and H_2O. Some laboratory studies have shown that stomatal activity can be important in controlling the surface uptake of iodine at low humidity levels also (Barry and Chamberlain, 1963; Garland and Cox, 1984).

Gas uptake on moist surfaces by dissolution is difficult to generalise. It is feasible that re-emission could occur in order to maintain the equilibrium constant if atmospheric concentrations fall (i.e. after the passage of a main plume) or if evaporation of the surface moisture occurs. However, this will depend on the aqueous chemistry that occurs soon after deposition and whether the deposited material reacts with the underlying surface.

4.2. Physical factors affecting deposition velocity

The physical characteristics of an underlying surface and the prevailing meteorological conditions are both important factors in determining the dry deposition velocity. Turbulence in the free atmosphere increases with increasing surface roughness and with increasing height of the individual surface elements. The increased turbulence can be accompanied by a decrease in the laminar sub-layer thickness also (e.g. Garland, 1977), as well as the more frequent penetration of turbulent eddies into this layer. In addition, long surface elements can penetrate further into the free turbulent atmosphere, and deposition of particles via interception or impaction can occur, with both of these processes becoming increasingly important with increasing wind speed.

Generally, these effects work towards higher deposition rates being associated with rough surfaces for both particles (e.g. Sehmel, 1980a) and gases (e.g. Garland, 1977). However, very large surface elements can cause areas of sheltering in which dry deposition might be negligible. This is likely to be true on both a large scale (e.g. involving areas of buildings, trees, etc.), as well as on a small scale (e.g. in a grassed area). Consequently, deposition velocity needs to be used with caution when predicting contamination effects for specific surface locations or surface components. There remain significant uncertainties associated with deposition to large or complex surfaces, such as buildings or trees.

Increasing wind speed is associated with increasing turbulence (i.e. turbulent diffusion), enabling a more rapid transport through the turbulent surface layer (Garland, 1977). In addition, an increase in wind speed can result in a decrease in the depth of the laminar sub-layer (including the more frequent penetration of turbulent eddies through this layer), as well as an increase in the efficiency of inertial and interception mechanisms for particle deposition. These effects result in a decrease in the laminar surface layer resistance and higher overall deposition velocities, notably for particles in the range 0.1–1.0 μm aerodynamic diameter.

Consequently, most modelling studies predict an increase in deposition velocity with increasing wind speed or surface roughness. This is especially true for atmospheric particles (e.g. Ruijgrok et al., 1995), but it might be less important for gases, as surface uptake processes can often determine the overall deposition rate. Moreover, since gas deposition can vary over a relatively short period, according to variations in levels of surface moisture or stomatal activity, there are inherent errors in assuming average deposition velocities for gases in the case of a single short-duration contamination event (e.g. an accidental release).

Other environmental factors can also affect the dry deposition of particles, although these are poorly understood. It has been proposed that atmospheric thermal stability is an important factor (e.g. Wesely et al., 1988), and other suggestions have included phoretic effects (which depend on temperature and humidity gradients in the close vicinity of a surface; see, e.g. Sievering, 1982).

4.3. Dry deposition in the urban environment

Dry deposition in an urban environment might be considered as an overall effect or as deposition to individual surface types (see Figure 3). Underwood (1987) cautioned against the use of a generalised deposition velocity for urban areas noting the likely regional variations according to surface type. Another problem in predicting dry deposition rates in the urban environment is that the variety of surface types and structures means that boundary layers are seldom allowed to develop fully and edge effects can dominate. An example of where this might be apparent is when considering deposition to the walls of buildings, as it would be highly probable that a significant variation in the deposition levels would occur according to location on those walls. Consequently, the determination of an appropriate area-averaged deposition velocity is a difficult modelling or empirical task.

There are difficulties in determining deposition rates to individual surface elements also, since this depends on the characteristics of the upwind fetch. In particular, deposition rates to vegetation types such as hedges or trees are difficult to predict, as they are likely to have a filtering effect on the free air stream. Nevertheless, these types of vegetation have been shown to be efficient collectors of large particles and modelling suggests that they are efficient at collecting small particles also (Tiwary et al., 2006).

Figure 3 Representation of dry deposition in an urban environment.

Underwood proposed that an estimate of an integrated deposition velocity for an urban area might be obtained by summing the surface types for a given type of location. The main difficulty with this approach is the general shortage of reliable experimental data. In addition, the potential for other factors to affect the local deposition velocity (i.e. sheltering or edge effects) means that it is difficult to measure a representative deposition velocity according to surface type and experimental results are difficult to generalise.

Roed (1987a) and Nicholson (1987) carried out separate studies on the deposition levels of ^{134}Cs and ^{137}Cs shortly after the Chernobyl accident (in Denmark and the UK, respectively). There was no rainfall at their study sites during the passage of the main plume from Chernobyl and this enabled the dry deposition velocity to be inferred for several types of surface (see Table 1). Both Roed and Nicholson found that values of the deposition velocity for the walls of buildings were of the order of 0.001 cm s^{-1}. The values for the roofs were a little over an order of magnitude higher and in the range 0.03–0.09 cm s^{-1}. Both authors assumed that the Cs would have been present in association with particles predominantly in the aerodynamic diameter range 0.1–1.0 μm.

Nicholson et al. (1993) reported a series of field studies involving the release of various sizes of particles between 2 and 22 μm diameter and the measurement of their subsequent deposition on a building downwind from the source. They reported that on average, deposition velocities were around four times higher for the roof of the building than for the walls. Nicholson et al. replicated these results partly for building surfaces at different orientations

Table 1 Examples of deposition velocities measured following the Chernobyl accident.

Radionuclide	Reference	Surface	Deposition velocity (cm s^{-1})
^{134}Cs	Nicholson (1987a)	Walls (sheltered)	<0.0005
		Walls (average)	<0.004
		Roofs	0.05
		Flat roof	0.08
	Roed (1987a)	Roofs	0.03
		Trees	0.03–0.13
		Asphalt	0.005–0.11
		Concrete	0.005–0.008
		Grass	0.02–0.10
^{137}Cs	Roed (1987a)	Walls	0.001
		Grass	0.02–0.09
^{131}I	Roed (1987a)	Roofs	0.3–0.4
		Trees	0.3–1.0
		Asphalt	0.03–0.07
		Concrete	0.03–0.07
		Grass	0.2–1.2
		Walls	0.03
		Soil	0.1–0.4

in a wind tunnel, although the differences were much less marked than in the field. These results are consistent with those on the Chernobyl plume, in that deposition to roofs was apparently higher than to walls. It is worth noting however, that despite the relatively low deposition velocities found for walls, Andersson et al. (2002) have found that the weathering of contamination from them can be extremely slow (see Chapter 5).

Roed (1987a) measured the dry deposition of Chernobyl-derived ^{134}Cs and ^{137}Cs to grass (see Table 1) noting values for the deposition velocity in the range 0.02–0.1 cm s^{-1}. These values are similar to those reported in many micrometeorological studies that have investigated the deposition of adventitious trace species, assumed to be in the diameter range 0.1 to 1.0 μm, on grass (e.g. Garland and Cox, 1982; Nicholson and Davies, 1987; Allen et al., 1991). In addition, Roed reported that values of the deposition velocity for roads were intermediate to those found for walls and roofs, and that values for isolated trees were comparable to, but slightly higher than, those found for roofs. Roed noted deposition velocities for a range of other particulate radionuclides that were significantly larger than those he reported for Cs. This is a possible indication of the inherent uncertainties in the assumption that all radionuclides originating from an accident approximate to the natural background aerosol size distribution.

The studies on Chernobyl-derived radionuclides are useful in providing estimates of deposition velocity under certain conditions, but there are inherent assumptions in the calculations. These originate because it was possible to determine only the averages of the air concentration and deposition flux, and both would be expected to vary with time even given the relatively short duration of exposure. Nevertheless, the studies provide still some of the best measurements available for estimating dry deposition to urban surfaces.

Using estimates of the dry deposition velocity for different urban surfaces, Roed and Jacob (1990) calculated an integrated (bulk) deposition velocity for an urban area equal to about $0.03\ cm\ s^{-1}$ for $0.4\ \mu m$ diameter particles. However, it must be stressed that there remain huge uncertainties regarding the values of deposition velocities for particulate materials that might be appropriate for inclusion in such a calculation.

The dry deposition rate of gases is highly dependent on their reactivity. Noble gases are assumed generally to be non-depositing. Of the remaining radionuclides, it is iodine that has received most attention, although not specifically for urban areas.

An important consideration when evaluating the dry deposition of iodine will be its chemical nature, and gaseous radioiodine has been reported in both reactive (i.e. HI and I_2) and organic (CH_3I) forms. Chamberlain (1991) summarised deposition data for reactive iodine and found values that were around or greater than $1\ cm\ s^{-1}$ for vegetational surfaces. In reality, this would be likely to represent quite a low surface resistance and the rates of turbulent diffusion in the free atmosphere and molecular diffusion through the laminar sub-layer would be the most dominant controlling factors in such a case. Nevertheless, some laboratory experiments have shown that stomatal activity can be important in controlling surface uptake at low humidity levels (Barry and Chamberlain, 1963; Garland and Cox, 1984). Consequently, the dry deposition of reactive iodine to vegetation might be expected to follow a diurnal cycle, with deposition velocities being at a maximum during the daytime when stomate are open and photosynthesis is taking place. Chamberlain also reported a very small value of the deposition velocity for methyl iodide ($0.01\ cm\ s^{-1}$) as reported by Atkins et al. (1967).

Generally, reported modelling studies have assumed a deposition velocity of $1\ cm\ s^{-1}$ for inorganic gaseous iodine (e.g. Jenkin, 1992; McFiggans et al., 2000), and these have been assumed often to be typical for water-soluble gaseous molecules of moderate molecular weight. Likewise, much lower deposition velocities (typically 0.001 or $0.01\ cm\ s^{-1}$) have been assumed for methyl iodide in most modelling studies (Simmonds et al., 1995).

It is possible that deposited iodine could re-volatilise. However, based on the few data available, it seems that this might be of limited importance.

In an experimental study, Chamberlain and Chadwick (1966) reported that only 0.16% of deposited iodine was volatilised from grass over a period of 14 days. Amiro and Johnston (1989) made a similar observation when investigating the uptake and loss of ^{125}I for bean plants. In this case, the authors reported a loss of 0.07% per day and noted that the iodine was not lost in elemental form.

Roed (1987a) measured the dry deposition of Chernobyl-derived ^{131}I to various urban surfaces (see Table 1). He reported deposition velocities in the range 0.2–1.2 $cm\,s^{-1}$ for grass and this is consistent with many of the above-mentioned assumptions. Roed found that ^{131}I deposition velocities for surfaces were about an order of magnitude larger than those of Cs. Nevertheless, there remain significant uncertainties over iodine deposition rates to most types of urban surface and this is an area in which further study is required.

5. Wet Deposition

Wet deposition is the removal of material from the atmosphere in falling rain or snow and it can be broken down into either in-cloud or below-cloud scavenging processes. Usually, the term rainout is used to describe in-cloud scavenging and the term washout is used for below-cloud scavenging. However, various authors have used the terms differently and these descriptions have become confused.

The relative importance of in-cloud and below-cloud scavenging depends on the distance from the source and the extent to which a released material is mixed within rain clouds. Below-cloud scavenging would be the only process expected normally at short distances from a source, where it is unusual for released material to reach cloud-base level. There may be exceptions to this however, especially in hilly or mountainous terrain.

Measurements of wet deposition have been expressed often in terms of a scavenging ratio or scavenging factor (W).

$$W = \frac{\text{Concentration in rain (Bq g}^{-1})}{\text{Concentration in air (Bq g}^{-1})} \qquad (5)$$

W has been termed the washout ratio or washout factor also by some authors, although it clearly relates to scavenging that comprises both in-cloud and below-cloud processes. The measured values of W calculated as in Equation (5) have been reported usually in the range of several hundred to several thousand for particulate materials. Typical values for gases are often more difficult to generalise, however. This is because the complexities associated with in-cloud chemistry and aqueous reactions subsequent to collection mean that it is can be difficult to distinguish whether the depositing species was scavenged in a particulate or a gaseous form.

The use of W to describe contamination from a single episode is likely to be of limited use, since ground-level and in-cloud air concentrations are likely to vary significantly according to meteorological conditions. This was illustrated in the case of the Chernobyl plume, when there were significant levels of wet deposition in some areas that did not record measurable air concentrations at ground level (Cambray et al., 1987). The explanation for this observation was that the rain was due to a frontal system and the ground-level air mass had a different origin to the in-cloud air mass where scavenging was taking place. Nevertheless, contamination in many areas was related to the level of rainfall received during the passage of the main plume, so that W may have some use in predicting deposition levels in so far as rainfall amounts can be predicted. Its use otherwise might be restricted to estimating long-term deposition levels from routine emissions, however.

Below-cloud scavenging is usually expressed in terms of a washout (or scavenging) coefficient (or rate), Λ (e.g. Nicholson et al., 1991).

$$\Lambda(\mathrm{s}^{-1}) = \text{Fraction of material removed per second} \tag{6}$$

Λ can be used to calculate an atmospheric concentration (c) that has been reduced from an initial air concentration (c_0) after time t(s), due to below-cloud scavenging as:

$$c = c_0 \exp(-\Lambda t) \tag{7}$$

Although Λ is ideal for inclusion in transport modelling calculations close to the source, where scavenging is restricted to below-cloud processes, there are significant uncertainties over the adoption of appropriate values. This is due mostly to the practical difficulties associated with undertaking quantitative experimental studies.

5.1. The below-cloud scavenging of particulate materials

The mechanisms that affect the collection of particulate material by falling raindrops (i.e. below-cloud scavenging) are similar to those that are important in determining their dry deposition rate (i.e. Brownian diffusion, interception and impaction). In addition, phoretic effects (Slinn and Hales, 1971) and electrical charge effects (Byrne and Jennings, 1993) have been thought to play an important role for particles in the size range 0.1–1.0 μm aerodynamic diameter. As well as the size of the particles, the scavenging rate depends on the size (and speed) of the falling raindrops (Andronache, 2003; Loosmore and Cederwall, 2004). As in the case of dry deposition, there is an expected minimum in the collection efficiency curve for particles somewhere in the size range from 0.1 to 1.0 μm aerodynamic diameter (often called the 'Greenfield Gap'; Flossman, 1998). However, there has been some uncertainty over the possible depth of this minimum.

Λ can be calculated for raindrops and particles of a given size (Stensland, 1977) as:

$$\Lambda = NV_{\mathrm{T}}E\pi R^2 \tag{8}$$

In Equation (8) N is the number of raindrops per unit volume and E, V_{T} and R are the collection efficiency, terminal velocity and radius of the raindrops. For a spectrum of raindrop sizes (R_{s}) and particle radius a, Equation (8) can be modified according to Dana and Hales (1974) as:

$$\Lambda(a, R_{\mathrm{S}}) = \int_0^\infty \pi R^2 V_{\mathrm{T}}(R)E(a, R)N(R)\mathrm{d}R \tag{9}$$

There have been various formulations for determining raindrop size distributions (e.g. Best, 1950). However, Slinn (1977) proposed the formulation of Mason (1971) for the mass mean radius of raindrops (R_{m}) in steady rain with a precipitation intensity of J (in mm h^{-1}):

$$R_{\mathrm{m}} = 3.5 \times 10^{-4}\left(\frac{J}{1\ \mathrm{mm\ h^{-1}}}\right) \tag{10}$$

Equation (10) eliminates the need for a detailed size distribution of raindrops and it can be shown from Equation (8) that:

$$\Lambda = \frac{0.75J}{R} \tag{11}$$

since

$$J = \frac{4}{3}\pi R^3 NV_{\mathrm{T}} \tag{12}$$

Slinn (1977) used R_{m} to approximate:

$$\Lambda = \frac{cJE(a, R_{\mathrm{m}})}{R_{\mathrm{m}}} \tag{13}$$

where c is a constant that might have a value equal to 0.5 for frontal rain (Slinn, 1984).

Some experimental measurements of the below-cloud scavenging of particulate material have involved the release of trace materials and the measurement of their subsequent deposition in rain (e.g. May, 1958; Slinn et al., 1979; Nicholson et al., 1991; Nicholson and Branson, 1993). The problem with many of these experiments relates to the difficulty in releasing particles of a fixed and known size distribution. Other experiments have measured the decrease in atmospheric concentration of an adventitious trace species (e.g. Graedel and Franey, 1977; Davenport and Peters, 1978) with time during a rainfall event. Unfortunately, these experiments suffer the disadvantage that usually the trace species covers a broad size spectrum and there is uncertainty that any change in particle concentrations might be explained partly by reasons other than rainfall (i.e. a change in origin of the air mass). However, the problems of a broad size spectrum have been partly overcome in some recent studies that have utilised real-time particle size analysers that have analysed atmospheric concentrations according to quite narrow size bands (e.g. Chate and Pranesha, 2004).

Despite the uncertainties associated with ascribing appropriate washout coefficients for modelling the wet deposition of particles, it is generally assumed that removal rate increases with increasing rainfall intensity. Some predictive models for below-cloud scavenging have adopted values of Λ/J to represent below-cloud scavenging and values equal to about 10^{-4} (s^{-1}/mm h^{-1}) have been adopted typically. Nevertheless, there is ample scope for further investigation into appropriate below-cloud scavenging coefficients for rain, especially regarding light rain or drizzle.

There are very few reported measurements of below-cloud scavenging of particles by snow. This is because field experiments are difficult to undertake due to the unpredictability of snowfall events. Furthermore, the complex shapes of snowflakes mean that scavenging by them is very difficult to model. A number of studies have suggested that Λ for particles during snow is greater than during rain, although there are a number of complicating factors that can make an interpretation the results difficult. Notably, riming (the interception of cloud droplets by falling snow which freeze on contact) has been associated with high concentrations in snow (e.g. Murakami et al., 1981; Davidson et al., 1987; Mitchell and Lamb, 1989), because of high concentrations within the cloud droplets and the efficient interception of these by the falling snowflakes. Consequently, it is difficult to distinguish between in-cloud and below-cloud scavenging in these studies. However, Sparmacher et al. (1993) undertook field experiments with the specific aim of determining below-cloud scavenging rate of particles by snow and concluded that this might be around five times higher than that for rain.

In-cloud scavenging involves nucleation processes, in which droplets or ice crystals form around pre-existing atmospheric particles. In order for this to happen, the particles must be well mixed within the condensation layer. The efficiency of the nuclei depends on their chemical and physical properties and it is difficult to describe these generally for material resulting from a contamination event. However, the high levels of Chernobyl deposition related to local rainfall (notably Cs), at distances more than 1,000 km from the source (e.g. Cambray et al., 1987; Clark and Smith, 1988), demonstrate the potential importance of in-cloud scavenging for some particulate radionuclides at least.

5.2. The wet deposition of gases

It is assumed often that Henry's law, as defined in Equation (14), controls the wet deposition of gases:

$$p = k^* c \tag{14}$$

where k^* is the Henry's law constant; and p and c are the respective concentrations of the gas in water and air when at equilibrium. Consequently, the potential for deposition would depend largely on the solubility of the gas.

There is little information on the scavenging of vapour-phase iodine by rain (Chamberlain, 1991), although Whitehead (1984) measured washout factors (defined as the ratio of the concentrations in unit masses of rain and air) equal to about 30 for stable iodine. These measurements related to iodine vapour, presumably in elemental form, and Chamberlain (1991) calculated that the removal rate was no greater than the equivalent dry deposition velocity.

Pasquill and Smith (1983) quoted the results of Ranz and Marshall (unreferenced) in which the wet deposition of iodine vapour was consistent with a scavenging coefficient equal to around $10^{-4}\,s^{-1}$. They noted also that this rate of removal was comparable to the dry deposition rate. Pasquill and Smith went on to criticise the frequently adopted use of the Henry's law constant for calculating equilibrium concentrations in rainfall and suggested that it might be more appropriate to consider the rates at which soluble gases dissolve in water. Consequently, when considering wet deposition of iodine it might be very important to consider whether uptake into cloud elements is important (i.e. whether any released material has become mixed within rain clouds or whether it has been restricted to below-cloud level).

There are likely to be some differences in the wet deposition rates of the different chemical species of atmospheric iodine; however, these are difficult to measure and there has been little work carried out in this area. It might be expected that iodine in organic form (e.g. methyl iodide) will have a low wet deposition rate in comparison to reactive forms of iodine, as observed in the case of dry deposition.

There is little known also about the deposition of iodine vapour in snow, although there is some evidence that the below-cloud scavenging of other gases by snow might be higher than otherwise expected.

5.3. Retention of deposited material by surfaces

A major factor when considering wet deposition in urban areas is the impact of run-off and how much deposited material is likely to be retained on surfaces (see Figure 4). While certain types of surface, such as areas of

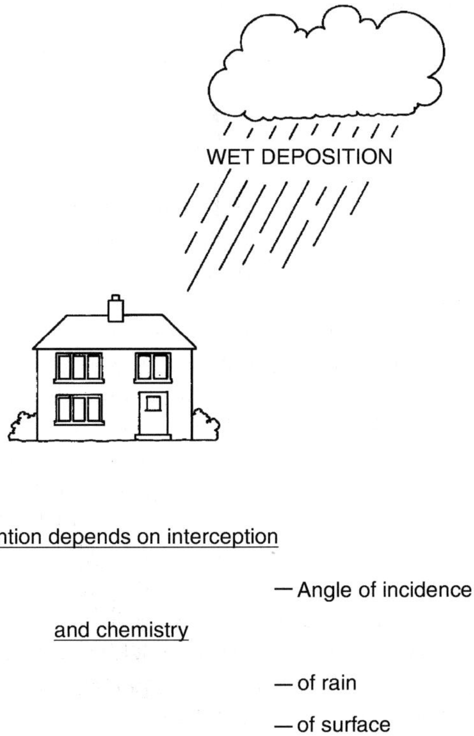

Figure 4 Summary of factors affecting retention of material deposited in rain.

grass, are not prone to run-off, drainage from buildings and roads is usually significant.

Roed (1987b) was undertaking experiments on run-off at the time of the Chernobyl accident and used these experiments to evaluate the interception of wet-deposited radionuclides by roofing materials. For those roofing materials that had not been treated with silicone, 60–90% of deposited Cs was retained (see Section 4.1 of Chapter 5 for details). Similar observations were noted for Ru, La and Ba also, but Roed found that iodine was retained only by red roof tiles (at about 50%). Jacob et al. (1987) undertook measurements in Germany following the Chernobyl accident and they found results similar to Roed's. They noted also that retention was found to decrease as the roof slope increased.

5.4. Deposition in fog or cloud

The deposition of material in fog droplets (sometimes termed 'occult deposition') can be important in certain circumstances (Dollard et al., 1983). Fog deposition incorporates aspects of both wet deposition in terms of nucleation processes and dry deposition in terms of aerodynamic transfer. The mechanisms for the deposition of fog droplets can be regarded as similar to those for the dry deposition of large particles, and the nucleation processes involved in their formation mean that the concentration of contaminants within the droplets can be much higher than in falling raindrops.

Fog droplets might be scavenged by falling rain also and this process has been termed the 'seeder–feeder' mechanism (Carruthers and Choularton, 1983). This process has been recognised as being especially important in enhancing the deposition of particulate pollutants in upland regions that are exposed to wind-driven cloud (Dore et al., 1992).

The relevance of fog deposition in inhabited areas will depend on the frequency of occurrence of fog. Fog can occur for various reasons; it can be due to the effects of advection (e.g. cooling due to mixing or by an air mass encountering a cooler surface), the evaporation of falling raindrops or by radiative heat transfer. Fogs are often assumed to be more frequent in towns and cities than in the surrounding countryside, because of the larger number of condensation nuclei available. However, the definition of fog in terms of visibility often confuses the issue and often it is only the visibility that is lower in urban areas (Oke, 1978). In addition, some studies have reported a decrease in the frequency of fog as building density increases (Sachweh and Koepke, 1995) due to the urban heat island effect. Consequently, the likelihood of fogs occurring in a given urban area is highly site specific. Moreover, the uncertainties associated with deposition fluxes when a fog does occur will be at least as great as those associated with dry deposition.

6. The Resuspension of Deposited Material

Material deposited on a surface can become airborne again and this process is termed resuspension (or re-entrainment). It is important to note that the term resuspension is used generally to describe the process in which any surface contaminant becomes airborne and it has not been restricted to materials that have been deposited via atmospheric mechanisms only. Resuspension has been regarded as potentially important in leading to both an inhalation dose and a spread of contamination. The relative importance of these effects will depend on the size of the particles resuspended, with small particles being dominant in causing an inhalation risk and large particles being most important in the spread of contamination.

Resuspension is usually modelled in terms of the resuspension rate (R_R) or the resuspension factor (K). Where:

$$R_R = \frac{\text{Resuspension flux (Bq m}^{-2}\text{ s}^{-1})}{\text{Surface deposit (Bq m}^{-2})} \tag{15}$$

and

$$K = \frac{\text{Air concentration (Bq m}^{-3})}{\text{Surface deposit (Bq m}^{-2})} \tag{16}$$

R_R is ideal for use in modelling studies, since it enables the deposited material to be considered as a secondary source and allows the prediction of both the inhalation dose and the downwind transport of deposited material. However, it is difficult to measure experimentally. K has been used commonly to predict inhalation dose and it is easier to measure than R_R. Its use is quite restricted however, since it does not take into consideration airborne material advected from upwind sources (including that originating from resuspension). In practice, neither should be ruled out completely for modelling purposes and both K and R_R have their uses.

Resuspension is quite difficult to measure and enormous ranges of values of K and R_R have been reported from a relatively small number of studies (Sehmel, 1980b, 1984; Nicholson, 1988b). Parts of the ranges might be explained by the different conditions in individual studies, although it seems likely that there might be significant errors associated with some studies also (Nicholson, 1988b), reflecting the experimental difficulties.

6.1. Factors affecting resuspension

Resuspended contamination has been found to be associated usually with host particles that were previously resident on the underlying surface (e.g. Sehmel, 1976a; Nicholson, 1988b). Consequently, the ease at which a contaminant can become resuspended in the atmosphere is related often to the characteristics of the resident surface material, rather than the depositing species.

Even the largest of particles can lift off from or move across a surface. However, gravitational effects may restrict the material from becoming resuspended properly in the atmosphere and it might be deposited very soon after becoming airborne. In this respect, thresholds for entrainment have been equated in terms of the settling velocity of the particle and the friction velocity (Gillette et al., 1974), which is a measure of the tangential velocity of the turbulent eddies in the free air. In practice, this means that only particles less than around 100 μm diameter can become fully resuspended in the atmosphere at normal wind speeds, while particles in the diameter range ~100–500 μm diameter are restricted to skipping across the surface (termed saltating). Larger particles still are restricted to rolling or creeping across the surface (Nicholson, 1988b).

Saltation and surface creep are likely to cause a local spread of contamination only, as was evident in Goiania following the rupture of a ^{137}Cs source (Pires do Rio et al., 1994). However, Nicholson and Fulker

(1994) demonstrated the potential for large resuspended particles to reach significant heights in the turbulent surface layer at Sellafield, UK. Their results demonstrate the potential for the spread of contamination via resuspension and this might be of concern when considering the local recontamination of cleaned areas. It is only the truly suspended smaller material that is likely to present the potential for an inhalation hazard however, not least because it is inhaled more readily.

Particulate material on a surface might be attached to it due to a number of attractive forces. These might include electrostatic and surface tension effects, and van der Waal's forces (e.g. Hinds, 1986). There are various lift forces that can overcome these attractive forces, including those originating from wind action and mechanical effects (Garland, 1979). Such mechanical effects might include vehicle or pedestrian movement, cleaning and construction activity. In addition, they might include collisions involving wind-generated saltating particles (Gillette, 1976), which in turn could lead to a cascade effect. In the urban environment, vehicular activity has been regarded as being possibly the most important mechanical action leading to resuspension in many circumstances.

In the case of wind-generated resuspension, lift forces might occur due to the penetration of turbulent eddies through the laminar sub-layer (e.g. Corn and Stein, 1965; Wilmarth and Lu, 1972). The shape of the particle is important in determining the area on which the lift forces can act and for particles of the same mass, the more irregularly shaped ones would have the largest effective areas. Consequently, they would be expected to resuspend more easily than more symmetrically shaped ones. Likewise, area increases with particle size and particles will become more easily resuspended with increasing size until gravitational restraints become an issue (see above). In this respect, the potential for resuspension is likely to increase with decreasing particle density for particles of the same mass (i.e. as the particles become correspondingly larger).

In particle deposition studies and elsewhere, an aerodynamic diameter is used often. This is equal to the diameter of a spherical particle of unit density that would have the same sedimentation (i.e. terminal settling) velocity. In this respect, sedimentation velocity decreases with increasing particle irregularity and decreasing particle density, for particles of the same mass. However, these two parameters counteract one another when considering resuspension, so that there are inherent problems in using the aerodynamic diameter in such cases. This is something that has been overlooked frequently in resuspension studies.

6.2. Wind-generated resuspension

Most experimental studies are in general agreement that resuspension fluxes increase with increasing wind speed (e.g. Nicholson, 1993; Giess et al., 1994).

The precise relationships differ, although most relate the resuspension flux to the wind speed raised to some power, usually in the range 1–6 and mostly >3 (Sehmel, 1984). Threshold wind speeds have been reported in soil erosion studies (e.g. Gillette, 1983), although such thresholds are less apparent for particle resuspension (e.g. Garland, 1983). One possible explanation is that soil erosion tends to depend on the occurrence of saltating particles, which impart momentum on to the surface material and help to erode the underlying surface (Nicholson, 1988b). On the other hand, resuspension involves particles that are less tightly bound to the surface and these do not require saltating material to enable them to become airborne.

In nearly all experimental studies, wind-generated resuspension has been found to decrease with increasing time after deposition or contamination (e.g. Garland, 1979; Nicholson, 1993). One explanation for this time dependence is that the most easily resuspended particles are removed first and this leaves those that are more tightly bound to the surface. Consequently, there are greater uplift forces required to dislodge the more tightly bound particles and statistically these occur less frequently as the required force becomes greater.

The time dependence of resuspension requires careful consideration, since under even quite constant conditions of wind flow (including those replicated in a wind tunnel), it is evident that there must be fluctuations in the factors determining whether a particle becomes resuspended or not. Some models have described the resuspension rate as being proportional to the reciprocal of time (e.g. Loosmore, 2003) or as a negative exponential (e.g. Linsley, 1978). Time dependence has been described in various theoretical models in terms of statistical distributions in the surface-particle interactive forces and the lift forces necessary for resuspension. One model (Reeks et al., 1988; Reeks and Hall, 2001) has described particles as being resident in potential energy wells with fluctuating surface-particle adhesion forces. This has been combined with a statistical distribution of turbulent eddies and a dependence of the resuspension rate on the reciprocal of time has been determined.

A number of wind tunnel studies have indicated a decline in the resuspension rate in proportion to the reciprocal of time also (Garland, 1979; Nicholson, 1993; Giess et al., 1994, 1997). Some of these (e.g. Nicholson, 1993) have found this decline to be rapid even over a period of seconds to minutes and that a significant fraction of material (several percent) might become resuspended during this time. How this relates to deposition following a contamination event is debatable, but it is possible that there could be a period following a deposition episode when optimum conditions for resuspension occur and this could lead to a short-term highly elevated air concentration.

Nicholson (1993) considered resuspension from two different surfaces (concrete and grass). The results are very variable but there was some indication of a more rapid decline in the resuspension rate from the concrete surface. This would be consistent with the smoother surface having fewer sheltered sites necessary to maintain resuspension over time, although such an explanation is speculative and a better understanding of resuspension from different types of surfaces is needed still.

It is useful to note that over much longer periods, Garland and Playford (1991) found that resuspension originating from Chernobyl declined in proportion to the reciprocal of time rather than as a negative exponential. Whilst it is tempting to relate this observation to the above-mentioned modelling studies, there were probably a number of factors causing this observation and a clear understanding of the processes leading to a long-term decline in the resuspension rate has yet to be achieved.

It is unlikely to be appropriate to ascribe any single resuspension rate or factor when determining the effects of an accidental contamination episode, since resuspension is highly dependent on the time since contamination. Moreover, certain conditions (e.g. high wind speeds) can result in temporarily elevated air concentrations that cannot be described by a resuspension rate that decreases steadily with time. Consequently, a range of values might be needed for modelling purposes according to the application, although there remains a need to understand the mechanisms controlling resuspension better if this range is to be narrowed.

6.3. Resuspension from roads

Vehicular activity is a potentially important resuspension mechanism for particulate materials (e.g. Nicholson, 1988b) and many early studies in this area were driven by the need to evaluate potential inhalation doses following a radioactive contamination event (e.g. Sehmel, 1973). Traffic-generated resuspension studies have been undertaken because of the potential for a dust nuisance to occur also, although research in this area has been concerned primarily with unpaved (i.e. dirt) roads (e.g. Dyck and Stukel, 1976). More recently, resuspension from roads has received attention due to the potential health effects of respirable particulate material (i.e. PM_{10}; Venkatram, 2000). Unfortunately, many of these later studies have focussed on the overall effect of vehicular activity on airborne particle concentrations and there has not been any clear distinction between resuspension and exhaust emissions. Consequently, such studies have provided only a limited insight into the physical mechanisms that lead to resuspension.

The potential for material to become resuspended from paved roads will depend on whether the road surface is wet or not. In the case where the road surface is covered in water, then it is reasonable to assume that the resuspension of surface material might be via tyre spray. Tyres are designed to displace water, and in achieving this, quite large droplets are generated either directly or when the displaced water hits the wheel arch. Because of their size, it is expected that these droplets would be deposited very close to the roadway (or on the vehicle itself). This has been demonstrated by vegetation damage, caused by road salt applied as a de-icing agent, being restricted to the close vicinity of roads (e.g. Smith, 1970). Nicholson and Branson (1990) reported measurements of respirable particles containing sodium or chloride in urban air, but were unable to relate these to applications of road salt (used as a de-icing agent). Their results support the assumption that the generation of tyre spray droplets would be unlikely to be a source of small airborne particles.

In dry conditions, resuspension from roads will depend on parameters relating to the road surface, the nature of the particulate material (i.e. size and shape) and the speed and type of vehicle being driven. The passage of the vehicle can cause resuspension via either induced turbulence or the shearing action of the tyres on the road surface (Sehmel, 1973).

As in the case of wind-generated resuspension, the resuspension of large particles by vehicle-induced turbulence would be expected to be faster than for small particles. The rate of removal of fluorescent particles from a paved road surface due to vehicle-induced turbulence was measured by Nicholson et al. (1989) and Nicholson and Branson (1990). They noted that the removal of material was not instantaneous, but continued with subsequent vehicle passes and this is analogous to the time dependence of wind-generated resuspension. Whilst a precise relationship between the amount of material removed from the road surface and the number of vehicle passes cannot be determined from their results, it is apparent that up to around 60% was removed during the first vehicle pass in some of their experiments. Nicholson and Branson found that this amount was higher for larger particles and greater vehicle speeds, as might be expected.

Sehmel (1973) estimated resuspension from road surfaces also by measuring downwind air concentrations and calculating resuspension fluxes using dispersion–deposition equations. He reported levels of resuspension up to around 1% per vehicle pass for a small truck and a decline in resuspension with time after deposition. The difference between Sehmel's results and those of Nicholson and Branson might be due to material in his experiments being deposited before reaching the air samplers or it becoming entrained in the vehicle's wake. Nevertheless, these studies confirm the importance of vehicle-generated turbulence as a potential mechanism for resuspension.

The presence of any surface moisture on an otherwise apparently dry road could be important in affecting the levels of material resuspended, as it is possible that surface tension effects could increase particle adhesion. This effect was noted by Nicholson et al. (1989) in their measurements using fluorescent particles. It might be expected also that material would be removed more easily from smoother surfaces and Sehmel (1976b) found this to some extent by looking at resuspension by traffic on a grass-covered area. In addition, Sehmel reported that a truck resuspended more material than a car, which he attributed to the greater levels of turbulence associated with the larger vehicle.

Tyre shear could cause resuspension due to the momentum imparted on the surface material. However, the rotation of the wheel of a vehicle generates turbulence also as air is squeezed out from under the tyre when the vehicle moves forward. In practice, it is difficult to separate these two effects. Nicholson and Branson (1990) reported that resuspension rates were slightly higher only for small particles (4 μm diameter), when a car was driven through an area of tracer. Sehmel (1973) however reported higher differences in his series of experiments. It is difficult to generalise on the relative importance of tyre shear and induced turbulence in initiating resuspension from roads, but it would seem possible that either could dominate according to the conditions.

Recent studies undertaken to calculate roadways as sources of PM_{10} have included such techniques as using a dual non-depositing tracer (Clairborn et al., 1995) or measuring concentration gradients both upwind and downwind of a roadway (Venkatram et al., 1999). Unfortunately, these experiments have evaluated only total PM_{10} emissions, which include exhaust emissions. However, Etyemezian et al. (2003a) used a lidar-based measurement system (TRAKER: Testing Re-entrained Aerosol Kinetic Emissions from Roads; as described by Kuhns et al., 2001; Etyemezian et al., 2003b) to survey vehicle-generated PM_{10} resuspension from paved roads in northwestern USA (see Table 2). The technique involved estimating

Table 2 Emission factors for paved roads (Etyemezian et al., 2003a).

	Emission factor ($g\,km^{-1}$ vehicle travelled)
Summer	
Residential roads (~11 $m\,s^{-1}$)	5.2
Interstate highways (~25 $m\,s^{-1}$)	3.0
Winter	
Residential roads (~11 $m\,s^{-1}$)	5.9
Interstate highways (~25 $m\,s^{-1}$)	4.9
	Emission factor/vehicle speed ($g\,km^{-1}/m\,s^{-1}$)
Summer (residential roads)	0.71
Summer (interstate highways)	0.13
Winter (residential roads)	0.78
Winter (interstate highways)	0.21

resuspension fluxes from measurements made behind the front wheels in comparison to others made in front of the bumper.

Etyemezian et al. surveyed over 400 km of paved roads in southwestern Idaho and found that their results spanned a narrow range with measured emission factors ranging from 3.0 to 8.0 g (vehicle) km^{-1} travelled. They found that emission factors were higher in winter than in summer; were higher in urban areas than in rural ones; and were lowest for the roads with the highest vehicle speeds (see Table 2). Interpretation of Etyemezian et al.'s results is difficult however, since as the authors acknowledge, road dust levels, vehicle speed and traffic density are interdependent. Kuhns et al. (2003) used the same TRAKER technique also and found that cleaning had little effect on reducing the resuspension of PM_{10} from paved roads, indicating the difficulty of removing small (i.e. respirable) particles from the road surface.

While the relatively few experimental data available on traffic-induced resuspension can be explained in terms of expected mechanisms, there remains a need for a resuspension model that can provide a quantitative prediction of effects. Part of the problem is that although vehicle wakes have been studied in the past, most of the work has been undertaken by aerodynamicists with the aim of determining only the effects of vehicle drag. Huggins et al. (2003) have gone some way towards characterising vehicle-generated turbulence by undertaking measurements using a sonic anemometer and they noted regimes of turbulence for a passing vehicle that were similar to those for high-speed trains. However, the wide range of road surfaces, and vehicle types and speeds, means that any generalisation for modelling purposes would need a much larger experimental data set.

As in the case of wind-generated resuspension, it is not possible to include a single value to allow for the effects of vehicle-generated resuspension. This reflects partly the lack of knowledge regarding the mechanisms that control resuspension, but also that it is likely to vary enormously according to traffic speed and density. Nevertheless, resuspension from roads is a potentially important contributor to dose in the aftermath of a contamination incident, especially in an urban area.

REFERENCES

Allen, A. G., R. M. Harrison, and K. W. Nicholson. (1991). Dry deposition of fine aerosol to a short grass surface. *Atmospheric Environment*, 25A, 2671–2676.

Amiro, B. D., and F. L. Johnston. (1989). Volatilization of iodine from vegetation. *Atmospheric Environment*, 23, 533–538.

Andersson, K. G., J. Roed, and C. L. Fogh. (2002). Weathering of radiocaesium contamination on urban streets, walls and roofs. *Journal of Environmental Radioactivity*, 62, 49–60.

Andronache, C. (2003). Estimated variability of below-cloud aerosol removal by rainfall for observed aerosol size distributions. *Atmospheric Chemistry and Physics*, 3, 131–143.

Astrup, P., S. Thykier-Nielsen, and T. Mikkelsen (2005). *In-Town Dispersion Calculations with RIMPUFF and UDM.* Risø-R-1539(EN). Risø National Laboratory, Roskilde, Denmark.

Atkins, D. H. F., R. C. Chadwick, and A. C. Chamberlain. (1967). Deposition of radioactive methyl iodide to vegetation. *Health Physics*, 13, 91–92.

Barry, P. J., and A. C. Chamberlain. (1963). Deposition of iodine onto plant leaves from the air. *Health Physics*, 9, 1149–1157.

Berkowicz, R. (2000). OSPM-A parameterised street pollution model. *Environmental Monitoring and Assessment*, 65, 323–331.

Best, A. C. (1950). The size distribution of raindrops. *Quarterly Journal of the Royal Meteorological Society*, 76, 16–36.

Brandt, J., T. Mikkelsen, S. Thykier-Nielsen, and Z. Zlatev. (1996). The Danish Rimpuff and Eulerian accidental release model (the DREAM). *Physics and Chemistry of the Earth*, 21, 441–444.

Byrne, M. A., and S. G. Jennings. (1993). Scavenging of sub-micrometer aerosol particles by water drops. *Atmospheric Environment*, 27A, 2099–2105.

Cambray, R. S., P. A. Cawse, J. A. B. Gibson, P. Johnson, G. N. J. Lewis, D. Newton, L. Salmon, and B. O. Wade. (1987). Observations on radioactivity from the Chernobyl accident. *Nuclear Energy*, 26, 77–101.

Carruthers, D. J., and T. W. Choularton. (1983). A model of the feeder-seeder mechanism of orographic rain including stratification and wind-drift effects. *Quarterly Journal of the Royal Meteorological Society*, 109, 575–588.

Chamberlain, A. C. (1953). *Aspects of Travel and Deposition of Aerosol and Vapour Clouds.* AERE HP/R 1261. HMSO, London.

Chamberlain, A. C. (1991). *Radioactive Aerosols.* Cambridge Environmental Chemistry Series 3, Cambridge University Press, Cambridge, UK.

Chamberlain, A. C., and R. C. Chadwick. (1966). Transport of iodine from atmosphere to ground. *Tellus*, 17, 226–239.

Chate, D. M., and T. S. Pranesha. (2004). Field studies of scavenging by rain events. *Journal of Aerosol Science*, 35, 695–706.

Clairborn, C., A. Mitra, G. Adams, L. Bamesberger, G. Allwine, R. Kantamaneni, B. Lamb, and H. Westberg. (1995). Evaluation of PM_{10} emission rates from paved roads and unpaved roads using tracer techniques. *Atmospheric Environment*, 29, 1075–1089.

Clark, M. J., and F. B. Smith. (1988). Wet and dry deposition of Chernobyl releases. *Nature*, 332, 245–249.

Clarke, R. E. (1979). *A Model for Short and Medium Range Dispersion of Radionuclides Released to the Atmosphere.* Report NRPB-R91. National Radiological Protection Board, Harwell, Didcot, Oxon, UK.

Corn, M., and F. Stein. (1965). Re-entrainment of particles from a plane surface. *American Industrial Hygiene Association Journal*, 26, 325–336.

Coulter, C. T., and P. A. Eckhoff. (1998). *Comparison of CALPUFF with ISC3.* Technical Report EPA-45/R-98/020. US Environmental Protection Agency, Research Triangle Park, NC.

Dana, M. T., and J. M. Hales. (1974). Washout coefficients for polydisperse aerosols. In: *Precipitation Scavenging. Proceedings of a Conference at Champaign, Illinois, 14–18 October 1974*, US Department of Energy Technical Information Service, Springfield, VA, pp. 247–257.

Davenport, H. M., and L. K. Peters. (1978). Field studies of atmospheric particulate concentration changes during precipitation. *Atmospheric Environment*, 12, 997–1008.

Davidson, C. I., R. E. Honrath, J. B. Kidane, R. S. Tsay, P. A. Magewsk, W. B. Lyons, and N. Z. Heidam. (1987). The scavenging of atmospheric sulphate by Arctic snow. *Atmospheric Environment*, 21, 871–882.

Dollard, G. J., M. H. Unsworth, and M. J. Harve. (1983). Pollutant transfer in upland regions by occult precipitation. *Nature*, 302, 241–242.

Dore, A. J., T. W. Choularton, and D. Fowler. (1992). An improved wet deposition map of the United Kingdom incorporating the seeder-feeder effect over mountainous terrain. *Atmospheric Environment – Part A: General Topics*, 26A, 1375–1381.

Dyck, R. I. J., and J. J. Stukel. (1976). Fugitive dust emissions from trucks on unpaved roads. *Environmental Science and Technology*, 10, 1046–1048.

Etyemezian, V., H. Kuhns, J. Gillies, J. Chow, M. Green, K. Hendrickson, M. McGown, and M. Pitchford. (2003a). Vehicle based road dust emissions measurement (III): Effect of speed, traffic volume, location, and season on PM_{10} road dust emissions. *Atmospheric Environment*, 36, 4583–4593.

Etyemezian, V., H. Kuhns, J. Gillies, M. Green, M. Pitchford, and J. Watson. (2003b). Vehicle based road dust emissions measurements (I): Methods and calibration. *Atmospheric Environment*, 36, 4559–4571.

Flossman, A. I. (1998). Clouds and pollution. *Pure and Applied Chemistry*, 70, 1345–1352.

Garland, J. A. (1977). The dry deposition of sulphur dioxide to land and water surfaces. *Proceedings of the Royal Society (London)*, A354, 245–268.

Garland, J. A. (1979). *Resuspension of Particulate Material from Grass and Soil.* AERE-R9452. HMSO, London.

Garland, J. A. (1983). Some recent studies of the resuspension of deposited material from soil and grass. In: *Precipitation Scavenging, Dry Deposition and Resuspension* (Eds H. R. Pruppacher, R. G. Semonin, and W. G. N. Slinn). Elsevier, Amsterdam, pp. 1087–1097.

Garland, J. A., and L. C. Cox. (1982). Deposition of small particles to grass. *Atmospheric Environment*, 16, 2699–2702.

Garland, J. A., and L. C. Cox. (1984). The uptake of elemental iodine by bean leaves. *Atmospheric Environment*, 18, 199–204.

Garland, J. A., and K. W. Nicholson. (1991). A review of methods for sampling large airborne particles and associated radioactivity. *Journal of Aerosol Science*, 22, 479–499.

Garland, J. A., and K. Playford. (1991). Resuspension following the Chernobyl accident. In: *Precipitation Scavenging and Atmosphere–Surface Exchange* (Eds S. E. Schwartz and W. G. N. Slinn). Hemisphere, Washington, DC, pp. 1605–1614.

Giess, P., A. J. H. Goddard, and G. Shaw. (1997). Factors affecting particle resuspension from grass swards. *Journal of Aerosol Science*, 28, 1331–1349.

Giess, P., A. J. H. Goddard, G. Shaw, and D. Allen. (1994). Resuspension of monodisperse particles from short grass swards: A wind tunnel study. *Journal of Aerosol Science*, 25, 843–857.

Gillette, D. A. (1976). Production of fine dust by wind erosion of soil: Effect of wind and soil texture. *Proceedings of the Atmosphere–Surface Exchange of Particulate and Gaseous Pollutants, 4–6 September 1974*, Richland, Washington, DC, pp. 591–609. Energy Research and Development Administration Symposium Series. CONF740921, National Technical Information Service, Springfield, VA.

Gillette, D. A. (1983). Threshold velocities for wind erosion on natural terrestrial arid surfaces (a summary). In: *Precipitation Scavenging, Dry Deposition and Resuspension* (Eds H. R. Pruppacher, R. G. Semonin, and W. G. N. Slinn). Elsevier, Amsterdam, pp. 1047–1058.

Gillette, D. A., I. H. Blifford, and D. W. Fryrear. (1974). The influence of wind velocity on the size distributions of aerosols generated by the wind erosion of soils. *Journal of Geophysical Research*, 79, 4068–4075.

Graedel, T. E., and J. P. Franey. (1977). Field measurements of submicron aerosol washout by rain. In: *Precipitation Scavenging. Proceedings of a Conference at Champaign, Illinois, 14–18 October 1974*, US Department of Energy Technical Information Service, Springfield, VA, pp. 503–523.

Hall, D. J., A. M. Spanton, F. Dunkerley, M. Bennett, and R. F. Griffiths. (2000). *A review of dispersion model inter-comparison studies using ISC, R91, AERMOD and ADMS.* R&D Technical Report P353. UK Environment Agency, Swindon, Wilts, UK, ISBN 1 85705 276 5.

Hosker, R. P., and S. E. Lindberg. (1982). Review: Atmospheric deposition and plant assimilation of gases and particles. *Atmospheric Environment*, 16, 889–910.

Hinds, W. C. (1986). *Aerosol Technology*. Wiley Interscience, New York.

Huggins, T., R. Kinnersley, and C. Baker. (2003). *Personal Communication*. Department of Environmental Health Resources Management, University of Birmingham, UK.

Jacob, P., R. Meckbach, and H. M. Muller. (1987). Reduction of external exposure from deposited Chernobyl activity by run-off, weathering, street cleaning and migration in the soil. *Radiation Protection Dosimetry*, 21, 51–57.

Jost, P. T., H. W. Gaggler, U. Baltensperger, B. Zinder, and P. Haller. (1986). Chernobyl fallout in size fractionated aerosol. *Nature*, 323, 22–23.

Jenkin, M. E. (1992). *The Photochemistry of Iodine-Containing Compounds in the Marine Boundary Layer*. AEA Technology Report Number AEA-EE-0405. Harwell, UK.

Kauppinen, E. I., R. E. Hillamo, S. H. Aaltonen, and T. S. Sinkko. (1986). Radioactive size distributions of ambient aerosols in Helsinki, Finland during May 1986 after the Chernobyl accident. *Environmental Science and Technology*, 10, 1257–1259.

Kuhns, H., V. Etyemezian, M. Green, K. Hendrickson, M. McGown, K. Barton, and M. Pitchford. (2003). Vehicle based road dust emissions measurements – Part II: Effect of precipitation, wintertime road sanding and street sweepers on inferred PM_{10} emission potentials from paved and unpaved roads. *Atmospheric Environment*, 36, 4572–4582.

Kuhns, H., V. Etyemezian, D. Landwehr, C. MacDougall, M. Pitchford, and M. Green. (2001). Testing re-entrained aerosol kinetic emissions from roads (TRAKER): A new approach to infer silt loading on roadways. *Atmospheric Environment*, 35, 2815–2825.

Linsley, G. S. (1978). *Resuspension of the Transuranium Elements – A Review of Existing Data.* NRPB-R75. HMSO, London.

Loosmore, G. A. (2003). Evaluation and development of models for resuspension of aerosols at short times after deposition. *Atmospheric Environment*, 37, 639–647.

Loosmore, G. A., and R. T. Cederwall. (2004). Precipitation scavenging of atmospheric aerosols for emergency response applications: Testing an updated model with real-time data. *Atmospheric Environment*, 38, 993–1003.

Mason, B. J. (1971). *The Physics of Clouds*. Clarendon Press, Oxford, UK.

May, F. G. (1958). The washout of lycopodium spores by rain. *Quarterly Journal of the Royal Meteorological Society*, 84, 451–458.

Mitchell, D. L., and D. Lamb. (1989). Influence of riming on the chemical composition of snow in winter orographic storms. *Journal of Geophysical Research*, 94, 14831–14840.

McFiggans, G. J., J. M. C. Plane, B. J. Allan, L. J. Carpenter, H. Coe, and C. O'Dowd. (2000). A modelling study of iodine in the marine boundary layer. *Journal of Geophysical Research*, 105, 14371–14385.

Murakami, M., H. Chikashi, and M. Choji. (1981). Observation of aerosol scavenging by falling snow crystals at two sites of different heights. *Journal of the Meteorological Society of Japan*, 59, 763–771.

Nicholson, K. W. (1987). Deposition to caesium to surfaces of buildings. *Radiation Protection Dosimetry*, 21, 37–42.

Nicholson, K. W. (1988a). The dry deposition of small particles: A review of experimental measurements. *Atmospheric Environment*, 22, 2653–2666.

Nicholson, K. W. (1988b). A review of particle resuspension. *Atmospheric Environment*, 22, 2639–2651.

Nicholson, K. W. (1993). Wind tunnel measurements on the resuspension of particulate material. *Atmospheric Environment*, 27A, 181–188.

Nicholson, K. W. (1995). Physical aspects of bioaerosol sampling and deposition. In: *Bioaerosols Handbook* (Eds C. M. Wathes and C. S. Cox). CRC, London. Chapter 4.

Nicholson, K. W., and J. R. Branson. (1990). Factors affecting resuspension by road traffic. *The Science of the Total Environment*, 93, 349–358.

Nicholson, K. W., and J. R. Branson. (1993). *Field Measurements of the Below-Cloud Scavenging of Particulate Material*. AEA Technology Report Number AEA-EE-0515. Harwell, UK.

Nicholson, K. W., J. R. Branson, and P. Giess. (1991). Field measurements of the below-cloud scavenging of particulate material. *Atmospheric Environment*, 25A, 771–777.

Nicholson, K. W., J. R. Branson, P. Giess, and R. J. Cannell. (1989). The effects of vehicle activity on particle resuspension. *Journal of Aerosol Science*, 20, 1425–1428.

Nicholson, K. W., J. R. Branson, J. B. Hedgecock, I. R. Pomeroy, and J. D. Watterson. (1993). *The Dry Deposition of Particulate Materials in an Urban Environment*. AEA Technology Report Number AEA-EE-0424. Harwell, UK.

Nicholson, K. W., and T. D. Davies. (1987). Field measurements of the dry deposition of particulate sulphate. *Atmospheric Environment*, 21, 1561–1571.

Nicholson, K. W., and M. J. Fulker. (1994). The atmosphere–surface exchange of radionuclides at the Sellafield reprocessing plant. *Journal of Aerosol Science*, 25, 807–820.

Oke, T. R. (1978). *Boundary Layer Climates*. Wiley, London.

Pasquill, F., and F. B. Smith. (1983). *Atmospheric Diffusion*. Wiley, Chichester, UK.

Pires do Rio, M. A., E. C. S. Amaral, and H. G. Paretzke. (1994). Resuspension and redeposition of ^{137}Cs in an urban area: The experience after the Goiania accident. *Journal of Aerosol Science*, 25, 821–831.

Reeks, M. W., and D. Hall. (2001). Kinetic models for particle resuspension in turbulent flows: Theory and measurement. *Journal of Aerosol Science*, 32, 1–31.

Reeks, M. W., J. Reed, and D. Hall. (1988). On the resuspension of small particles by a turbulent flow. *Journal of Physics D*, 21, 574–589.

Roed, J. (1987a). Dry deposition in rural and urban areas in Denmark. *Radiation Protection Dosimetry*, 21, 33–36.

Roed, J. (1987b). Run-off from and weathering of roof material. *Radiation Protection Dosimetry*, 21, 59–63.

Roed, J., and P. Jacob. (1990). Deposition on urban surfaces and subsequent weathering. *Proceedings of a CEC Seminar on Methods and Codes for Assessing the Off-site Consequences of Nuclear Accidents*, EUR-13013/1, CEC, Luxembourg, pp. 335–356.

Ruijgrok, W., C. I. Davidson, and K. W. Nicholson. (1995). Dry deposition of particles. *Tellus*, 47B, 587–601.

Sachweh, M., and P. Koepke. (1995). Radiation fog and urban climate. *Geophysical Research Letters*, 22, 1073–1076.

SCOPE 50. (1993). Radioecology after Chernobyl. In: *Biogeochemical Pathways of Artificial Radionuclides* (Eds F. Warner and R. M. Harrison). Wiley, Chichester, UK.

Sehmel, G. A. (1973). Particle resuspension from an asphalt road caused by car and truck traffic. *Atmospheric Environment*, 7, 291–309.

Sehmel, G. A. (1976a). *^{238}Pu and ^{239}Pu Associated with Larger Than 'Respirable' Resuspended Particles at Rocky Flats during July 1973*. BNWL-2119. Pacific Northwest Laboratory, Richland, Washington, DC, pp. 96–98.

Sehmel, G. A. (1976b). *Particle Resuspension from Truck Traffic in a Cheat Grass Area*. BNWL-2000 – Part 3. Pacific Northwest Laboratory, Richland, Washington, DC, pp. 96–98.

Sehmel, G. A. (1980a). Particle and gas deposition: A review. *Atmospheric Environment*, 14, 983–1011.

Sehmel, G. A. (1980b). Particle resuspension: A review. *Environment International*, 4, 107–127.

Sehmel, G. A. (1984). Deposition and resuspension. In: *Atmospheric Science and Power Production* (Ed. D. Randerson). US Department of Energy Technical Information Service, Springfield, VA, pp. 533–583.

Sievering, H. (1982). Profile measurements of particle dry deposition velocity at an air–land interface. *Atmospheric Environment*, 16, 301–306.

Simmonds, J. R., G. Lawson, and A. Mayall. (1995). *Radiation Protection 72. Methodology for Assessing the Radiological Consequences of Routine Releases of Radionuclides to the Environment.* EUR 15760EN. CEC, Brussels.

Slinn, W. G. N. (1977). Precipitation scavenging: Some problems, approximate solutions and suggestions for future research. In: *Precipitation Scavenging. Proceedings of a Conference at Champaign, Illinois, 14–18 October 1974*, US Department of Energy Technical Information Service, Springfield, VA, pp. 1–60.

Slinn, W. G. N. (1984). Precipitation scavenging. In: *Atmospheric Science and Power Production* (Ed. D. Randerson). US Department of Energy Technical Information Service, Springfield, VA, pp. 466–532.

Slinn, W. G. N., and J. M. Hales. (1971). A re-evaluation of the role of thermophoresis as a mechanism of in- and below-cloud scavenging. *Journal of the Atmospheric Sciences*, 28, 1465–1471.

Slinn, W. G. N., P. C. Katen, M. A. Wolf, W. D. Loveland, L. F. Radke, E. L. Miller, L. J. Ghannam, B. W. Reynolds, and D. Vickers. (1979). *Wet and Dry Deposition and Resuspension of AFCT/TFCT Fuel Processing Radionuclides.* SR09810-10. Oregon State University, Corvallis. Available from the National Technical Information Service, Springfield, VA.

Smith, W. H. (1970). Salt contamination of white pine planted adjacent to an interstart highway. *Plant Disease Reporter*, 54, 1021–1025.

Sparmacher, H., K. Fulber, and H. Bonka. (1993). Below-cloud scavenging of aerosol particles: Particle-bound radionuclides – Experimental. *Atmospheric Environment – Part A: General Topics*, 27A, 605–618.

Stensland, G. S. (1977). Modelling rainwater contaminant concentrations due to hygroscopic particle washout. In: *Precipitation Scavenging. Proceedings of a Conference at Champaign, Illinois, 14–18 October 1974*, US Department of Energy Technical Information Service, Springfield, VA, pp. 794–812.

Thykier-Nielsen, S., and T. Mikkelsen. (1991). *RIMPUFF users guide. Version 30 (PC version).* Risø National Laboratory, Roskilde, Denmark.

Tiwary, A., H. P. Morvan, and J. J. Colls. (2006). Modelling the size-dependent collection efficiency of hedgerows for ambient aerosols. *Journal of Aerosol Science*, 37, 990–1015.

Underwood, B. (1987). Dry deposition to an urban complex. *Radiation Protection Dosimetry*, 21, 21–32.

US EPA. (2007). AERMOD Implementation Guide, US Environmental Protection Agency, Air Quality Assessment Division, Research Triangle Park, NC, www.epa.gov/scram001/7thconf/aermod/aermod_implmtn_guide_09jan2008.pdf

Venkatram, A. (2000). A critique of empirical emission factor models: A case study of the AP-42 model for estimating PM_{10} emissions from paved roads. *Atmospheric Environment*, 34, 1–11.

Venkatram, A., D. Fitz, K. Bumiller, S. Du, M. Boeck, and C. Ganguly. (1999). Using a dispersion model to estimate emission rates of particulate matter for paved roads. *Atmospheric Environment*, 33, 1093–1102.

Wesely, M. L., D. R. Cook, R. L. Hart, R. E. Speer, K. W. Nicholson, and T. D. Davies. (1988). Field measurements of the dry deposition of particulate sulphur – Discussion. *Atmospheric Environment*, 22, 198–199.

Whitehead, D. C. (1984). The distributions and transformations of iodine in the atmosphere. *Environment International*, 10, 321–339.

Wilmarth, W. W., and S. S. Lu. (1972). Structure of the Reynolds stress near the wall. *Journal of Fluid Mechanics*, 55, 65–92.

CHAPTER 3

Airborne Contamination Inside Dwellings

Miriam A. Byrne*

Contents

1. Introduction

It is well known that in the developed world (and, to an increasing extent, in the developing world), more than 80% of people's lives are spent indoors, whether it is in the home, office, school, workplace or public building (Simoni et al., 2003). Even time spent within an enclosed vehicle can be considered to be in an 'indoor' environment. For this reason, it is important that any assessment of radiation dose received by an individual or population following an accidental or deliberate outdoor release must

*Corresponding author. Tel.: +353 91 524411; Fax: +353 91 494 584
E-mail address: miriam.byrne@nuigalway.ie

School of Physics, National University of Ireland, Galway, Ireland

Radioactivity in the Environment, Volume 15
ISSN 1569-4860, DOI 10.1016/S1569-4860(09)00403-3

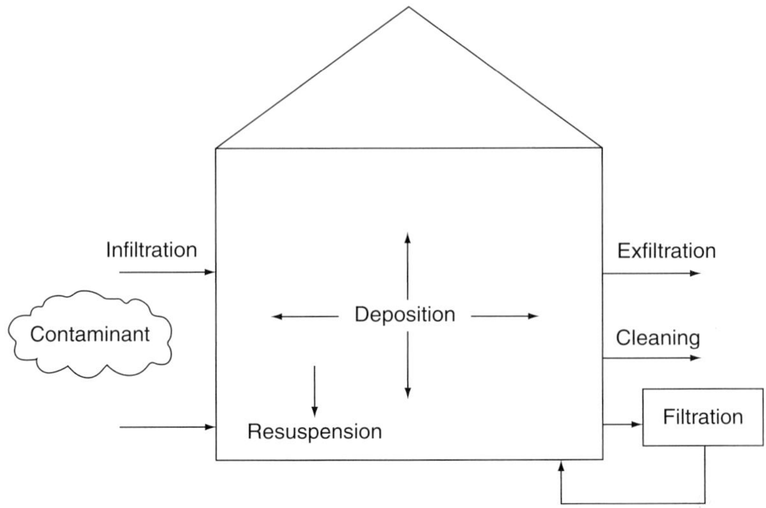

Figure 1 Schematic diagram of the routes of entry and fate of airborne pollutants within a dwelling (no indoor sources shown).

consider the ingression of the radioactive material into buildings. This chapter presents evidence of the occurrence of ingression, but as indicated in Figure 1, the actual indoor airborne concentration at any time is influenced by a number of time-dependent factors, such as contaminant deposition, contaminant removal and resuspension. Explanations are given of the parameters that can be used to describe these processes, and based on published studies, likely numerical values for these parameters are presented. Most of the data reviewed and discussed in this chapter are not derived from radiation-related studies, but are nevertheless of great significance in this context, as the physical characteristics of the pollutants, which in many of the cases cited are comparable to the characteristics of radioactive contaminants, are often the major determinant of their fate indoors.

2. Ingression of Contaminants into Dwellings

The building fabric presents a barrier to outdoor airborne pollutants, the effectiveness of which is determined by the permeability of the structure and the nature of particulate matter. Lange (1995) has demonstrated that the size distribution of ingressed particulates may be modified by the filtration process.

In addition to infiltration through crevices in the building structure, airborne particles may also enter a building through natural ventilation systems (i.e. windows and doors). The air exchange rate between indoors and outdoors, described by the parameter λ_a, is influenced by meteorological factors and dimensions of building openings, and governs the rate of infiltration. Air exchange rates in naturally ventilated buildings span a significant range, depending on geographically determined ambient temperature variations, as well as stringency of building regulations.

A measure of the indoor/outdoor concentration ratio for a specific pollutant is a useful tool for assessing the degree of protection against exposure afforded by the building shield, provided that the pollutant can positively be identified as being of outdoor origin. Numerous datasets for indoor/outdoor pollutant concentration ratios exist for many different types of enclosures, including vehicles (see Table 1 for a selected summary).

Generally, the data indicate that indoor concentrations of outdoor-originating pollutants are lower than concentrations observed outdoors, and several decades ago initiated a belief that the building fabric provided a 'shield' against outdoor pollution. However, Alzona et al. (1978) showed that, in the case of a single-room dwelling, and constant air exchange driving the passage of an outdoor pollutant cloud, the ratio of the time-integrated indoor and outdoor concentrations, C_i and C_o, respectively, is

$$C_i/C_o = f\lambda_a/(\lambda_a + \lambda_d) \qquad (1)$$

Table 1 A selected summary of indoor/outdoor air pollution concentration ratio measurements.

	Enclosure type (and number)	Air pollutant	Indoor/outdoor air pollutant concentration ratio
Andersen (1972)	Educational (1)	Sulphate aerosol	0.31
Alzona et al. (1978)	Residential (3)	Metal dusts (no indoor sources)	0.42 (fine aerosol fraction)
	Educational (5)		0.24 (coarse fraction)
	Vehicular (2)		
Roed et al. (1985)	Residential (17)	Cosmogenic Be-7, Ru-103 (weapons tests)	0.33 (average)
Roed and Cannell (1987)	Residential (1)	Chernobyl aerosol, Cosmogenic Be-7	0.78 (Be-7), 0.31 (Zr-95), 0.80 (Cs-137)
Engelmann et al. (1992)	Vehicular (5)	Surrogate 3 μm aerosol particles	0.013–0.705

where λ_d is the rate constant for aerosol deposition on indoor surfaces, and *f*, the building fabric filtration factor (which has a value of unity when the passage of a pollutant through the building fabric is uninhibited). The presentation of this relationship acknowledged for the first time the importance of the indoor deposition process in reducing the potential for indoor inhalation of outdoor-originating contamination. Further, Roed and Cannell (1987) determined values for the building fabric filtration factor, by studying the ingression of Chernobyl aerosol into a house, and deduced a value of unity for all particle sizes, demonstrating the importance of deposition, in combination with air exchange, in determining indoor pollutant concentrations.

The current state of knowledge regarding indoor deposition is discussed in the next section of this chapter. Later sections discuss pollutant transport processes which do not occur in every situation and are strongly dependent on human intervention, namely removal of pollutants from surfaces by cleaning or resuspension.

3. Deposition and Removal of Contaminants on Indoor Surfaces

3.1. Deposition

3.1.1. Background theory

Deposition is the process by which airborne contaminant concentrations are reduced, thereby reducing inhalation exposure. Therefore, in the case of air pollutants that are only of health concern when inhaled, deposition of pollutants onto surfaces can be of potential benefit. However, as discussed in Chapter 6, for radioactive contaminants, it has been shown by Andersson et al. (2002) that significant contributions to the whole-body dose may arise from radionuclides deposited on indoor surfaces and on the human body, and therefore in this context, the process of deposition is not necessarily a beneficial one in terms of exposure reduction.

The concept of a rate constant for aerosol deposition was introduced in the previous section. Generally, as described by Byrne et al. (1995), a value for this rate constant is experimentally determined by injecting and mixing a tracer aerosol in a room or test chamber and monitoring its exponential decay rate over time. Many published studies quote values for the deposition rate constant, but this has little value for intercomparison of data, as it is a value that is specific to an individual room or chamber. To allow comparison between data derived for different enclosures, it is necessary to introduce the concept of an indoor deposition velocity, V_{dep}. The rate constant for deposition has been linked to V_{dep}, by the following definition attributed to Kocher (1980), Nazaroff and Cass (1986) and Porstendorfer and Reineking (1992):

$$V_{\text{dep}} = V\lambda_{\text{d}}/S \tag{2}$$

where V is the volume and S the surface area of the room or building, including the ceiling.

3.1.2. A review of experimental research to establish likely ranges of deposition parameter values

An extensive review of the process of indoor deposition, describing background theory, experimental studies and simulation efforts, has been presented by Lai (2002). Based on the data reviewed in his paper, Table 2 has been generated to show a number of deposition studies carried out in buildings and the range of deposition velocities calculated. These studies encompass a wide range of laboratory aerosol generation and detection techniques, and also include studies that incorporate natural aerosol sources. It should be noted that Table 2 only makes reference to studies where deposition velocities have been reported, as distinct from particle loss rate coefficients.

The data presented in Table 2 show aerosol deposition velocities which span the range 10^{-7}–$10^{-2}\,\text{m}\,\text{s}^{-1}$. An additional review paper, by Porstendorfer and Reineking (1992), focuses entirely on radon progeny and quotes deposition velocity values of the order of $10^{-4}\,\text{m}\,\text{s}^{-1}$ for the unattached fraction and $10^{-2}\,\text{m}\,\text{s}^{-1}$ for the fraction that is attached to aerosol particles.

3.1.3. A review of experimental work to establish influencing factors

In order to examine the influencing factors on aerosol deposition, a review of aerosol deposition measurements in laboratory test chambers is presented, as these studies generally permit greater control of parameters than is possible in the case of those carried out in real buildings, and additionally allow one room surface to be studied in isolation. A summary of these measurements is shown in Table 3. Following the review procedure of Lai (2002), only those studies involving rectangular or cubic chambers (i.e. room-like enclosures) are included, and only those with dimensions comparable to that of a small room. In this section, reference is also made to a limited number of 'real building' cases in which individual parameters were studied.

3.1.3.1. Particle size. Figure 4 from the review paper of Lai (2002) is reproduced here as Figure 2 and presents a compilation of aerosol deposition velocity measurements as a function of particle size, based on various experimental studies that are summarised in Table 2 and Table 3.

Table 2 A summary of indoor deposition velocity measurements carried out in rooms.

	Building type	Ventilation	Active/passive sampling?	Particle composition	Particle size range	Deposition velocity range ($m\,s^{-1}$)
Raunemaa et al. (1989)	Residential and commercial (30)	Both	Both	Fe, Cl, S, K, Ca	Coarse: >1.5 μm, Fine: >0.2 μm	2×10^{-3}–5×10^{-3}
Nazaroff et al. (1990)	Institutional (5)	Mechanical	Both	Soot, Na, S, Pb, miscellaneous	0.05–0.211 μm	10^{-7}–2×10^{-5}
Roed and Cannell (1987)	Residential (1)	Overpressurised	Active	Cs, I, Ru, Rh, Cu, Cl, Zr, Nb (from Chernobyl)	0.4–1.4 μm	6.4×10^{-5}–1.9×10^{-4}
Roed and Cannell (1987)	Residential (1)	Overpressurised	Active	Be-7 (cosmogenic)	0.5 μm	7.1×10^{-5}
Sinclair et al. (1990)	Commercial (4)	Mechanical	Both	SO_4^{--}, Cl^-, Na^+, NH_4^+, K^+, Mg^{++}, Ca^{++}	Fine: >2.5 μm, Coarse: 2.5–15 μm	6×10^{-5}–2.2×10^{-2}
Fogh et al. (1997)	Residential (4)	Natural	Both	Tracer labelled silica and nebulised indium	0.5, 0.7, 2.5, 4.5, 5.5 μm, monodisperse	0.53×10^{-4}–3.89×10^{-4}
Thatcher and Layton (1995)	Residential (1)	Natural	Active	House dust	0.3–>25 μm	1.6×10^{-4}–2.7×10^{-3} [a]
Xu et al. (1994)	Residential (1)	Natural	Active	Tobacco smoke	0.09–1.25 μm	1.7×10^{-6}–2.9×10^{-5}

[a]Loss rates determined, and deposition velocities deduced by Thatcher and Layton (1995), based on an assumption that all particles fell to the floor.

Table 3 A summary of indoor deposition studies carried out in room-sized test chambers.

	Chamber size, and surface/volume ratio (m^{-1})	Particle size range (μm)	Air circulation type	Interior surface characteristics
Abadie et al. (2001)	Cube, side 0.6 m, S/V = 10	0.7, 1, 5	Single mixing fan	Glazing, linoleum, wood, wallpaper, carpet
Byrne et al. (1995)	Cube, side 2 m, S/V = 3	0.7, 2.5, 4.5, 5.4	Single mixing fan (22–43% turbulence)	Aluminium, wallpaper, carpet, astroturf
Harrison (1979)	Volume 0.2 m^3, S/V = 10.3	0.2–2.0	Natural convection	Smooth and sandpaper
Lai et al. (2002)	Cube, side 2 m, S/V = 3	0.7, 2.5, 4.5, 5.4	Three mixing fan speeds	Smooth aluminium, and pseudo 3D roughness elements
Morawska and Jamriska (1996)	Cube, side 1.44 m, S/V = 4.2	0.12–0.5 (radon)	Continuous flow, 0.37 and 0.74 air changes per hour	Smooth
Nomura et al. (1997)	0.75 × 0.75 × 1.8 m, S/V = 6.4	0.015, 0.03, 0.08, 0.3, 2.0	Continuous flow, 0.8, 1.7 and 2.4 air changes per hour	Smooth
Thatcher et al. (2002)	Test room, 2.2 × 2.7 × 2.4 m, S/V = 2.48	0.5–10	Four fan-generated air speeds, range <0.5–1.9 m s^{-1}	Textured walls and metal floor (a); textured dry walls and carpeted floor (b); (c) = (b) + furniture

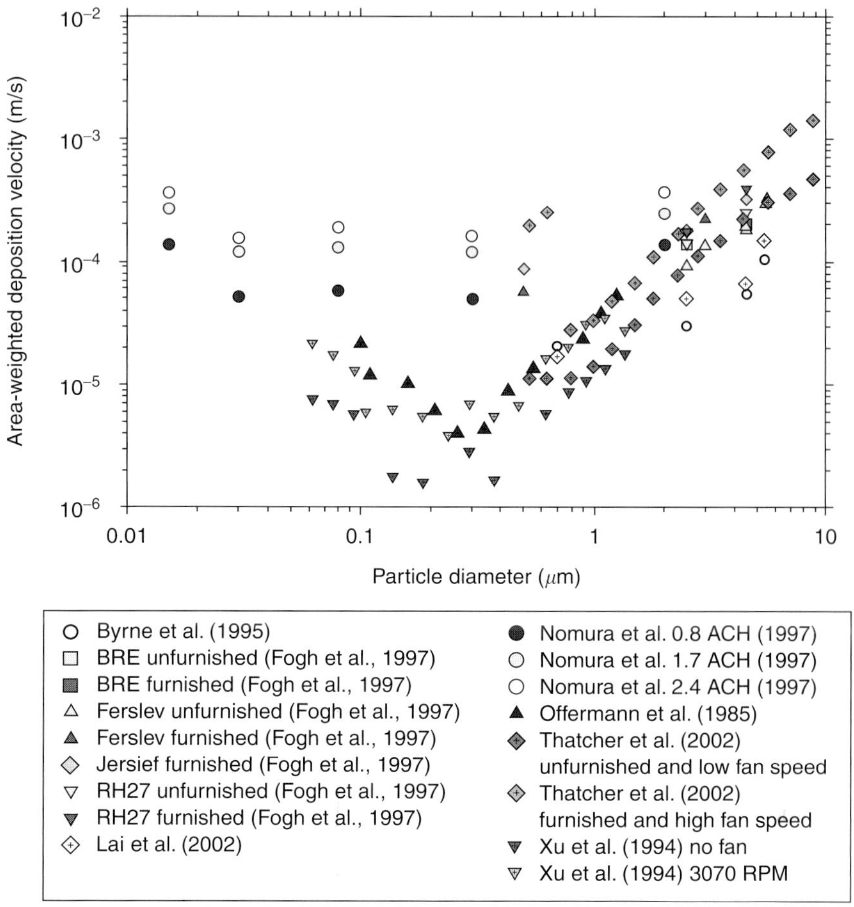

Figure 2 Measured indoor aerosol deposition velocities as a function of particle size, derived from studies carried out in rooms and room-sized test chambers (Adapted from Lai, 2002 with permission from Elsevier).

Although there is a considerable spread in the data, due to the wide range of experimental conditions considered, the figure shows the characteristic 'V' shape, which indicates that for very small particles (<0.1 μm), Brownian and turbulent diffusion are dominant deposition mechanisms, while for particles larger than 1 μm, gravitational settling is dominant. For particles between 0.1 μm and 1 μm, neither mechanism is dominant and therefore particles in this size range pose the greatest inhalation exposure risk as they have the longest residence time in air.

3.1.3.2. Surface orientation. The data shown in Figure 2 are largely derived from aerosol concentration decay curves, and are therefore

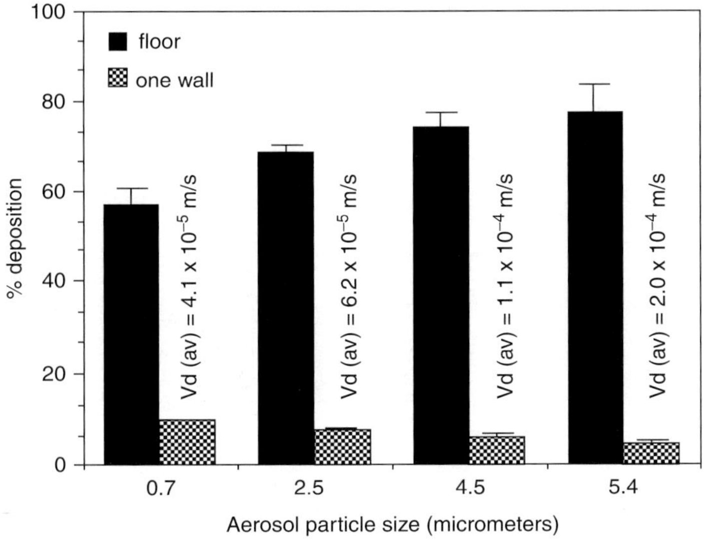

Figure 3 Observed proportion of the total aerosol particle mass deposited on vertical and horizontal surfaces for four particle sizes. Vd (av) denotes the average aerosol deposition velocity to all the chamber surfaces, and this is equivalent to the term V_{dep} used elsewhere in this chapter (from Byrne et al., 1995).

averaged over surfaces of various orientations (i.e. walls vs. floors vs. ceilings). The measurements of Byrne et al. (1995), through the use of tracer particles, permitted direct measurement of aerosol deposition on individual surfaces, and therefore the effect of surface orientation can be observed. Figure 3 is based on this work, and shows that, as might be expected, on horizontal upward-facing surfaces, the effect of gravity is strongest for largest particles. However, a reverse in this trend is seen for vertical surfaces.

3.1.3.3. Enhancement of surface area, through surface roughness or addition of furnishings. A number of studies, notably Fogh et al. (1997) and Thatcher et al. (2002), have demonstrated that aerosol deposition velocities are greater in furnished rooms than in unfurnished rooms. A number of factors contribute to this observation: first, on a macroscopic scale, the presence of furnishings simply constitutes extra surface area in a room and provides an additional impaction medium for particles, especially in the supra-micron diameter range. On a microscopic scale, the distorting effect of the rough surface on the air–surface boundary layer has an effect on the behaviour of particles within that layer.

From a particle's perspective, most surfaces, with the exception of glass and metal sheeting, appear rough, but nevertheless, there have been few systematic studies that have examined the effect of surface roughness on

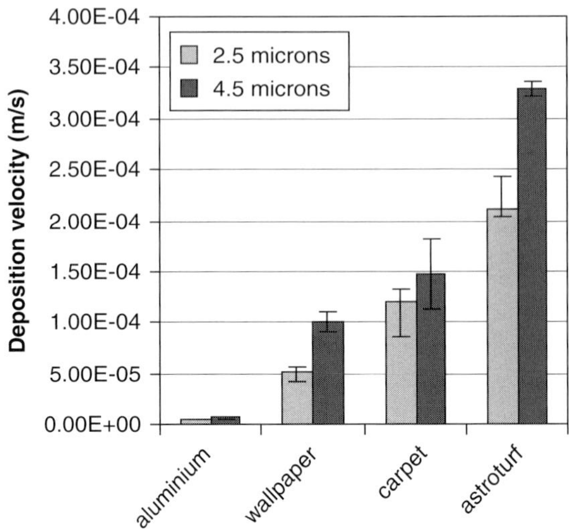

Figure 4 Observed variation, by Byrne et al. (1995), of aerosol deposition velocities on vertical aluminium, wallpaper, carpet and astroturf surfaces. Data are shown for 2.5 and 4.5 μm particles.

aerosol deposition. This may be due to analytical limitations, which prevent direct measurement of particle fluxes to rough surfaces, and therefore data can only be derived from loss rate coefficients. The most significant studies are cited in Table 3. Early work by Harrison (1979) employed five monodisperse particle sizes (ranging from 0.2 μm to 2 μm) in a 0.2 m^3 chamber, in which the surface was roughened by the application of latex paint. No discernible difference in aerosol deposition rate was noted, relative to smooth surfaces, but this may have been due to a limited analytical capability to detect the effect of a relatively small degree of roughness. Byrne et al. (1995) carried out measurements in a 8 m^3 test chamber with four monodisperse tracer particle sizes in the range 0.7–5.4 μm, and aluminium, wallpaper, carpet and astroturf surfaces, so that direct measurements on the rough surfaces were permitted; Figure 4 uses data derived from this work and shows a histogram of measured deposition velocities on rough vertical surfaces for 2.5 and 4.5 μm particles. Similar observations were made, using comparable particle sizes, on wallpaper, carpet and wood surfaces, by Abadie et al. (2001) albeit in a smaller chamber (0.216 m^3).

3.1.3.4. Circulation of air. A difficulty in intercomparing the studies cited above arises from the variability in flow conditions employed. Lai et al. (2002) highlight the complex interaction between surface effects and circulation effects in a study incorporating pseudo-3D roughness elements on the walls of the same chamber used by Byrne et al. (1995), and with the same particle sizes. A number of mixing fans were used to generate turbulence, and in some cases, reduced deposition velocity was observed on rough surfaces relative to smooth surfaces. It was postulated that this was due to the formation of a stable cavity vortex between the protrusion elements that prevented the interaction of the turbulent eddy current from the well-mixed core from reaching the roughness elements, and therefore deposition was reduced.

On a more macroscopic scale, in real-room experiments, where deposition velocities were merely derived from aerosol concentration decay curves, the presence of mixing fans has been seen to enhance aerosol deposition. In measurements reported by Thatcher et al. (2002), which employed polydisperse particles in the size range 0.5–10 μm, an enhancement in deposition loss rate by a factor ranging from 1.3 to 2.4 was observed, with the fan-generated airspeed in the room varying from $<5\,\mathrm{cm\,s^{-1}}$ to $19\,\mathrm{cm\,s^{-1}}$. Abu-Jarad and Sextro (1988) also report a significant reduction in the airborne concentration of radon progeny when a mixing fan is employed in a test house. In their study, they computed that deposition on the fan blades constituted a minor factor (less than 1%), but in other studies (e.g. Thatcher et al., 2002), 9–13% of aerosol deposition was attributed to this factor.

*3.1.3.5. **Electrostatic effects.*** Most of the data on aerosol deposition that has already been reported has been explained in terms of the effects of diffusion (Brownian and turbulent) and also gravitational settling. Limited studies have examined the effects of surface characteristics such as electrostatics on deposition. Liebermann and Rosinski (1962) lined the walls of test chambers with electrostatic materials, and McMurry and Rader (1985) studied the loss rate of aerosol particles in Teflon film bags. The outcome of both studies suggested that electrical forces might contribute significantly to particle loss, as singly charged particles exceeded loss rates of neutral particles by several orders of magnitude. Also, Shimada et al. (1989) carried out experiments with monodisperse particles in the size range 0.02–2 μm in a small metal tank to which an electrical field was applied, and observed that in the presence of the field, deposition rate increased by up to an order of magnitude.

3.2. Removal of contaminants from indoor surfaces

Traditionally, research on the effectiveness of cleaning of particulate deposits has concentrated on industrial clean rooms, pharmaceutical facilities and hospitals. However, with the existence in recent decades of threats of accidental and deliberate releases of radioactive and biochemical materials, research has been extended to other areas, including the domestic environment. As was stated in the previous section, from an aerosol particle's perspective, the vast majority of surfaces on which it deposits appear rough, and therefore the most efficient means of *in situ* removal of particulate from surfaces is through the generation of a controlled lift force, that is, through vacuum cleaning. While considerable research contributions have been made (notably Fogh et al., 1999) on recommending cleaning strategies, including vacuuming, for outdoor areas, following a radioactive contamination incident such as Chernobyl, similar studies indoors are scarce. One exception is the work of Cannell et al. (1987), who conducted experiments on the decontamination of carpet tiles impregnated with a 0.6 μm fluorescent tracer, using a light industrial dry vacuum cleaner. They found that for carpets vacuumed for just $20\,s\,m^{-2}$, 61% of the tracer was removed from carpets which had not been trodden on, while only 37% could be removed from those which were traversed 60 times. Furthermore, intensive decontamination (5 min) removed totals of 72% and 52% for the two carpeted surfaces.

Although quantitative data on intentional removal of particles from indoor surfaces is scarce, the important factors relating to the particles, surfaces and cleaning devices can be deduced from research in the occupational sector. Byrne (2000) has carried out an extensive review of studies incorporating suction devices, although the objective of these studies was not surface cleaning, but surface sampling, to assess the concentration of deposited material. Numerous factors were found to have significance; Table 4 summarises those research efforts that have investigated various vacuuming parameters and indicates whether significant differences in cleaning potential was observed when parameter values were varied. Parameters include dust loading, suction rate, relative humidity and electrostatic forces, particle size, dryness of contaminated surface, duration of cleaning, residence time of contamination and operator variability.

In the use of vacuum systems to remove particulate from surfaces, it is important to note that the exact contact mechanism between the device and the surface will determine whether particles are directed into the device for collection (the success of which depends on the filtration system within, as even the best high-filtration paper bags will have an efficiency of no greater than 75% at 0.3 μm, compared to 99.97% at 0.3 μm for HEPA filter media (D. E. Ristenbatt, Ristenbatt Vacuum Cleaner Service, Inc.)), or whether the agitation of the surface by the vacuum cleaning head results in resuspension into the room air. The latter as a possible enhanced route of exposure to radioactive dust was highlighted by Goddard et al. (1986) in the course of dust measurements in houses in Cumbria, UK. For this reason, in the context of removal of radioactive materials from surfaces, the application of a solid clay suspension to the surface (as described by Belfiore et al., 1984) might enhance the efficient removal of contamination by suction.

Table 4 A summary of experimental studies that investigated the efficiency of removal, by suction, of particulates from surfaces.

Parameter	Investigator(s)	Significant effect observed if parameter value varied? (Yes/No)
Dust loading	Royster and Fish (1964)	No
	Reynolds et al. (1997)	Yes
Vacuum cleaner suction rate	Royster and Fish (1964)	Yes
	Que Hee et al. (1985)	Yes
Relative humidity and electrostatic forces	Wang et al. (1995)	Yes
Number of sample passes	Que Hee et al. (1985)	Yes
Particle size	Byrne et al. (1998)	No
Dryness/greasiness of sampled surface	Wheeler and Stancliffe (1998)	Yes
	Royster and Fish (1964)	Yes
Residence time of contamination on surface	Kildesø et al. (1999)	Yes
Sample area	Royster and Fish (1964)	No
Operator variability	Royster and Fish (1964)	Yes
Sample surface type	Royster and Fish (1964)	Yes
	Leese et al. (1997)	Yes
	Kildesø et al. (1998)	Yes

4. Resuspension

4.1. Background theory

All deposited particles are potential sources for resuspension, when the surface on which they are deposited is disturbed, either mechanically or through enhanced airflow over the surface. Typical mechanical disturbances might be due to occupants walking or engaging in surface cleaning activities. In the indoor environment, air currents caused by normal room ventilation are expected to be a lesser source of particulate resuspension (Beyeler et al., 1999).

Extensive computational modelling of the process of particle resuspension from surfaces has been carried out by a number of researchers, but most notably Michael Reeks and colleagues; one example of their numerous publications is Reeks and Hall (2001). However, in purely analytical terms, resuspension can be described by a resuspension *factor* K and a *resuspension rate* R_R, as defined in Section 6 of Chapter 2 for outdoor conditions. Numerically, resuspension rates are orders of magnitude lower than deposition rates, but the actual amount of dust resuspended from a surface can contribute significantly to air concentration.

According to Biwer et al. (2002), a relationship between K and R_R can be derived as follows:

Consider that the variation with time in indoor airborne particle concentration (C_i) in a room of volume V (where V is the product of the room's surface area A and height H) is due to resuspension (which increases concentration), deposition and ventilation (which both decrease concentration), that is

$$V_{dep} C_i/dt = C_s \times A \times R_R - C_i \times A \times V_{dep} - C_i \times V \times \lambda_a \quad (3)$$

where C_s is the surface concentration of particulate material, λ_a the air exchange (ventilation) rate and V_{dep} the particle deposition velocity.

If both sides of Equation (3) are divided by V, the room volume, and steady-state conditions are assumed (i.e. $dC_i/dt = 0$), then

$$\frac{C_s \times R_R}{H} - \frac{C_i \times V_{dep}}{H} - C_i \times \lambda_a = 0 \quad (4)$$

and as $K = C_i/C_s$, Equation (4) can be rearranged to give

$$R_R = K(V_{dep} + (\lambda_a \times H)) \quad (5)$$

4.2. A summary of selected studies to establish likely ranges of resuspension parameter values

Biwer et al. (2002) present an extensive review of resuspension factor measurements, mostly relevant to the nuclear industry, and also present resuspension rates derived from the resuspension factors (based on Equation 5). These data are summarised in Table 5. Maximum and minimum values for resuspension rate are shown in each case; these are derived by assigning likely maximum and minimum values to the parameter V_{dep}, the indoor aerosol deposition velocity, in Equation 3.

Table 5 A summary of data, indicating the likely range of resuspension factor and resuspension rate values under various conditions in nuclear installations.

	Resuspension factor (m^{-1})	Range of resuspension rates (s^{-1})	Conditions
Brunskill (1967)	1.2×10^{-4}	7.7×10^{-7}–1.1×10^{-6}	4–6 persons walking
Fish et al. (1967)	1.9×10^{-4}	5.1×10^{-10}–5.1×10^{-7}	Vigorous sweeping
	3.9×10^{-5}	1.1×10^{-10}–1.1×10^{-7}	Vigorous walking
	7.1×10^{-4}	1.9×10^{-9}–1.9×10^{-6}	Light sweeping, fans operating
Ikezawa et al. (1980)	4×10^{-6}–2×10^{-5}	3.3×10^{-8}–2.2×10^{-7}	Spill clean-up activity
Jones and Pond (1967)	2×10^{-8}	1.9×10^{-10}–2.5×10^{-10}	No movement
	1×10^{-5}	9.5×10^{-8}–1.2×10^{-7}	Walking at 14 steps per minute
	5×10^{-5}	4.8×10^{-7}–6.1×10^{-7}	Walking at 36 steps per minute
Kvostov and Kostyakov (1969)	2.5×10^{-6}	4.2×10^{-8}–4.8×10^{-8}	No work
	2.2×10^{-4}	3.7×10^{-6}–4.3×10^{-6}	Floor scrubbing
Tagg (1966)	2.2×10^{-4} (personal sampling)	1.4×10^{-6}–2×10^{-6}	Walking at 100 steps per minute
	5.9×10^{-5} (area sampling)	3.8×10^{-7}–5.4×10^{-7}	Walking at 100 steps per minute

Note: Data are selected from a larger compilation presented by Biwer et al. (2002).

4.3. A review of published work to establish the influencing factors on aerosol resuspension

4.3.1. Human activity

'Human Activity' is a term that encompasses the mere presence of persons in a room, their potential to disturb surfaces through walking or simply through enhancement of air circulation, and also the deliberate disturbance of surfaces through procedures such as vacuuming.

The influence of human activity on resuspension factor and resuspension rates, due to the presence of persons and also their active cleaning of surfaces, is evident in Table 5; the fourth column documents the type of activity associated with each resuspension value recorded. In the domestic sector, many studies have shown the effect of human activity on resuspension of particles; resuspension factors estimated by Goddard et al. (1986) in Cumbrian houses in the north of England were a factor of 2 higher during the day ($2 \times 10^{-5}\,m^{-1}$) than at night, indicating the significance of human activity as a promoter of resuspension. Goddard et al. (1986) also recorded significant 'spikes' in real-time aerosol concentration data during vacuuming events in the houses, and similar observations were recorded by Raunemaa et al. (1989) in Finnish houses.

A striking representation of the influence of human activity of resuspension is presented by Thatcher and Layton (1995), who carried out measurements in a California dwelling. Figure 7 from their paper is reproduced below as Figure 5. The horizontal line represents the 'undisturbed' aerosol concentration, and the enhancement of that concentration for various particle size fractions and various human activity patterns is very clear.

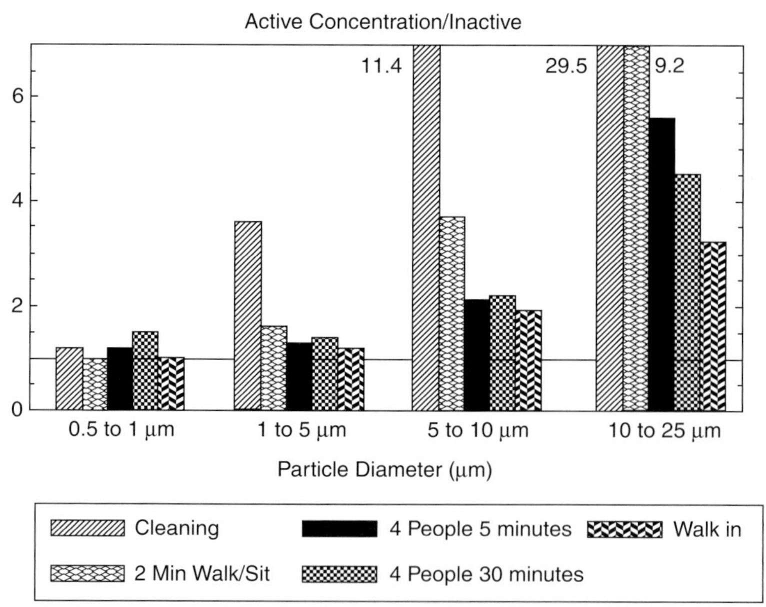

Figure 5 Depiction, by Thatcher and Layton (1995), of the elevation in aerosol concentration in a room as a result of the following activities: (1) vigorous vacuuming and housecleaning; (2) 2 min of continuous walking and sitting in the living area by one person; (3) 5 min of normal activity by four people; (4) 30 min of normal activity; and (5) one person walking in and out of the living area.

4.3.2. Particle size

An additional feature that is evident from the work of Thatcher and Layton (1995) is the influence of particle size on resuspension rate. In Figure 5, for all the activity patterns shown, the elevation in concentration for the largest particles (10–25 microns) is particularly striking, whereas the sub-micrometre particles appear to be essentially non-resuspendable. This observation, from which it can be concluded that resuspended particles are generally not of major respiratory health concern due to their relatively large size, can be explained by examining the theory of particle adhesion to surfaces. Particle adhesion has been extensively studied and it is generally well understood that the main adhesive forces are van der Waal's forces, electrostatic forces and/or surface tension forces of adsorbed liquid films (Hinds, 1982). These forces are influenced by particulate shape, size and composition, by ambient temperature and humidity, and by surface roughness (see Section 6 of Chapter 2). In general, the magnitude of adhesive forces is inversely proportional to particle diameter, but removal forces are proportional to a second or third power of diameter. This implies that as particle size decreases, it becomes increasingly difficult to remove particles from surfaces.

The work of Raunemaa et al. (1989) included one of the earliest observations of a shift in the size distribution of resuspended particles relative to the size distribution of depositing particles. More recent measurements by Andersson et al. (2004) involved the release of tracer particles with two different size distributions (2.5 and 0.7 μm, respectively) into test rooms, and the re-analysis of the room air (after the room surfaces had been agitated with a filter-less vacuum cleaner) after 24 h, 2 days and 5 days. For the particles that had an original mean diameter of 2.5 μm, no major change in size distribution was recorded during any of the resuspension events, but for the particles originally centred around 0.7 μm, their mean size evolved to 2.8 μm after 24 h and to 5.6 μm after 5 days.

4.3.3. Particle residence time on surfaces

The work of Raunemaa et al. (1989) notes a linear dependence of resuspension rate on surface accumulation time of aerosol material. In addition, the measurements of Andersson et al. (2004), described in Section 4.3.2, indicate that in the case of particles that were originally of mean diameter 2.5 μm, 60% of the 'resuspendable' fraction at 24 h was available after 48 h from a linoleum surface, and the same percentage was determined when attempting resuspension after 5 days from an acrylic carpet surface.

The above observations may, to some extent, be explained by Vasistat (1992), who states that the adhesion of particles to surfaces increases with time owing to the gradual increase in the contact area in which the attractive forces are operative. Although in the case of rigid particles depositing from air, the area of contact should reach its final value instantaneously, if the air contains moisture, the area increases over time due to capillary condensation which reinforces the adhesion of the particles to the wall surface.

4.3.4. Surface type: roughness and orientation

In test chamber measurements by Andersson et al. (2004), wood, plastic and carpet surfaces were 'contaminated' with polydisperse silica particles and subjected to disturbance via the dropping of weights (chosen to provide a comparable level of disturbance to a human footfall on a surface). Results indicated greater resuspension factors from wood and plastic than from wool, and were consistent with observations of Braun et al. (2002) that the concentration of particles in the zone close to the surface (and therefore available for resuspension) for a hard surface is greater than for an equally contaminated rough, but soft, surface (where the particles can penetrate into complex layers and become embedded).

A doctoral thesis published by ETH Zurich (Lengweiler, 2000) discussed an experimental and computational study of particle deposition and resuspension. Resuspension rates were evaluated in this work by contaminating surfaces with talcum powder particles (mean diameter 4.5 μm) and then subjecting the surfaces to aggressive vacuuming. The resuspension rates calculated were $2.9 \times 10^{-5}\,s^{-1}$ for floor surfaces, $6.6 \times 10^{-5}\,s^{-1}$ for wall surfaces and $8.2 \times 10^{-5}\,s^{-1}$ for ceiling surfaces, indicating that if all surfaces are subject to the same disturbance, the gravitational force acts most favourably for particles deposited on a downward-facing horizontal surface. This is, of course, not representative of the way in which 'real-world' surfaces are generally agitated.

REFERENCES

Abadie, M., K. Limam, and F. Allard. (2001). Indoor particle pollution: Effect of wall textures on particle deposition. *Building and Environment*, 36, 821–827.

Abu-Jarad, F., and R. G. Sextro. (1988). Reduction of radon progeny concentration in ordinary room due to a mixing fan. *Radiation Protection Dosimetry*, 24(1/4), 507–511.

Alzona, J., B. L. Cohen, H. Rudolph, H. N. Jow, and J. O. Frohliger. (1978). Indoor–outdoor relationships for airborne particulate matter of outdoor origin. *Atmospheric Environment*, 13, 55–60.

Andersen, I. (1972). Relationships between outdoor and indoor air pollution. *Atmospheric Environment*, 6(4), 275–278.

Andersson, K. G., C. L. Fogh, M. A. Byrne, J. Roed, A. J. H. Goddard, and S. A. M. Hotchkiss. (2002). Radiation dose implications of airborne contaminant deposition to humans. *Health Physics*, 82(2), 226–232.

Andersson, K. G., J. Roed, M. A. Byrne, H. Hession, P. Clark, E. Elahi, A. Byskov, X. L. Hou, H. Prip, S. K. Olsen, and T. Roed. (2004). *Airborne Contamination in the Indoor Environment and its Implications for Dose.* R-1462 (EN). Riso National Laboratory, Roskilde, Denmark.

Belfiore, A., G. Panciatici, and A. Lomoro. (1984). Removal of thorium and uranium from surfaces by attapulgus clay suspensions. *Inorganica Chimica Acta*, 94(1–3), 159–160.

Beyeler, W. E., W. A. Hareland, F. A. Duran, T. J. Brown, E. Kalinina, D. P. Gallegos, and P. A. Davis. (1999). *Residual Radioactive Contamination After Decommissioning: Parameter Analysis.* NUREG/CR-5512. United States Nuclear Regulatory Commission.

Biwer, B. M., S. Kamboj, J. Arnish, C. Yu, and S. Y. Chen. (2002). *Technical Basis for Calculating Radiation Doses for the Building Occupancy Scenario Using the Probabilistic RESRAD-BUILD 3.0 Code.* NUREG/CR-7655. ANL/EAD/TM/02-1. United States Nuclear Regulatory Commission.

Braun, W. X., B. A. Cicciarelli, D. L. Davidson, E. H. Hart, A. Luedtke, K. McIntosh, and P. R. Peoples. (2002). Indoor pollutant measurement and modelling, comparing impact of surface characteristics. *Proceedings of Indoor Air 2002, the 9th International Conference on Indoor Air Quality and Climate*, July 2002, Monterey, CA, pp. 885–890.

Brunskill, R. T. (1967). The relationship between surface and airborne contamination. In: *Surface Contamination* (Ed. B. R. Fish). Pergamon Press, New York, pp. 93–105.

Byrne, M. A. (2000). Suction methods for assessing contamination on surfaces. *Annals of Occupational Hygiene*, 44, 523–528.

Byrne, M. A., K. F. Bell, C. L. Fogh, D. V. Vollmair, J. Roed, and A. J. H. Goddard. (1998). Evaluation of sampling techniques for aerosol exposed skin. *Journal of Aerosol Science*, 29(S1), S1225–S1226.

Byrne, M. A., A. J. H. Goddard, C. Lange, and J. Roed. (1995). Stable tracer aerosol deposition measurements in a test chamber. *Journal of Aerosol Science*, 26(4), 645–653.

Cannell, R. J., A. J. H. Goddard, and H. M. ApSimon. (1987). Contamination of dwellings by particulate matter: Ingress and distribution within the dwelling. *Radiation Protection Dosimetry*, 21, 111–116.

Engelmann, R. J., W. R. Pendergrass, J. R. White, and M. E. Hall. (1992). The effectiveness of stationary automobiles as shelters in accidental releases of toxic materials. *Atmospheric Environment*, 26A, 3117–3125.

Fish, B. R., et al. (1967). Redispersion of settled particulates. In: *Surface Contamination* (Ed. B. R. Fish). Pergamon Press, New York, pp. 75–81.

Fogh, C. L., K. G. Andersson, A. N. Barkovsky, A. S. Mishine, A. V. Ponamarjov, V. P. Ramzaev, and J. Roed. (1999). Decontamination in a Russian settlement. *Health Physics*, 76(4), 421–430.

Fogh, C. L., M. A. Byrne, J. Roed, and A. J. H. Goddard. (1997). Size specific indoor aerosol deposition measurements and derived I/O concentration ratios. *Atmospheric Environment*, 31(15), 2193–2203.

Goddard, A. J. H., M. J. Minski, I. Thornton, and E. B. Culbard. (1986). *Household Particulate Survey (Radioactivity in Housedust in West Cumbria and Its Significance).* UK Department of the Environment Report DOE/PW/8703.

Harrison, A. W. (1979). Quiescent boundary layer thickness on aerosol enclosures under convective stirring conditions. *Journal of Colloid and Interfacial Science*, 69, 563–570.

Hinds, W. C. (1982). *Aerosol Technology*. Wiley, New York.

Ikezawa, Y., T. Okamoto, and A. Yabe. (1980). Experiences in monitoring radioactive contamination in JAERI. In: *Radiation Protection, a Systematic Approach to Safety Proceedings of the 5th International Radiation Protection Society Conference*. Pergamon Press, New York.

Jones, I. S., and S. F. Pond. (1967). Some experiments to determine the resuspension factor of plutonium from various surfaces. In: *Surface Contamination* (Ed. B. R. Fish). Pergamon Press, New York, pp. 83–92.

Kildesø, J., P. Vinzents, and T. Schneider. (1998). Measuring the potential resuspension of dust from carpets. *Journal of Aerosol Science*, 29(Suppl. 1), S287–S288.

Kildesø, J., P. Vinzents, and T. Schneider. (1999). A simple method for measuring the potential resuspension of dust from carpets in the indoor environment. *Textile Research Journal*, 69(3), 169–175.

Kocher, D. C. (1980). Effects of indoor residence on radiation doses from routine releases of radionuclides to the atmosphere. *Nuclear Technology*, 48(2), 171–179.

Kvostov, N. N., and M. S. Kostyakov. (1969). Hygienic significance of radioactive contamination of working surfaces. *Hygiene and Sanitation (English edition)*, 34, 43–48.

Lai, A. C. K. (2002). Particle deposition indoors: A review. *Indoor Air*, 12, 211–223.

Lai, A. C. K., M. A. Byrne, and A. J. H. Goddard. (2002). Experimental studies of the effect of rough surfaces and air speed on aerosol deposition in a test chamber. *Aerosol Science and Technology*, 36, 973–982.

Lange, C. (1995). Indoor deposition and the protective effect of houses against airborne pollution. Risø-R-780(EN). PhD Thesis, Risø National Laboratory, Roskilde, Denmark.

Leese, K. E., E. C. Cole, R. M. Hall, and M. A. Berry. (1997). Measurement of airborne and floor dusts in a nonproblem building. *American Industrial Hygiene Association Journal*, 58, 432–438.

Lengweiler, P. (2000). Modelling deposition and resuspension of particles on and from surfaces. No. 13734. Doctoral Thesis, Technische Wissenschaften ETH, Zurich, http://e-collection.ethbib.ethz.ch

Liebermann, A., and J. Rosinski. (1962). Behavior of an aerosol cloud in a plastic chamber. *Journal of Colloid Science*, 17, 814–822.

McMurry, P. H., and D. J. Rader. (1985). Aerosol wall losses in electrically charged chambers. *Aerosol Science and Technology*, 4, 249–268.

Morawska, L., and M. Jamriska. (1996). Deposition of radon progeny on indoor surfaces. *Journal of Aerosol Science*, 27(2), 305–312.

Nazaroff, W. W., and G. R. Cass. (1986). Mathematical modelling of indoor aerosol dynamics. *Environmental Science and Technology*, 23(2), 157–166.

Nazaroff, W. W., M. P. Ligocki, T. Ma, and G. R. Cass. (1990). Particle deposition in museums: Comparison of modelling and measurement results. *Aerosol Science and Technology*, 13, 332–348.

Nomura, Y., P. K. Hopke, B. Fitzgerald, and B. Mesbah. (1997). Deposition of particles in a chamber as a function of ventilation rate. *Aerosol Science and Technology*, 27, 62–72.

Porstendorfer, J., and A. Reineking. (1992). Indoor behaviour and characteristics of radon progeny. *Radiation Protection Dosimetry*, 45, 303–311.

Que Hee, S. S., B. Peace, C. S. Clark, J. R. Boyle, R. L. Bornschein, and P. B. Hammond. (1985). Evolution of efficient methods to sample lead sources, such as house dust and hand dust, in homes of children. *Environmental Research*, 38, 77–95.

Raunemaa, T., M. Kulmala, H. Saari, M. Olin, and M. H. Kulmala. (1989). Indoor air aerosol model: Transport indoors and deposition of fine and coarse particles. *Aerosol Science and Technology*, 11, 11–25.

Reeks, M. W., and D. Hall. (2001). Kinetic models for particle resuspension in turbulent flows: Theory and measurement. *Journal of Aerosol Science*, 32(1), 1–31.

Reynolds, S. J., L. Etre, P. S. Thorne, P. Whitten, M. Selim, and W. J. Popendorf. (1997). Laboratory comparison of vacuum, OSHA, and HUD sampling methods for lead in household dust. *American Industrial Hygiene Association Journal*, 58(6), 439–446.

Roed, J., and R. J. Cannell. (1987). Relationship between indoor and outdoor aerosol concentration following the Chernobyl accident. *Radiation Protection Dosimetry*, 21, 107–110.

Roed, J., H. L. Gjørup, and H. Prip. (1985). *Huses Beskyttende Virkning ved Luftforureningsuheld (in Danish)*. Risø-M-2484. Risoe National Laboratory, Roskilde, Denmark, ISBN 87-550-1171-3; ISSN 0418-6435.

Royster G. W., and B. R. Fish. (1964). Techniques for assessing "removable" surface contamination. *"Surface Contamination": Proceedings of a conference held in Galtinburg Tennessee* (published by Pergamon Press).

Shimada, M., K. Okuyama, Y. Kousaka, Y. Okuyama, and J. H. Seinfeld. (1989). Enhancement of Brownian and turbulent diffusive deposition of charged aerosol particles in the presence of an electrical field. *Journal of Colloid and Interfacial Science*, 128, 157–168.

Simoni, M., M. S. Jaakola, L. Carrozzi, S. Baldacci, F. Di Pede, and G. Veigi. (2003). Indoor air pollution and respiratory health in the elderly. *European Respiratory Journal*, 21(40), 15–20.

Sinclair, J. D., L. A. Psota-Kelty, C. J. Weschler, and H. C. Shields. (1990). Measurement and modelling of airborne concentrations and indoor surface accumulation rates of ionic substances at Neenah, Wisconsin. *Atmospheric Environment*, 24A(3), 627–638.

Tagg, B. (1966). Determination of the resuspension of plutonium arising from contaminated surfaces in a plutonium operational area. In: *International Symposium on the Radiological Protection of the Worker by the Design and Control of His Environment*. Society for Radiological Protection, Bournemouth, UK.

Thatcher, T. L., A. C. K. Lai, R. Morena-Jackson, R. G. Sextro, and W. W. Nazaroff. (2002). Effects of room furnishings and air speed on particle deposition rates indoors. *Atmospheric Environment*, 36, 1811–1819.

Thatcher, T. L., and D. W. Layton. (1995). Deposition, resuspension and penetration of particles within a residence. *Atmospheric Environment*, 29(13), 1487–1497.

Vasistat, N. T. (1992). Effect of adhesion time on the particle deposition: Re-entrainment and rolling. *Industrial and Engineering Chemistry Research*, 31, 1549–1554.

Wang, E., G. G. Rhoads, T. Wainman, and P. J. Lioy. (1995). Effects of environmental and carpet variables on vacuum sampler collection efficiency. *Applied Occupational and Environmental Hygiene*, 10, 111–119.

Wheeler, J. P., and J. D. Stancliffe. (1998). Comparison of methods for monitoring solid particulate surface contamination in the workplace. *Annals of Occupational Hygiene*, 42(7), 477–488.

Xu, M., M. Nematollahi, R. G. Sextro, A. J. Gadgil, and W. W. Nazaroff. (1994). Deposition of tobacco smoke particles in a low ventilation room. *Aerosol Science and Technology*, 20, 194–206.

CHAPTER 4

Contamination of Humans: In the Respiratory Tract and on Body Surfaces

Miriam A. Byrne*

Contents

1. Introduction

As discussed in Chapter 3, ingress of airborne contamination into dwellings is a process that occurs readily, and in the context of a radionuclide release into the atmosphere, indoor inhalation of contamination (while the material is airborne) and deposition of contamination onto

*Corresponding author. Tel.: +353 91 524411; Fax: 353 91 494584
E-mail address: miriam.byrne@nuigalway.ie

School of Physics, National University of Ireland, Galway, Ireland

Radioactivity in the Environment, Volume 15
ISSN 1569-4860, DOI 10.1016/S1569-4860(09)00404-5

the human body surface (skin, hair, clothing) may therefore represent important dose contributions. Historically, the latter contribution, that of deposited material, was largely overlooked, until Linsley et al. (1986) included an experimentally unvalidated value for the deposition velocity of particles to skin in order to derive emergency levels for counter-measure introduction following nuclear accidents. Since then, the topic of contaminant deposition on the human body surface has received considerable attention, and following a description in this chapter on respiratory tract structure and a discussion of the current state of knowledge regarding lung dosimetry, a description of the skin structure and a discussion of the mechanisms by which it can become contaminated is presented.

2. Biological Effects of Radiation on the Respiratory Tract

2.1. Epidemiological studies

The considerable radio-sensitivity of the human lung, together with the highly localised alpha doses in the bronchial and pulmonary regions from naturally occurring and man-enhanced radon decay products make the respiratory tract the most critical organ for cancer induction due to exposure to ionising radiation in our environment (Burkart, 1989). This is borne out by the results of epidemiological studies of radiation exposure among uranium miners and the general population. Frumkin and Samet (2001) provide an extensive review of the outcome of several studies on uranium miners, and refer to data amassed by Lubin et al. (1995) of studies of miners in former Czechoslovakia, France, Canada, Australia, Sweden and China, which included 68,000 miners and showed 2,700 lung cancer deaths. Although this review notes that several issues of uncertainty relate to the datasets (i.e. incomplete information available regarding smoking, medical treatments, exposures before the study period etc.), it is able to extract a number of important points. These are that at all but the highest levels, risk increases in a linear fashion with exposure, and at the lower end, there is no indication that there is a threshold. In those miners known to be non-smokers, a linear relationship between lung cancer risk and exposure also exists.

On the basis of considerable experience gained by studying health effects in uranium and other miners who worked in radon-rich environments, which identified radon progeny as a cause of lung cancer, and motivated by concerns among homeowners in the United States, the US Environmental Protection Agency commissioned a study of the biological effects of ionising radiation, with particular reference to radon; the outcomes of this study were published in 1988 (NRC, 1988). A scientific committee was formed to conduct a further phase of the study in 1994, and the findings of this group included the conclusion that among 11,000 never-smokers who died from lung cancer in 2005, between 2,100 and 2,900 (depending on the model used) had radon-related cancers (NRC, 1994).

2.2. The ICRP lung model

One of the most widely used and highly regarded bases for dosimetric calculations associated with inhalation of radioactive materials is the human respiratory tract model (HRTM) of The International Commission on Radiological Protection. This model is described in considerable detail in ICRP Publication 66 (ICRP, 1994), but a more accessible summary of various aspects of the model, together with advice on the selection of input parameters, is given by Valentin (2002). As discussed by Bailey (1994), the model is a general update of that described in ICRP Publication 30 (ICRP, 1979), but is of wider scope, as it takes account of the variability in radio-sensitivity among different respiratory tract tissues (the earlier model calculated average dose to the whole lung). It applies explicitly to workers and all members of the public, inhalation of gases, vapours and particles, evaluation of dose per unit intake or exposure and bioassay interpretation. The model can readily incorporate information related to the subject (i.e. age, activity, health status) or the exposure (i.e. particle size, chemical form). Various aspects of the model are described in the following sections.

2.2.1. Morphology and physiology

The morphometric model, as is described in Chapter 2 of ICRP Publication 66 (ICRP, 1994), treats the respiratory tract as two tissues: the thoracic airways (TH) and the extrathoracic airways (ET), which are, in turn, sub-divided as follows (Figure 1):

- The extrathoracic (ET) airways are divided into ET_1, the anterior nasal passage and ET_2, which consists of the posterior nasal and oral passages, the pharynx and larynx, together with associated lymphatic tissue (LN_{ET}).
- The thoracic regions are bronchial (BB: trachea, generation 0, and bronchi, airway generations 1–8), bronchiolar (bb: airway generations 9–15) and alveolar interstitial (AI: the gas exchange region, airway generations 16 and beyond), together with associated lymphatic tissue (LN_{TH}).

In the alveolar region (AI) and lymph nodes (LN_{TH} and LN_{ET}) the cells at risk are considered to be distributed throughout the region, and the average dose to the region is calculated. The whole mass of the region forms the target tissue. For the regions making up the conducting airways

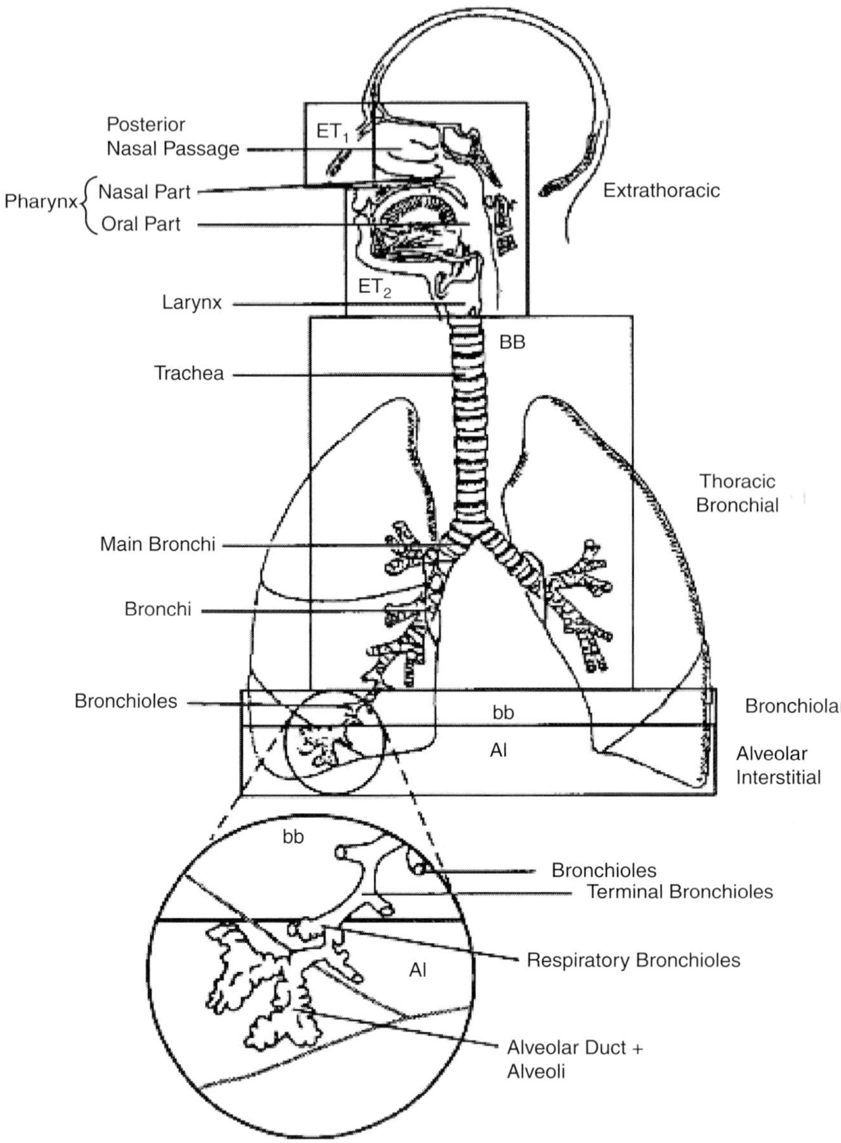

Figure 1 Respiratory tract regions defined in the HRTM. (From ICRP, 1994.)

(ET_1 and ET_2, bronchial, BB, and bronchiolar, bb) the target cells are considered to lie in a layer of tissue at a certain range of depths from the airway surface. In the extrathoracic regions (ET_1 and ET_2) the target cells are the basal cells; in the bronchial (BB) region they are both basal and secretory cells; and in the bronchiolar (bb) region they are secretory cells.

The dimensions of the airways and the ranges of depths of target cells define the masses of tissues containing radio-sensitive target cells.

For modelling purposes, morphological dimensions are assigned for a reference worker, and scaled to apply to members of the general population: 3-month-old infants, 1-, 5-, 10- and 15-year-old children, and adults. Some airway dimensions that define the volume and airflow for various age groups are shown in Table 1 (Harvey and Hamby, 2002).

Reference values of the masses of tissue containing target cells in each region, for each age, for use in dose calculations, are given in ICRP Publication 71 (ICRP, 1995). The mass of the AI region is assumed to be proportional to body mass, irrespective of age, gender or race (ICRP, 1994). The mass of the extra-thoracic lymph nodes in the reference adult male is assumed to be equal to that of the thoracic lymph nodes, and for other subjects, the masses of the lymph nodes are assumed to be proportional to the mass of the lungs.

The ICRP lung model is also cognisant of lung physiology, as lung volumes increase with the height of subject, and this affects regional deposition. Additionally, breathing rates are influenced by the subject's activities. Based on an exhaustive search of lung physiological parameters for various ethnic groups of both sexes and a range of ages, reference values have been defined for radiation worker's lung volumes and breathing rates (breathing rates being defined for a range of activities, varying from $0.45\,m^3\,h^{-1}$ during sleeping to $3.0\,m^3\,h^{-1}$ during heavy exercise), derived from data for working Caucasian males and females; these data are tabulated in Annex B of ICRP Publication 66. A second dataset is presented, intending to represent the general population, derived from data for non-working Caucasian males and females of various ages; for example, the total lung capacity for a 138 cm tall 10-year-old male of 33 kg is given as 2.87 l, compared with 6.98 l for a 73 kg adult male of height 176 cm. Data on variations in lung parameters for other ethnic groups are also presented to allow calculations to be adapted for particular populations.

2.2.2. Deposition model

The gaseous or particulate nature of a radioactive species is a critical consideration, as this will determine its deposition location and hence, its dosimetric potential; it is generally assumed that deposition of particles is primarily governed by particle size, and that deposition of gas and vapours is influenced most strongly by chemical form. Extensive studies, notably those of Dorrian and Bailey (1995) and Dorrian (1997) (an example from Dorrian (1997), showing measured size distributions of radioactive particles released to the environment in a range of conditions, is given in Table 2), allow an estimation of the likely particle sizes involved in a release. The chemical species involved in a release will indicate the likely existence of a

Table 1 Suggested lung model input parameters for the general population.

	Parameter (units)	Adult male	Adult female	15-yearold male	15-yearold female	10-year-olds	5-year-olds	1-year-olds	3-month-olds
d_0	Diameter of trachea (cm)	1.65 (0.067)	1.53 (0.45)	1.59 (0.068)	1.52 (0.065)	1.31 (0.06)	1.06 (0.049)	0.75 (0.028)	0.62 (0.027)
d_9	Diameter of airway at generation 9 (cm)	0.165 (0.067)	0.159 (0.006)	0.161 (0.007)	0.156 (0.007)	0.143 (0.007)	0.127 (0.006)	0.107 (0.004)	0.099 (0.004)
d_{16}	Diameter of airway at generation 16 (cm)	0.051 (0.002)	0.048 (0.002)	0.047 (0.002)	0.045 (0.002)	0.039 (0.002)	0.031 (0.001)	0.022 (0.001)	0.020 (0.001)
BR	Breathing rate ($m^3 h^{-1}$)	1.74 (0.67)	1.37 (0.45)	1.51 (0.51)	1.41 (0.45)	1.21 (0.41)	0.65 (0.19)	0.40 (0.11)	0.22 (0.06)
V_d	Anatomic dead space (ml)	146 (25.5)	124 (21.0)	130 (22)	114 (19)	78 (14.9)	46 (8.5)	20 (3.7)	14 (2.6)
FRC	Functional residual capacity (ml)	3301 (600)	2681 (500)	2677 (562)	2325 (488)	1484 (311)	767 (161)	244 (26)	148 (28)

Source: Adapted from Harvey and Hamby (2002) with permission.

Note: Parameters show mean (standard deviation) of normal distributions. A proposed modification of the breathing rate parameterisation is discussed in the text.

Table 2 Estimated particle size distributions resulting from various radioactive release types.

Type of aerosol	Number of measurements	Range of AMADs (μm)	Median (μm)	σ_g [a]
All artificial aerosols	125	0.3–18	1.5	2.9
Chernobyl fallout aerosols	78	0.3–4.8	0.64	1.5
Resuspension aerosols	16	1.025–15	6.0	1.9
Natural ^{7}Be aerosols	56	0.29–1.18	6.0	1.9

Source: Adapted from Dorrian (1997) with permission from HPA.
[a] σ_g is the geometric standard deviation obtained from the cumulative distribution of measured AMADs.

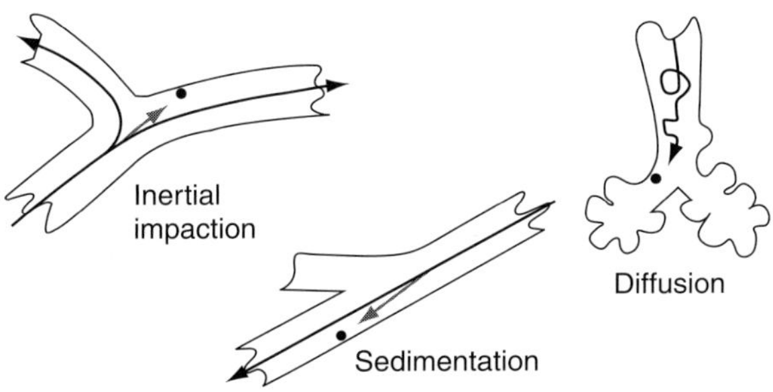

Figure 2 Main mechanisms of particle deposition in the respiratory tract. (Adapted from Yeh et al., 1976 with permission from *EHP*.)

gas/vapour phase: as discussed by Valentin (2002) gas/vapour phases are highly probable when noble gases and hydrogen are considered, and particulate phases are more common, for elements such as nickel, ruthenium and mercury. The ICRP lung model recommends a default of 50% particulate and 50% gas/vapour composition in the absence of full information, but Valentin (2002) illustrates the importance of a correct ratio assignment, by providing a dose coefficient calculation based on this default; the coefficients span the range of 0.65–2.2 times those that would correspond to a 100% particulate or 100% vapour release.

For particles, the main deposition loss mechanisms for inhaled particles, as depicted in Figure 2, are inertial impaction, sedimentation and diffusion, and the relative significance of these mechanisms depends upon fluid dynamics, airway geometries and the particle's physical characteristics. Inertial impaction occurs when particles have sufficient momentum to deviate from fluid streamlines and strike boundary airway surfaces. As momentum is a product of particle mass and particle velocity, inertial impaction is an important deposition mechanism for larger particles greater than 1 μm in diameter, and for particles travelling at higher velocities, and this occurs primarily in the upper airways. Inertial impaction as a particle deposition mechanism has been widely studied, for example by Martonen (1993) for both laminar and turbulent flow conditions.

Reduced flow velocities lead to increased particle residence times in airways and the occurrence of sedimentation. While particles as small as 0.1 μm may be affected, this mechanism increases in importance with increasing particle size (Stahlhofen et al., 1989).

Deposition via diffusion occurs when particles exhibiting random Brownian motion collide with the airway surface. While deposition of particles between 0.1 and 1 μm in size occurs by both diffusion and sedimentation, diffusion is the primary deposition mechanism for particles less than 0.1 μm in size.

For particulate matter, predicted values of total and regional lung deposition can be obtained by utilising the ICRP lung model, taking into account age-related differences in airway diameters and breathing rates. ICRP Publication 66 tabulates regional deposition for aerosols ranging in size from 0.006 to 0.2 μm activity median thermodynamic diameter (AMTD[1]) and for 0.5 to 20 μm activity median aerodynamic diameter (AMAD) for workers (normal nose breathers and habitual mouth breathers) and for members of the public (normal nose breathers) of each age group: 3-month-old infants, 1-, 5-, 10- and 15-year-old children, and adults, at the four reference levels of exercise, sleep, sitting, light exercise and heavy exercise: a graphical example of predicted regional deposition for a reference worker is shown in Figure 3. Deposition in individual and combination lung regions is shown in Figure 3(a) and (b), respectively.

When inhalation of gas molecules is considered, it is assumed that almost all inhaled gas molecules make contact with the airway surfaces but are returned to the air unless they dissolve in or react with the surface lining of the airways. The HRTM model assigns three classes to gases and vapours, according to their solubility/reactivity:

1. SR-0 refers to insoluble and non-reactive materials, which therefore undergo negligible deposition.
2. SR-1 refers to soluble or reactive materials, and for all such materials, the deposition is assumed to occur according to the following pattern: 10% ET_1, 20% ET_2, 10% BB, 20% bb, 40% AI.
3. SR-2 refers to highly soluble material which is assumed to deposit entirely in the extrathoracic ET_2 region.

[1] For radioactive aerosols, as measurements are usually made on the basis of radioactivity, it is convenient to refer the activity median diameter of the aerosol. AMAD is generally used in the region above 0.5 μm, where sedimentation and inertial impaction are dominant deposition mechanisms and it is also used when deposition depends mainly on diffusion, typically for particles with diameters less than 0.5 μm.

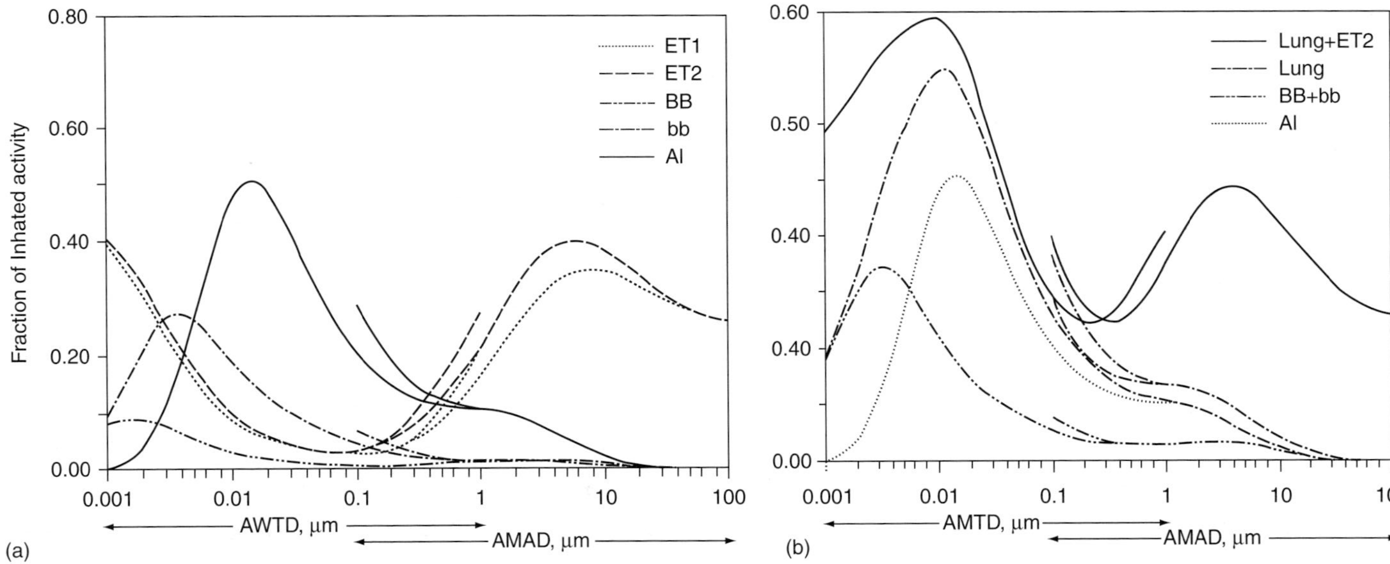

Figure 3 Deposition in (a; left) each respiratory tract region and (b; right) combinations of regions as predicted by the HRTM for a reference worker, as a function of aerosol median size. In the transition size range (0.1–1 µm) values are given for both AMAD and AMTD. (Adapted from Valentin, 2002 with permission from ICRP.)

2.2.3. Clearance model

There are three transfer mechanisms considered in the clearance model, and these are (i) removal of material by nose wiping and blowing, (ii) removal of material from respiratory tract by transfer to the gastrointestinal (GI) tract or lymph notes, and (iii) removal of material from the respiratory tract by absorption into body fluids (such as blood).

The simplifying assumptions made by the model regarding clearance are

(a) the rate of occurrence is independent of age and sex,
(b) independent clearance rates exist regarding particle transport and absorption to blood,
(c) the rate of removal of material from the nose and the rate of transfer to the GI tract are independent of the nature of the material, and
(d) absorption of material by body fluids is assumed to occur at the same rate throughout the regions of the respiratory tract (no absorption is assumed to occur in ET_1), but the rate of occurrence is strongly influenced by the material type.

Absorption is considered to be a two-stage process, with time-dependent clearance rates existing for both stages. Dissolution of materials is the first stage, and materials are considered to have a fraction f that dissolves rapidly and a fraction $(1-f)$ that dissolves slowly. The second stage, material uptake by body fluids, is assumed to occur instantaneously.

The model allows material-specific absorption rates to be used for materials for which reliable experimental data exist, and in other cases, default values are recommended. Materials can be classified as Type V, which are absorbed instantaneously, and therefore there is no need for a regional deposition assessment, Type F which are totally absorbed with a half-time of 10 min, and Type M for which 10% is absorbed with a half-time of 10 min and 90% with a half-time of 140 days.

2.2.4. Validation of the ICRP model with measurement data

Hisamatsu and Takizawa (2003) estimated intakes of the transuranic elements (associated with weapons testing) in the Japanese population, based on measurement data, in Tokyo and Akita, and compared these with the output of ICRP metabolic and lung models. Parameters in the ICRP-66 lung model that were specified only for Caucasians were corrected for Japanese heights and tissue weights. Estimates for inhaled amounts of plutonium were derived from measured fallout rates in Tokyo and Akita. Figure 4 shows a comparison of measured and modelled plutonium concentration values in lungs and indicates a good agreement between the two.

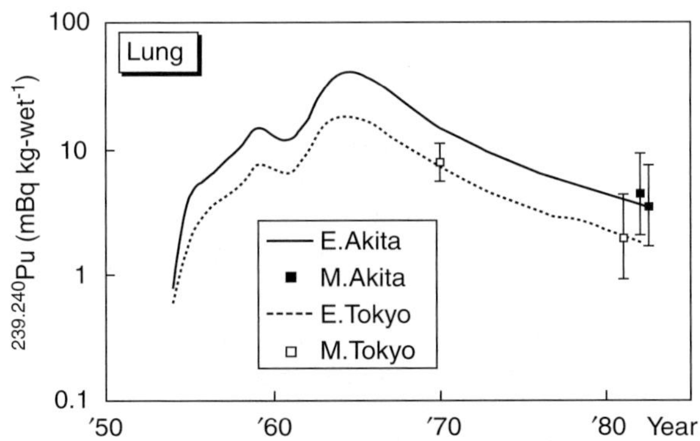

Figure 4 Measured lung Pu concentrations compared with those estimated with the ICRP 66 model. (Adapted from Hisamatsu and Takizawa, 2003 with permission from Wolters Kluwer Health.)

3. Biological Effects of Radiation on the Skin

3.1. The structure of the human skin

The skin is the largest organ in the human body; for a 70 kg man it weighs approximately 2.1 kg, covers an area of the order of 2 m^2, and has an average thickness of 3 mm. The skin has a complex structure and serves several vital functions. It forms a physical barrier between the body and the environment and also has a temperature regulation role, controlling heat loss by altering blood flow in the skin and modifying moisture evaporation. In addition, the skin has important sensory functions and also plays a role in the body's immune system. The skin can be divided into two discrete layers, which overlie the subcutaneous fat layer:

(i) The Epidermis is a thin outer layer of skin, itself divided into several layers. The outermost is the stratum corneum, which is composed of 15–20 layers of dead cells which are constantly being shed and replaced from the layers beneath. It is estimated that, on average, the stratum corneum is completely replaced every 2–3 weeks (Ness, 1994). The other layers of the epidermis are the stratum granulosum, stratum spinosum and stratum germinativum (or basal layer). The thickness of the epidermis varies from approximately 30 μm on the face to >200 μm on the fingertips.

(ii) The Dermis comprises the greater part of the total thickness of the skin and lies directly under the basal layer of the epidermis. It is largely composed of a network of collagen fibres and bundles and it is these which give the skin its mechanical strength. The dermis is highly complex in structure and contains blood vessels, nerves and lymphatic vessels as well as hair follicles and sebaceous glands.

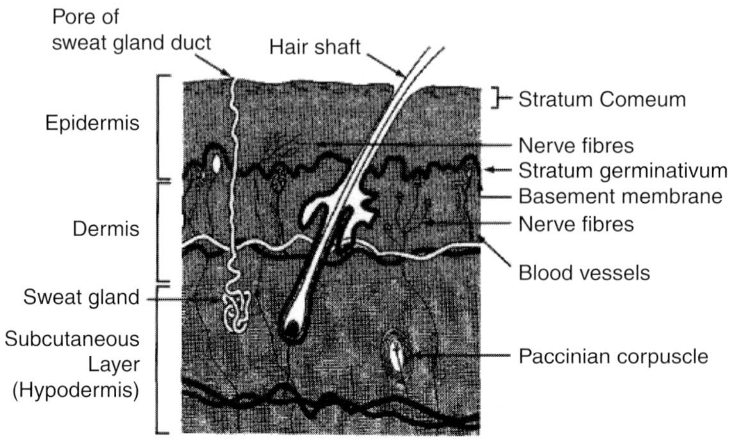

Figure 5 Schematic diagram of a cross section through human skin. (Adapted from WHO, 2006 with permission.)

Figure 5 is a schematic diagram of a cross-section through the skin, depicting the various features of the skin structure.

3.2. Dose implications of radioactive contamination of the skin

The effects of a potentially harmful substance, including a radioactive one, deposited on the skin may be felt on the skin itself, or, if the substance is capable of penetrating the stratum corneum, then the whole body. Fiserova-Bergerova et al. (1990) describe the mechanisms by which chemicals are absorbed through the skin and are taken up by the body, and ill-health effects associated with such absorption have been well documented for many chemical species. The factors which determine the dose received by the body from material deposited or adsorbed on its surfaces are (i) the area of contaminated surface, (ii) the quantity of contaminating material deposited and (iii) the time of exposure, that is, how long the contamination remains on the body.

Several incidences of radioactive contamination of the human body surface have been reported in the literature. Pires do Rio et al. (1994) describe how, at Goiania in Brazil in 1987, parts of the source assembly from an abandoned teletherapy unit (containing a ^{137}Cs source) were perceived to be attractive glowing trinkets by members of the public, who took them home, resulting in unwitting contamination of the hands. Following the Chernobyl accident, several cases of human body contamination have been reported, notably a study by Barabanova and Osanov (1990) of the conditions of exposure of 56 victims of the

Chernobyl accident who suffered radiation lesions in the skin, and the documentation by Beentjes et al. (1988) of the contamination of two young women vacationing in Kiev shortly after the accident. In the hospital context, Mackie et al. (1994) have reported how, in an aerosol ventilation study involving technetium, the bodies of hospital staff (particularly the hair) became contaminated.

In the context of a radiological incident, skin exposure is likely to be particulate or vapour in nature, and will often take the form of a uniform exposure to deposited particles of aerosol dimensions or to vapour molecules. Alternatively, an exposure to 'hot particles' may occur, these being physically small radioactive sources with linear dimensions up to approximately 1 mm. Their origin may be either associated with nuclear accidents or with improper handling of high-level radioactive waste. The response of skin to small highly radioactive particles (i.e. hot particles) on its surface has been experimentally investigated by several researchers. These include Dean et al. (1970), who exposed the skin of a human volunteer to U-235 enriched microspheres and Hopewell et al. (1986), who conducted a series of experiments using a range of radioactive sources on pig and mouse skin. The dose received from hot particles deposited on the skin has been estimated mathematically by Pöllanen and Toivonen (1995). In a review by Charles (2004) on the current state of understanding of the effects of irradiation of the skin, by uniform and non-uniform sources, the discrepancy of a factor of 10 which exists in the recent recommended dose limits for hot-particle exposures, as set by the International Commission on Radiological Protection (ICRP) and the National Council on Radiation Protection and Measurements (NCRP), is discussed. Charles (2004) concludes that this is due to differences in views on acceptability of different levels of deterministic effects.

In his review, Charles (2004) highlights the complexity of the topic of radioactive contamination of the skin, as the layered structure of the skin means that critical skin structures are located at depths which may be beyond the physical penetration depth capability of some radiation types. For this reason, for alpha or low-energy (i.e. 0.23 MeV approximately) beta radiation interacting with skin, dose is restricted to the superficial layers below approximately 70 μm, and the most significant doses from radioactive material deposited on the human body surface would arise from high-energy beta or low-energy gamma emitters. As discussed by Charles (2004), the categories of severe irradiation-related skin damage are defined by the ICRP and the NCRP as follows:

Erythema is a skin-reddening effect, produced by acute radiation doses of at least several grays to the superficial blood vessels of the upper dermis.

Moist desquamation is the result of death of the epidermal basal cells and damage to the superficial network of blood vessels. It has been found, by the ICRP (1991), that for pig skin, this effect occurs approximately 6 weeks after exposure to doses in excess of 12 grays. The target depth for this effect is approximately 150 μm (i.e. roughly the depth of the basal cell layer around the hair follicle).

Ulceration and Necrosis results from damage to the deep dermal structure, on a timescale of 2–12 weeks following exposure. Low-energy beta radiation does not produce this effect, as the target depth is greater than 1 mm. Following the Chernobyl accident, Barabanova and Osanov (1990) observed these effects in persons who received high-energy beta doses of 2–20 grays.

Dermal atrophy and damage to the deep vasculature becomes apparent on a timescale of months or years, following received doses of 30–40 grays.

4. Contaminant Exposure and Clearance on Humans

4.1. Airborne contaminant deposition on human skin, hair and clothing

While the process of exposure via inhalation is relatively straightforward, in so far as it involves a transfer of airborne contamination from an air-stream to the respiratory tract, the process of contamination of the body's skin, hair and clothing is considerably more complex, as transfers can occur between any of these three surface types, in either direction, as well as from an air-stream. The complexity of the process is very well described by Schneider et al. (1999), who have formulated a 'conceptual model' for assessment of dermal exposure; the zones involved in this model, and the possible inter-zone transfers (via contact, deposition etc.) are shown in Figure 6. Contact transfer of contaminants between inert surfaces and human body surfaces is discussed later in the chapter. In this section, experimental studies that have determined contaminant deposition rates to skin, hair and clothing are described, and the results are reviewed.

Prior to any deposition velocity measurements being made to the body surfaces of human subjects, several researchers measured aerosol concentrations in the vicinity of human phantoms. Notable among these were Parker et al. (1990), who measured aerosol concentrations close to the mouth and chest of a heated phantom in a test chamber, and Gudmundsson et al. (1992) who measured deposition velocities of particles in the size range 2–32 μm in a wind tunnel. Values in the range 10^{-4}–2.7×10^{-3} m s^{-1} onto the eyes of the phantom were reported, and later measurements using the same phantom, by Schneider et al. (1994) recorded deposition velocities onto the forehead in the range 6×10^{-4}–9.8×10^{-3} m s^{-1}.

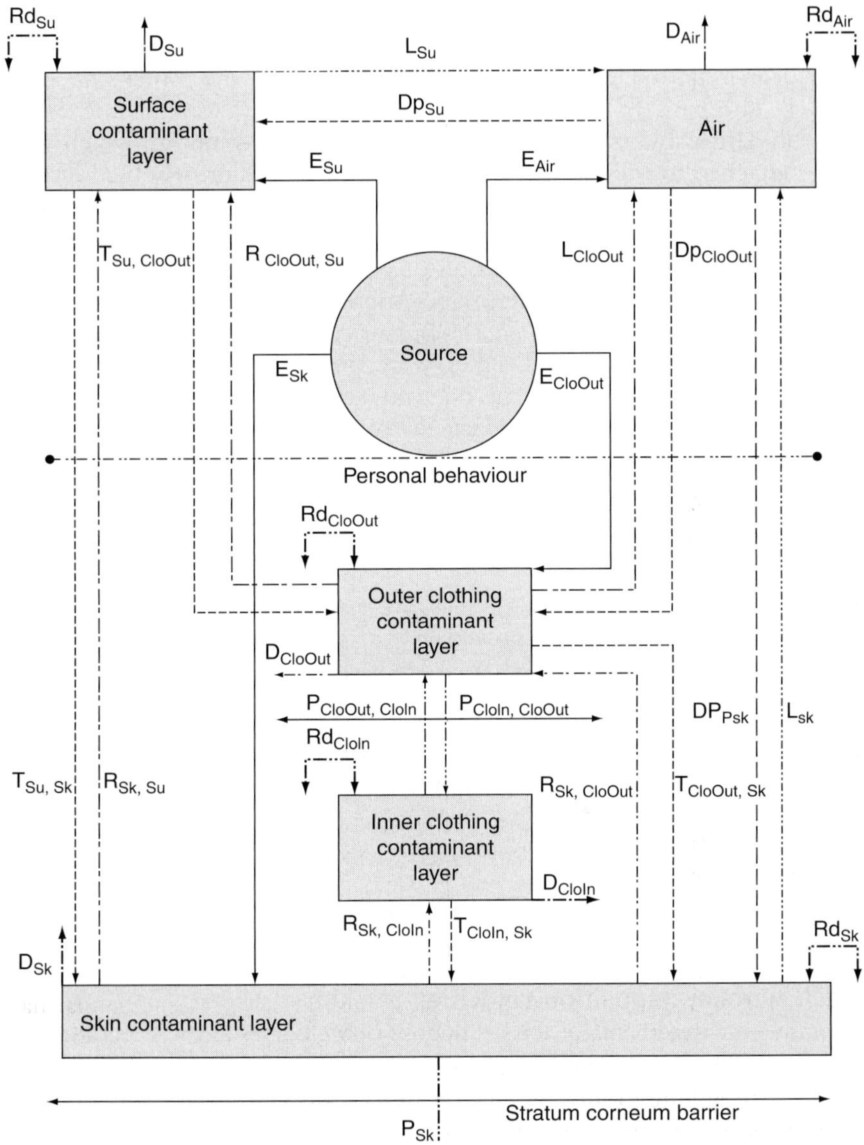

Figure 6 Overview of the conceptual model of Schneider et al. (1999) for dermal exposure. E, emission; Dp, deposition; L, resuspension or evaporation; T, transfer; R, removal; Rd, redistribution; D, decontamination; P, penetration and permeation.

In a series of projects funded by the European Commission, over the period 1995–2003, aerosol deposition velocities have been measured to skin, hair and clothing, with a view to obtaining data for radiological dosimetric assessment calculations. These projects were collaborative efforts between the Riso National Laboratory, Denmark, Imperial College London and the National University of Ireland, Galway, and the findings have been reported in detail in Fogh et al. (1999) and Andersson et al. (2004). The basis of all the experiments carried out was the use of the neutron-activatable tracers, indium, dysprosium or europium, which were either attached to monodisperse porous silica particles or nebulised directly to form an aerosol. The particles were then dispersed into a room or chamber in which volunteers were present. After a suitable deposition period, skin sampling (a comparison of particle recovery efficiencies from skin due to wiping, washing, tape-stripping and suction sampling is discussed in detail in Fogh et al., 1999), hair cutting and clothing sectioning were carried out, and the samples were placed in a nuclear reactor to activate the tracers. Subsequent neutron activation and gamma spectrometry, in comparison with standards, allowed a quantitative determination of tracer levels in each sample. These data, when combined with a measure of the average airborne tracer concentration in the room or chamber (which was also measured during each experiment), allowed a calculation of the deposition velocity onto the hair, skin or clothing surface, according to the following expression

$$V_{\mathrm{dep}} = \frac{\text{Tracer mass deposited per unit sample area}}{\text{Average airborne tracer concentration} \times \text{Exposure time} \times \text{Particle recovery efficiency}} \quad (1)$$

As discussed in Fogh et al. (1999), skin sampling was from a well-defined area, and clothing samples were bulk samples for which an area could be easily determined, but estimating the area of a hair sample was complicated, and several approaches were considered.

4.2. Aerosol deposition velocities to humans

4.2.1. Aerosol deposition velocities to skin

Extensive measurements of aerosol deposition velocities to the skin of human volunteers were made by Fogh et al. (1999) and Andersson et al. (2004) in a range of environments, including homes, offices and test chambers, for monodisperse aerosol particle sizes between 0.5 and 8 μm. Initially, the aim of the experiments was to determine the likely range of aerosol deposition velocity values to skin, compared with values for inert indoor surfaces and for outdoor surfaces, but later measurements focused on examining the factors influencing deposition velocities to the human body surface. As two examples of factors that were investigated, Figure 7 illustrates the observed enhancing effect of skin moisture on aerosol deposition, and Figure 8 demonstrates the effect of elevated human body movement (simulated through fan-induced enhanced air circulation) on aerosol deposition.

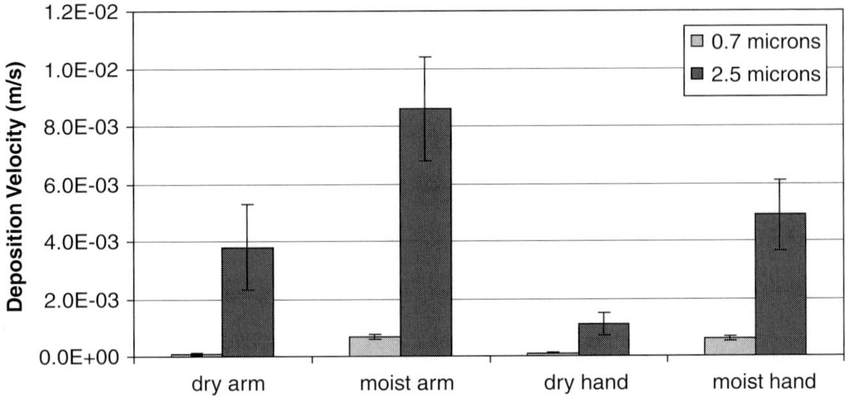

Figure 7 Comparison of the measured deposition velocities of two particle sizes on the moist and dry skin of two volunteers. Skin moisture levels were measured with a corneometer. (Adapted from Andersson et al., 2004 with permission.)

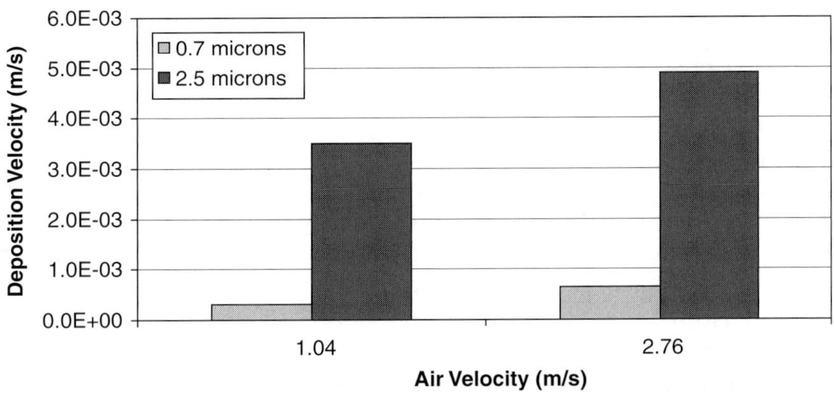

Figure 8 Comparison of the effect of enhanced air velocity (intended to simulate that caused by physical movement of the volunteer) on the measured aerosol deposition velocities of two particle sizes on a volunteer's skin. (Adapted from Andersson et al., 2004 with permission.)

As previously indicated, the motivation for making aerosol deposition velocity measurements to skin (and other body surfaces) was to obtain data for dosimetric assessment. Based on all the measurements recorded by Fogh et al. (1999) and Andersson et al. (2004), Andersson et al. (2002) defined suitable aerosol deposition velocities onto human skin for various particle size–specific radionuclides; a selection of these values are shown in Table 3 and are compared with aerosol deposition velocities to a grass surface, estimated by Roed (1990) shortly after the Chernobyl accident. It can be

Table 3 Selected particle size–specific aerosol deposition velocity values, measured by Fogh et al. (1999) and Andersson et al. (2004).

Particle size (μm)	Aerosol deposition velocity ($m\,s^{-1}$)	
	Skin	Cut lawn
0.5	5×10^{-3}	5×10^{-4}
0.7	1×10^{-3}	4×10^{-4}
4	1.2×10^{-2}	7.5×10^{-4}

Source: Adapted from Andersson et al. (2002) with permission.
Note: Values are compared with aerosol deposition velocities to a grass surface, estimated by Roed (1990) shortly after the Chernobyl accident.

seen that deposition velocities to the human body surface are at least an order of magnitude higher than to a stationary outdoor surface, highlighting the importance of considering radionuclide deposition on the body as a dose contribution following an accident or incident.

4.2.2. Aerosol deposition velocities to clothing

Following preliminary experiments which indicated an increased particle deposition velocity onto a rough surface compared with smooth, Fogh et al. (1999) conducted a comprehensive series of measurements of deposition velocities to different types of clothing, the results of which are indicated in Figure 9. Averaged over all clothing types, measured deposition velocities were $0.55 \pm 0.29 \times 10^{-4}\,m\,s^{-1}$ for 0.5 μm particles, $11.6 \pm 3.5 \times 10^{-4}\,m\,s^{-1}$ for 2.5 μm particles and $76 \pm 45 \times 10^{-4}\,m\,s^{-1}$ for 8 μm particles, indicating a clear increase in deposition velocity with particle size. The relationship between clothing roughness and deposition velocity is clearest for the smallest particles shown in Figure 9; a confounding factor might be triboelectric charging of fabrics containing synthetic materials compared to natural materials, although preliminary experiments reported by Andersson et al. (2004) have indicated that this effect is likely to be small.

4.2.3. Aerosol deposition velocities to hair

Early measurements carried out by Fogh et al. (1999) involved the use of exposure of synthetic wigs to tracer aerosol particles, and this possibly introduced electrostatic effects which might not be prevalent to the same extent when determining deposition velocities to real human hair. With this in mind, a new series of experiments was designed by Andersson et al. (2004), which not only used the natural hair of volunteers, but also recognised that deposition velocities to the distal and proximal parts of hair could differ due to differences in air movement. In relation to radiation dose, this was thought to be significant, as beta radiation from

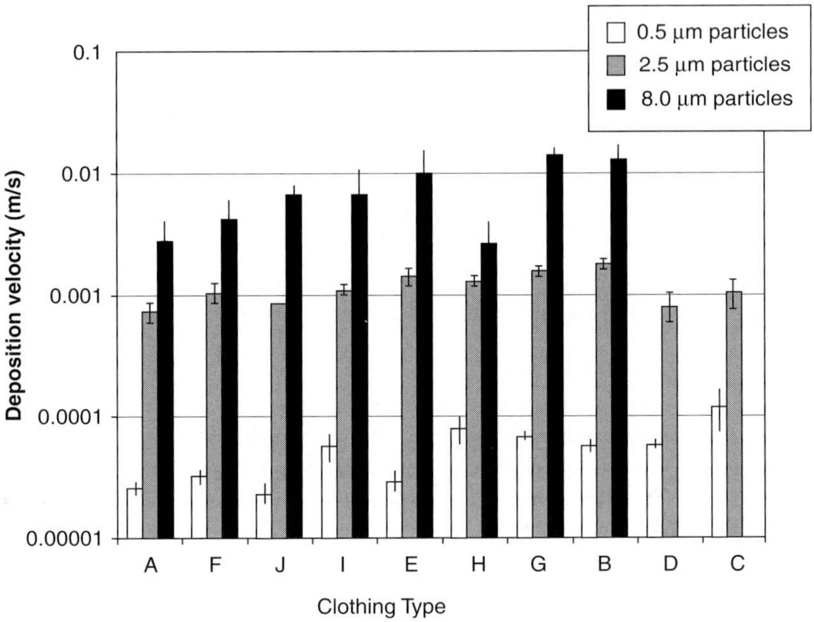

Figure 9 Measured deposition velocities for 3 particle sizes and 10 different clothing types. Each result is based on three samples. Samples are sorted left to right according to increasing roughness (A is smooth fabric, and C is highly textured). The percentage of synthetic material in each sample is: A (100%), F (65%), J (0%), I (35%), E (30%), H (50%), G (35%), B (0%), D (100%) and C (0%). (Adapted from Fogh et al., 1999 with permission.)

Table 4 Average aerosol deposition velocities to distal and proximal components of volunteers' hair, measured by Andersson et al. (2004) for two particle sizes.

Particle size (μm)	Average deposition velocity (distal) $\times 10^{-4}$ m s^{-1}	Average deposition velocity (proximal) $\times 10^{-4}$ m s^{-1}
0.7	2.6 ± 1.2 (30)	1.2 ± 0.8 (12)
2.5	18 ± 8 (22)	8 ± 5 (8)

Note: Standard deviations are shown, and number of samples is indicated in parentheses.

contamination on the outer hair layers would be attenuated through the mass of the inner layers. For these reasons, experiments were conducted whereby three volunteers were subjected to exposures of tracer particles, as described earlier, and strands from their hair were then cut, but sectioned into 'distal' and 'proximal' components prior to analysis. A summary of the results obtained is shown in Table 4 and unsurprisingly indicates higher deposition velocities to the more exposed distal component of the hair.

4.3. Contact transfer of contaminants to humans

Humans may become contaminated with radioactive material through unintentional contact with contaminated surfaces, such as those that occurred in the case of the Goiania incident.

Quantification of the efficiency of transfer of material to the skin, in an event such as that mentioned above, is important in calculating exposure. A percentage contact transfer efficiency, C, can be defined as

$$C = \frac{\text{Mass of contaminant per unit hand or glove area after contact}}{\text{Mass of contaminant per unit area of test surface prior to contact}} \times 100 \tag{2}$$

To date, studies of contaminant transfer efficiency to the skin have largely focused on liquid contamination in the horticultural sector (e.g. Aprea et al., 2001). One exception is the work of Brouwer et al. (1999) in which contact transfer efficiencies of powders from glass plates to volunteer's hands were assessed. This study used fluorescent-dyed particles, at loadings of 10–200 mg cm^{-2}, and between 1 and 12 hand presses. A fluorescence scanning system was used to demonstrate that only 4%–40% of the hand's palm area was in contact with the contaminated surface, and contact transfer efficiencies of the order of 2% were calculated, regardless of the number of hand presses. Figure 10 shows a typical image of the location of contamination on the palm during this study (i.e. the whitest areas). Further studies of particulate transfers, at lower surface loadings, were conducted by Fogh et al. (1999) and Andersson et al. (2004). The former study involved the contamination of wooden board, polycotton fabric, synthetic knitted fabric and polythene sheeting, with 5 and 10 μm neutron-activatable tracer labelled particles, and the percentage transferred to latex gloves during a single hand press was assessed by direct neutron activation analysis of the gloves. Experiments were carried out with dry and damp gloves. The latter study focused on sub-micrometre particles, and involved an assessment of the transfer of fluorescent polymer particles of 0.1, 0.5 and 0.026 μm from plastic sheeting, metal foil and wooden board during single and double hand presses; in this case, the gloves were rinsed and the rinse-water was analysed to determine particle uptake. Again, dry and damp gloves were studied. The results of both studies are summarised in Table 5.

The percentage transfer efficiencies shown in Table 5, although having a large variability, are significantly greater than values reported by Brouwer et al. (1999). Brouwer reports that the powder used in their study had a granular structure and was pounded by a household blender prior to use, so it is likely that the particle size was larger than in the later studies. The data in Table 5 indicate a general increase in transfer efficiency as particle size decreases, apart from the smallest size particles (0.026 μm), where it is likely that electrostatic effects bind those particles to the contaminated surface.

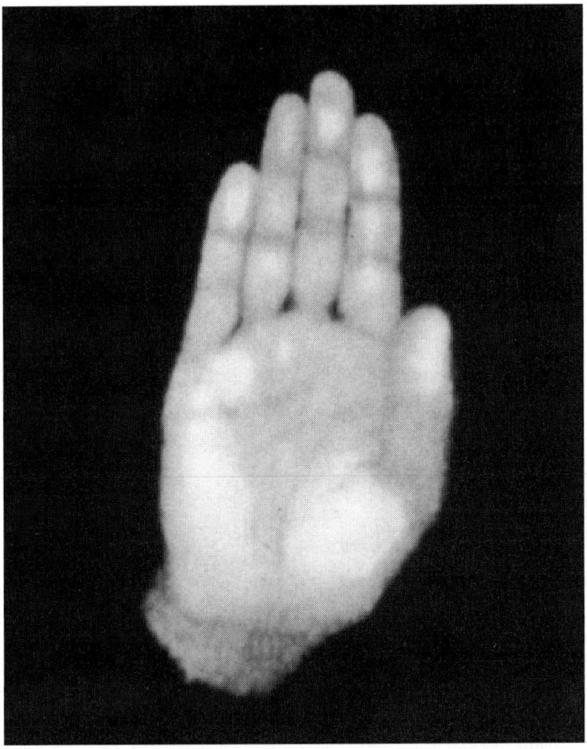

Figure 10 Image of location of fluorescent particles on the palm of the hand after repeated hand press trials. (Adapted from Brouwer et al., 1999, with permission from ACGIH.)

Transfer to the hand from smooth surfaces that were not electrostatically charged (metal foil) is generally higher than that from rough surfaces. It was also observed that surface moisture (i.e. damp gloves) had an enhancing effect on the degree of contact transfer.

In summary, it can be concluded that there is considerable potential for contamination of the human body through hand contact with contaminated surfaces, especially if the size distribution of the contamination tends towards the sub-micron particulate range, and the contaminated surface is relatively smooth.

Another form of contamination transfer that arises through contact involves the uptake of outdoor deposited contamination on footwear, and 'tracking' this contamination into the indoor environment. Evidence of this occurrence is presented by Al-Radady et al. (1994) and Thatcher and Layton (1995) for non-radioactive dust, and by Goddard et al. (1986) for radioactive deposits. In surveys of dust in sites close to the Sellafield nuclear fuel reprocessing plant in Cumbria, increases of up to 47% in airborne

Table 5 Percentage contact transfer efficiencies, for particulates, reported by Fogh et al. (1999) and Andersson et al. (2004).

	Cotton	Paper/wood	Knitted	Plastic	Metal foil
10	19.1% (11.9%, 9)	20.8% (11.9%, 4)	21.5% (11.4%, 5)	22.7% (11.1%, 9)	
5	16.9% (5.2%, 12)			24.3% (11.4%, 6)	
0.5 (2*)		59.5% (7.7%, 5)			63.6% (3.8%, 5)
0.1 (2)		15.73% (1)		21.3% (2.6%, 5)	49.3% (6.7%, 5)
1				35.2% (1.0%, 5)	
0.5		21.6% (4.5%, 3)		26.0% (4.8%, 3)	
0.026		22.9% (5.5%, 5)		16.8% (2.0%, 5)	
10^D	30.5% (16.7%, 8)	30.2% (17.1%, 4)	28.5% (15.2%, 4)	27.2% (15.0%, 6)	
0.1^D (2)		59.5% (2.7%, 5)		70.2% (15.1%, 5)	63.4% (15.7%, 6)
0.5^D (2)		87.0% (2.7%, 5)			65.5% (8.0%, 5)

Note: Standard deviations and number of replicates shown in parentheses. Superscript 'D' indicates that the surface was damp, and '*' indicates double hand presses.

activity concentration indoors were found by Goddard et al. (1986) in homes where, after coastal walking occupants did not remove outdoor footwear. In addition, a case study of indoor-lead contamination, carried out by Al-Radady et al. (1994) exhibited an increase in the efficiency of mechanical transport of dust indoors during wet weather conditions.

A number of laboratory studies, including the work of Cannell (1988), have attempted to define relationships between the efficiency of particle transfer from surfaces to footwear and particle size, surface type and footwear type.

4.4. Natural clearance of contaminants from humans

The residence time of particulate contamination on the human body is a factor that has an important impact on the accuracy of exposure assessment in the aftermath of an accidental release of radionuclides to the atmosphere. If a person is aware that their body surface has been contaminated, active steps such as washing may be taken towards decontamination. However, in the event of a person being unwittingly exposed, it is necessary to know the rate of 'natural' clearance of contamination in order to make an accurate exposure assessment.

The motivation for developing quantitative tools for determining removal rates of contamination from skin has arisen largely from the agricultural and horticultural sectors; in this context, active decontamination, through various washing procedures, is the issue of interest. The development of fluorescence scanning systems, for detecting deposits of fluorescent-tagged materials (e.g. fluorescors added to pesticides etc.) has afforded a means of investigating the efficiency of removal of contamination in a non-invasive manner. The critical components of such a system are (i) suitably filtered light to activate fluorescence in the fluorescor, (ii) a light-collecting device, such as a charge-coupled device (CCD) camera to detect the fluorescence (this device may need optical filtration also, to eliminate interference from the original illumination) and (iii) an image analysis system for quantification of the detected fluorescence.

The first quantitative fluorescence methods were developed in the early 1980s (Vo-Dinh and Gammage, 1981) and since then many systems have been developed. Fenske et al. (1986) were the first to use a fluorescent tracer and image analysis system (VITAE) to investigate quantitatively the dermal exposure of workers to pesticides. Roff (1994) later developed the FIVES (fluorescence interactive video exposure system) fluorescence monitor to measure dermal exposure to chemicals in the workplace. Most early systems employed ultraviolet illumination but the fluorescence of human skin itself under ultraviolet illumination caused interference with the fluorescence of the tracer particles. A more recent system developed by Bell (1998) overcame this difficulty by using visible illumination (in the

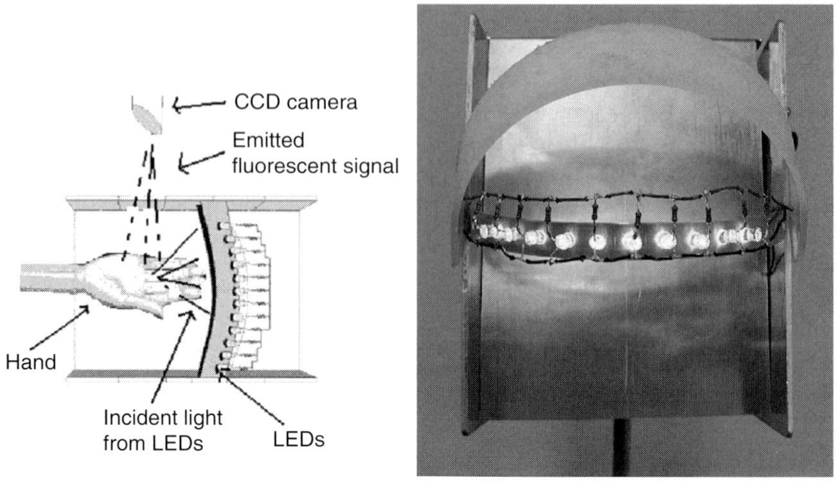

Figure 11 Schematic diagram of the system for illumination and detection of tracer particles on skin, designed by Hession et al. (2006). A photograph of the illuminating light-emitting diodes (LEDs) is also shown.

form of large security lights with optical filters attached). Since then, rapid advancements in CCD camera technology and illumination (i.e. small high-intensity light-emitting diodes are now widely available) has allowed the designing of systems that are more portable, compact and sensitive. Such a system is described by Hession et al. (2006), and the important components are presented in Figure 11. Using this system, a series of experiments were carried out to assess the natural clearance rate of tracer particles from the arms of volunteers, according to: (i) the amount of arm hair present, and (ii) the degree of movement in which the volunteer is involved.

Figure 12 shows the measured time variation in the concentration of 10 μm particles on the skin of two volunteers, one with little arm hair and another with significant hair. It can be seen that the arm hair plays a role in particulate retention. Figure 13 shows the measured time variation in the concentration of 10 μm particles on the skin of two volunteers, one who spent the experimental period seated indoors, and another who engaged in vigorous outdoor walking in dry, breezy conditions. In this case, the rate of clearance was seen to be greater from the more active volunteer. In this study, experiments were also carried out to measure clearance rates of 3 μm particles from skin, and averaged over all experimental scenarios, mean clearance rates of $6.40 \times 10^{-5}\,h^{-1}$ and $8.10 \times 10^{-5}\,h^{-1}$ were determined for 3 μm and 10 μm particles, respectively, suggesting that smaller particles adhere more effectively to hair and skin surfaces than larger particles (the physical processes involved are discussed in the section on resuspension in Chapter 3).

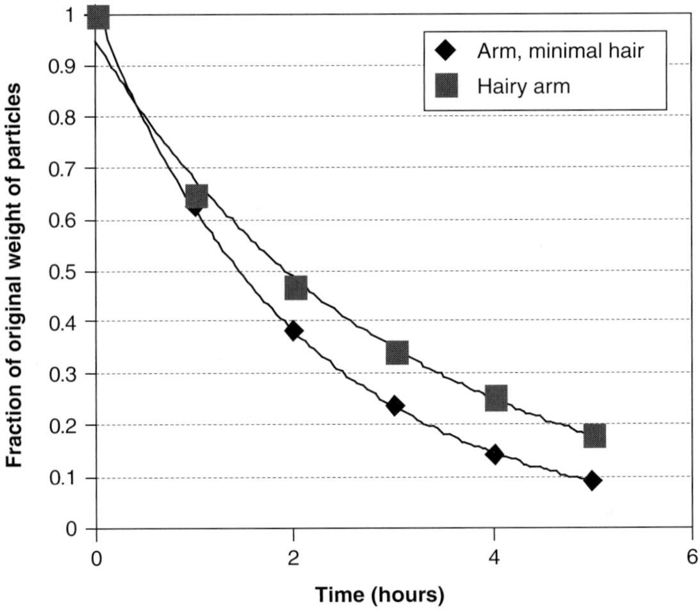

Figure 12 Measured time variation in the concentration of 10 μm particles on the skin of two volunteers with different levels of arm hair. Volunteer 1 is female with little arm hair; Volunteer 2 is male with significant arm hair (from Hession et al., 2006).

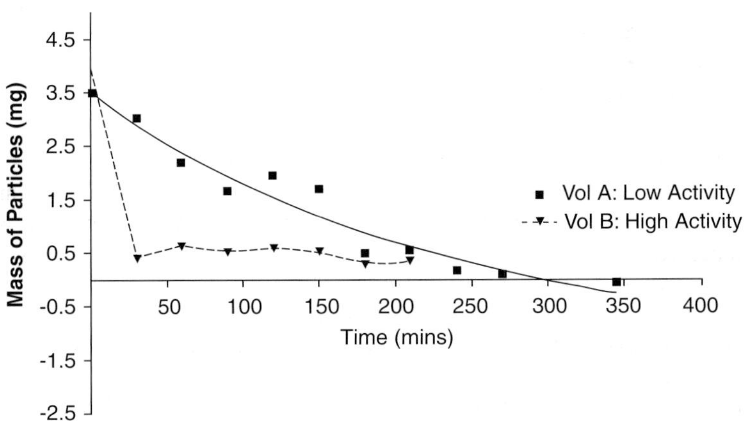

Figure 13 Measured time variation in the concentration of 10 μm particles on skin subjected to different degrees of movement (from Hession et al., 2006).

A number of issues must be considered when employing clearance rates, such as those quoted above, in exposure-assessment calculations. One factor is that a fluorescence scanning system permits the examination of a fluorescor on the skin until the time at which the concentration is below the limit of detection of the system, as distinct from the time at which all the fluorescor has been removed; as indicated earlier, technological developments in camera sensitivity and image analysis capability are occurring rapidly at the time of writing. For example, Bell (1998) quoted a detection limit for his system of 2.5 mg cm^{-2} of tracer particles, whereas the detection limit quoted by Hession et al. (2006) was 50 μg cm^{-2}. Another consideration should be the complexity of the skin surface from a particle's perspective; any particle that has penetrated into a skin crevice or hair follicle may become invisible to an optical analysis system. Experiments to examine the possibility of skin penetration by fine particulates have been carried out by many researchers. In a study described by Andersson et al. (2004), excised human skin was exposed to inert fluorescent microparticles of 0.1, 0.5 and 1.0 μm, respectively, for periods of up to 18 h, and the skin was then sectioned and analysed by confocal microscopy. Irrespective of particle size or exposure time, it was observed that particles remained within the stratum corneum region of the skin, but did not penetrate into the epidermis or dermis. However, it was observed that particles were capable of lodging in hair follicles in the space between the shafts of hairs and the surrounding stratum corneum. Similar observations were made by Cross et al. (2007) in experiments involving zinc oxide nanoparticles. It is likely that particles lodged in this way would be invisible to a fluorescence scanning system, meaning that particle clearance rates would be underestimated. Furthermore, due to their lodgement, it is likely that these particles would be more difficult to remove, either through natural clearance or active decontamination, than particles resting on the outer skin surface, and therefore they would have a more significant dose contribution.

REFERENCES

Al-Radady, A. S., B. E. Davies, and M. J. French. (1994). Distribution of lead inside the home: Case studies in the North of England. *The Science of the Total Environment*, 145, 143–156.

Andersson, K. G., C. L. Fogh, M. A. Byrne, J. Roed, A. J. H. Goddard, and S. A. M. Hotchkiss. (2002). Radiation dose implications of airborne contaminant deposition to humans. *Health Physics*, 82(2), 226–232.

Andersson, K. G., J. Roed, M. A. Byrne, H. Hession, P. Clark, E. Elahi, A. Byskov, X. L. Hou, H. Prip, S. K. Olsen, and T. Roed. (2004). *Airborne Contamination in the Indoor Environment and its Implications for Dose*. R-1462 (EN). Riso National Laboratory, Roskilde, Denmark.

Aprea, C., G. Sciarra, L. Lunghini, L. Centi, and F. Ceccarelli. (2001). Evaluation of respiratory and cutaneous doses and urinary excretion of alkylphosphates by workers in greenhouses treated with omethoate, fenitrothion, and tolclofos-methyl. *American Industrial Hygiene Association Journal*, 62(1), 87–95.

Bailey, M. R. (1994). The new ICRP model for the respiratory tract. *Radiation Protection Dosimetry*, 53, 107–114.

Barabanova, A., and D. P. Osanov. (1990). The dependence of skin lesions on the depth-dose distribution from beta-irradiation of people in the Chernobyl nuclear power plant accident. *International Journal of Radiation Biology*, 57(4), 775–782.

Beentjes, L. B., W. C. Buijs, F. H. Corstens, and J. Duijsings. (1988). Radioactive contamination of Kiev vacationers after the Chernobyl accident. Biological half-life of Cs. *Nuclear Medicine and Biology*, 15(2), 171–175.

Bell, K. F. (1998). An experimental study of aerosol deposition on skin, hair and clothing. PhD Thesis, Imperial College London.

Brouwer, D. H., R. Kroese, and J. J. V. Hemmen. (1999). Transfer of contaminants from surface to hands: Experimental assessment of linearity of the exposure process, adherence to the skin, and area exposed during fixed pressure and repeated contact with surfaces contaminated with a powder. *Applied Occupational and Environmental Hygiene*, 14(4), 231–239.

Burkart, W. (1989). Biological effects of radiation. *The Science of the Total Environment (Special Issue)*, 89(1–2), 1–230.

Cannell, R. J. (1988). Radiation exposure in an urban dwelling following an accidental release of radioactivity. PhD Thesis, Imperial College London.

Charles, M. W. (2004). The skin in radiological protection – recent advances and residual unresolved issues. *Radiation Protection Dosimetry*, 109(4), 323–330.

Cross, S. E., B. Innes, M. S. Roberts, T. Tsuzuki, T. A. Robertson, and P. McCormick. (2007). Human skin penetration of sunscreen Nanoparticles: *In-vitro* assessment of a novel micronized zinc oxide formulation. *Skin Pharmacology and Physiology*, 20(3), 148–154.

Dean, P. N., J. Langham, and L. M. Holland. (1970). Skin response to a point source of fissioned uranium 235 carbide. *Health Physics*, 19, 3–7.

Dorrian, M. D. (1997). Particle size distributions of radioactive aerosols in the environment. *Radiation Protection Dosimetry*, 69(2), 117–132.

Dorrian, M. D., and M. R. Bailey. (1995). Particle size distributions of radioactive aerosols measured in workplaces. *Radiation Protection Dosimetry*, 60(2), 119–133.

Fenske, R. A., J. T. Leffingwell, and R. C. Spear. (1986). A video imaging technique for assessing dermal exposure. I. Instrument design and testing. *Journal of the American Industrial Hygiene Association*, 41(2), 764–770.

Fiserova-Bergerova, V., J. T. Pierce, and P. O. Droz. (1990). Dermal absorption potential of industrial chemicals: Criteria for skin notation. *American Journal of Industrial Medicine*, 17(5), 617–635.

Fogh, C. L., M. A. Byrne, K. G. Andersson, K. F. Bell, J. Roed, A. J. H. Goddard, D. V. Vollmair, and S. A. M. Hotchkiss. (1999). *Quantitative Measurement of Aerosol Deposition on Skin, Hair and Clothing*. R-1075 (EN). Riso National Laboratory, Roskilde, Denmark.

Frumkin, H., and J. M. Samet. (2001). Radon. *CA: A Cancer Journal for Clinicians*, 51, 337.

Goddard, A. J. H., M. J. Minski, I. Thornton, and E. B. Culbard. (1986). *Household Particulate Survey (Radioactivity in Housedust in West Cumbria and its Significance)*. UK Department of the Environment Report DOE/PW/8703.

Gudmundsson, A., T. Schneider, O. H. Petersen, P. S. Vinzents, M. Bohgard, and K. R. Akelsson. (1992). Determination of particle deposition velocity onto the human eye. *Journal of Aerosol Science*, 23(S1), 563–566.

Harvey, R. P., and D. M. Hamby. (2002). Age-specific uncertainty in particulate deposition for 1 μm AMAD particles using the ICRP 66 lung model. *Health Physics*, 82(6), 807–816.

Hession, H., M. Byrne, S. Cleary, K. G. Andersson, and J. Roed. (2006). Measurement of contaminant removal from skin using a portable fluorescence scanning system. *Journal of Environmental Radioactivity*, 85(2–3), 196–204.

Hisamatsu, S., and Y. Takizawa. (2003). Validation of ICRP metabolic models for the transuranics in a Japanese population. *Health Physics*, 85(6), 701–708.

Hopewell, J. W., J. E. Coggle, J. Wells, R. Hamlet, J. P. Williams, and M. W. Charles. (1986). Contrib Title: The acute effects of different energy beta emitters on pig and mouse skin. *British Journal of Radiology*, Suppl. 19, 47–51.

ICRP. (1994). *Human Respiratory Tract Model for Radiological Protection*. ICRP Publication 66. *Annals of the ICRP*, 24(1-3), pp. 1–482.

ICRP. (1995). *Age-Dependent Doses to Members of the Public from Intake of Radionuclides: Part 4 Inhalation Dose Coefficients*. ICRP Publication 71. *Annals of the ICRP*, 25(3-4).

ICRP. (1979). *Limits for Intakes of Radionuclides by Workers*. ICRP Publication 30, Part 1. *Annals of the ICRP*, 2(3-4).

ICRP. (1991). *The Biological Basis for Skin Dose Limitation*. ICRP Publication 59. *Annals of the ICRP*, 22(2). Pergamon Press, Oxford.

Linsley, G. S., M. J. Crick, J. R. Simmonds and S. M. Haywood. (1986). *Derived Emergency Reference Levels for the Introduction of Countermeasures in the Early to Intermediate Stages of Emergencies Involving Release of Radioactive Materials to the Atmosphere*. NRPB Report DL10. National Radiological Protection Board, UK.

Lubin, J. H., J. D. Boice, Jr., C. Edling, R. W. Hornung, G. R. Howe, E. Kunz, R. A. Kusiak, H. I. Morrison, E. P. Radford, J. M. Samet, M. Tirmarche, A. Woodward, S. Xiang Yao, and D. A. Pierce. (1995). Lung cancer in radon-exposed miners and estimation of risk from indoor exposure. *Journal of the National Cancer Institute*, 87(11), 817–827.

Mackie, A., G. C. Hart, D. A. Ibbet, and R. J. S. Whitehead. (1994). Airborne radioactive contamination following aerosol ventilation studies. *Nuclear Medicine Communications*, 15, 161–167.

Martonen, T. B. (1993). Mathematical model for the selective deposition of inhaled pharmaceuticals. *Journal of Pharmaceutical Sciences*, 82, 1191.

Ness, S. A. (1994). *Surface and Dermal Monitoring for Toxic Exposures*. Wiley, Canada.

NRC (National Research Council). (1994). *Committee on Health Effects of Exposure to Radon (BEIR VI), and Commission on Life Sciences. Health Effects of Exposure to Radon: Time for Reassessment?* National Academy Press, Washington, DC.

NRC (National Research Council). (1988). *Committee on the Biological Effects of Ionizing Radiations. Health Risks of Radon and Other Internally Deposited Alpha-Emitters: BEIR IV*. National Academy Press, Washington, DC.

Parker, R. C., R. K. Bull, D. C. Stevens, and M. Marshall. (1990). Studies of aerosol distributions in a small laboratory containing a heated phantom. *Annals of Occupational Hygiene*, 34(1), 35–44.

Pires do Rio, M. A., E. C. S. Amaral, and H. G. Paretzke. (1994). The resuspension and redeposition of ^{137}Cs in an urban area: The experience after the Goiânia accident. *Journal of Aerosol Science*, 25, 821–831.

Pöllanen, R., and H. Toivonen. (1995). Skin dose calculations for uranium fuel particles below 500 μm in diameter. *Health Physics*, 68(3), 401–405.

Roed, J. (1990). *Deposition and Removal of Radioactive Substances in an Urban Area*. Final Report of the Project NKA AKTU-245. Nordic Liaison Committee for Atomic Energy, Roskilde, Denmark.

Roff, M. W. (1994). A novel lighting system for the measurement of dermal exposure using a fluorescent dye and an image processor. *Annals of Occupational Hygiene*, 30(6), 903–919.

Schneider, T., M. Bohgard, and Gudmundsson. (1994). A semi-empirical model for particle deposition onto facial skin and eyes. Role of air currents and electric fields. *Journal of Aerosol Science*, 25(3), 583–593.

Schneider, T., R. Vermeulen, D. H. Brouwer, J. W. Cherrie, H. Kromhout, and C. L. Fogh. (1999). Conceptual model for assessment of dermal exposure. *Occupational and Environmental Medicine*, 56, 765–773.

Stahlhofen, W., G. Rudolf, and A. C. James. (1989). Intercomparison of experimental regional deposition data. *Journal of Aerosol Medicine*, 2, 285.

Thatcher, T. L., and D. W. Layton. (1995). Deposition, resuspension and penetration of particles within a residence. *Atmospheric Environment*, 29(13), 1487–1497.

Valentin, J. (2002). *Guide for the Practical Application of the ICRP Human Respiratory Tract Model: Supporting Guidance 3. Annals of the ICRP*, 322(1-2).

Vo-Dinh, T., and R. B. Gammage. (1981). The lightpipe luminoscope for monitoring occupational skin contamination. *American Industrial Hygiene Association Journal*, 42(2), 112–120.

World Health Organization. (2006). *Dermal Absorption*. Environmental Health Criteria 235, International Programme for Chemical Safety (IPCS), ISBN 9241572353, p. 11, 197pp.

Yeh, H. C., R. F. Phalen, and O. G. Raabe. (1976). Factors influencing the deposition of inhaled particles. *Environmental Health Perspectives*, 15, 147–156.

CHAPTER 5

Migration of Radionuclides on Outdoor Surfaces

Kasper G. Andersson*

Contents

1. Introduction

An important lesson learned from the Chernobyl accident is that airborne radioactive contaminants deposited over an inhabited area may give rise to significant external doses to the inhabitants (see, e.g. Andersson and Roed, 2006; Fogh et al., 1999). To adequately protect the public in such

*Corresponding author. Tel.: +45 46 77 41 73; Fax: +45 4677 5330
E-mail address: kasper.andersson@risoe.dk

Risoe National Laboratory for Sustainable Energy, Technical University of Denmark, DK-4000 Roskilde, Denmark

Radioactivity in the Environment, Volume 15
ISSN 1569-4860, DOI 10.1016/S1569-4860(09)00405-7

situations it is important to be able to estimate the doses that they would receive over time by living in a given contaminated area. The inhabited environment consists of a complexity of different horizontal and vertical surfaces of different materials and shapes, which may receive very different levels of contamination. Proper estimation of doses thus requires the use of sophisticated models, as discussed in Chapter 6. A crucial factor in this context is the migration of the contamination on the different surfaces in the environment. Contaminants may penetrate into permeable surfaces, which can shield against the radiation, or they can migrate from one surface to another, which may reduce or increase the hazards they pose to people living in the area. The migration depends on the characteristics of the deposited contaminants as well as on the environment in which they deposited. This chapter is aimed at shedding light on the mechanisms responsible for the migration and retention of radioactive contaminants deposited in the inhabited environment, and describing the dynamic processes that influence dose. These processes also have implications for inhalation doses received from resuspended material (see Chapters 2 and 6), as they determine the amount of contaminant material available for resuspension from the different surfaces. It should be noted that in addition to the 'naturally' occurring processes described in this chapter, such factors as anthropogeneous mixing, renewal, demolition and replacement of elements in the environment may play a highly important, though not generically predictable, role in the changes of contamination levels in inhabited areas.

2. Influence of Initial Physico-Chemical Forms of Deposited Contaminants

The migration in the inhabited environment of radionuclides deposited after an airborne release will depend on the initial physico-chemical form of the contaminants in relation to the materials, orientations and conditions of the surfaces in the environment on which they are deposited, as well as on the weather and various physical and chemical disturbances. This section describes examples of different physico-chemical forms that might be envisaged in connection with an emergency and outlines some implications of these.

The Chernobyl accident demonstrated that essentially a single contaminating incident can generate a range of airborne contaminants of widely differing characteristics governing their environmental behaviour. The released fuel fractionation particles could be divided into two categories according to the processes that created them (Kashparov et al., 1996):

- non-oxidised fuel particles formed by the mechanical impact of the explosion on 26 April 1986;
- oxidised fuel particles formed over the period from 26 April to 5 May, as a result of the fire in the reactor.

Of these, the non-oxidised fuel particles were generally found to have higher chemical stability in the environment than the oxidised fuel particles (Kashparov et al., 2004). The fuel particles had, according to the processes of their formation, been subject to various high levels of heating over different periods of time. This led to varying degrees of depletion with respect to the more volatile radionuclides, which partially evaporated. The most important volatile contaminant elements from the Chernobyl accident were found to comprise caesium, ruthenium and iodine (Baltensperger et al., 1987; Bondietti et al., 1988; Horn et al., 1987, Jost et al., 1986). The reason why the ruthenium was found to be rather volatile in spite of its very high elemental boiling point of 2,700°C is that it is oxidised at considerably lower temperatures (700–1,200°C) into volatile forms, provided sufficient oxygen is present (Kashparov et al., 1996; Hunt et al., 1994). The very volatile iodine appeared in air masses in inhabited areas in different chemical forms, including the highly reactive elemental iodine gas, as well as caesium iodide, adsorbed on ambient particles, and in volatile organic compounds such as the inert methyl iodide. The airborne radioiodine transport is believed mainly to have taken place in the form of gas at equilibrium with local ambient aerosols (Lange, 1995). Other volatile elements released in a major reactor accident are, following evaporation, converted from gas to particles through chemical reactions, nucleation and condensation as well as coagulation, thus changing physico-chemical and aerodynamic characteristics during or immediately after the atmospheric release event. At Chernobyl this typically formed small contaminant particles with an aerodynamic diameter in the range of 1 μm or slightly less, which dispersed over large distances (Rulik et al., 1989; Tschiersch and Georgi, 1987; Winkelmann et al., 1987; Jost et al., 1986). In contrast, the larger particles (in the range from a few to ca. 150 μm; Sandalls et al., 1993) carrying radionuclides of less volatile elements from the Chernobyl release predominantly deposited at relatively short distances from the Chernobyl power plant, due to gravitational settling, impaction and interception. Although significant concentrations of more 'refractory' element radionuclides (e.g. ^{89}Sr, ^{90}Sr, ^{95}Zr, ^{140}Ba, ^{141}Ce and ^{144}Ce) were recorded several hundreds of kilometres away from Chernobyl (Kryshev, 1996), it was generally found that the ratio between concentrations of ^{90}Sr and ^{137}Cs was some 0.2–0.3 near the reactor, but only about 0.01–0.02 at a distance of 1,000–2,000 km (Aarkrog et al., 1992; Devell, 1987). A similar decline in concentrations relative to radiocaesium was found for the other 'refractory' element radionuclides (Mück et al., 2002), demonstrating the importance of physical characteristics.

Since scavenging by falling rain greatly influenced particle deposition, and meteorological conditions frequently changed over the 'critical' days after the Chernobyl accident, the deposition of the different contaminants often formed rather irregular and unpredictable patterns. An example of very small-scale inhomogeneity in the deposition process is that in the town of Pripyat, less than 4 km from the Chernobyl power plant, east-facing walls were measured to have more than twice as high contamination as west-facing walls (Roed and Andersson, 1996). This might possibly reflect the influence of inertial impaction of large particles.

The various physico-chemical species of the contaminants were found to have different solubility. Very near the reactor, for instance caesium was to a considerable extent in the form of large fuel particles, with low solubility, whereas the caesium that had travelled as far as Prague had a range of different degrees of solubility (Tomasek et al., 1995; Rulik et al., 1989). Although the overall solubility of the particles in Prague and other places at similar distances was much higher than that of the particles deposited near Chernobyl, part of the radiocaesium in Prague was however reported not to be in readily water-soluble form (Tomasek et al., 1995).

Over some years, a slow environmental weathering process will release radionuclides from the low-solubility fuel particles, depending on, for example matrix composition, oxidation states and pH, and thus increase their mobility in the environment (Salbu et al., 1998). However, a considerable difference was still recorded in 1993 between the binding strength of caesium to surfaces in the town of Pripyat and that in the village of Vladimirovka, at a distance of some 65 km from the power plant. Spraying of inert water solutions on similar limestone walls removed two-thirds of the caesium contamination in Pripyat, but only about one-fifth of the caesium contamination in Vladimirovka (Roed and Andersson, 1996).

The dissolution of deposited contaminant particles with high chemical stability depends on the environmental conditions. For instance, a comprehensive study in the strongly contaminated areas near Chernobyl (Kashparov et al., 2004) demonstrated a clear correlation between the fraction of undissolved fuel particles and soil acidity. It has been suggested that the particle dissolution process in soil can be described by a simple first-order kinetics equation (Kashparov et al., 2004; Petryaev et al., 1991; Konoplev et al., 1992). On the background of measurements, the dissolution half-life was estimated to be of the order of 14 years at pH 7. However, as demonstrated by the model results of Kashparov et al. (2004) (see Figure 1), which are verified by measurements made by different workers (e.g. Krouglov et al., 1998; Petryaev et al., 1991), the dissolution half-life is considerably shorter at low pH. As the fuel particles are generally comparatively inert, a slow process of particle dissolution will delay the migration and fixation of the contaminants in the environment as well as the uptake by living organisms.

It should be noted that in connection with other types of airborne releases, contaminants with significantly different physico-chemical properties should be expected, depending on, for example radionuclide composition, temperature and pressure at release. For instance, the shape of the uranium fuel particles originating from the low-temperature pre-fire

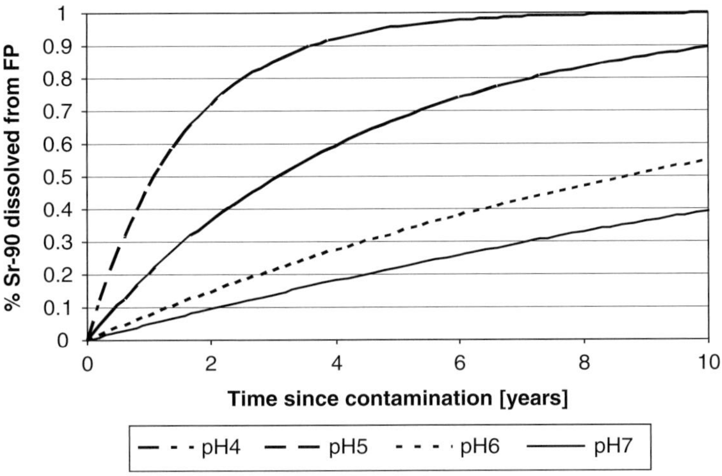

Figure 1 Time-dependence of the fraction of ^{90}Sr dissolved from fuel particles in the areas contaminated by the Chernobyl accident, according to an empirical Ukrainian model (Kashparov et al., 2004). Model curves are shown for different soil pH values.

Windscale releases was reported to be flake-like, indicating different releasing processes (Salbu et al., 1998). It has been reported that these particles could not be dissolved in 1 M HCl (Salbu, 2000). Also plutonium particles spread over the ice after the crash in 1968 near Thule air base in Greenland of a B52 aeroplane carrying nuclear weapons were found to be virtually insoluble (Eriksson, 2002). Livens and Baxter (1988) and Bunzl et al. (1989) found that radiocaesium originating from Chernobyl was in a form that was at that time more labile in mineral soil than that from the early weapons tests. This was attributed to either differences in physico-chemical form or the time dependence of the 'locking up' of caesium in soil. In support of this, Hilton et al. (1992) reported that in aerosol samples collected in the UK, the water-soluble fraction of the radiocaesium released from Chernobyl was some 70%, whereas it was less than 10% in the fallout particles from weapons testing. Dorrian and Bailey (1995) compiled results from 52 different publications covering a wide range of industries and workplaces, where radioactive aerosols have been measured to have been released. It was found that the overall particle size distribution could be adequately described by a log-normal distribution with a median value of 4.4 μm. Many of the reported cases were related to largely insoluble plutonium particles, but also a wide range of other, much more readily soluble, radioactive substances was recorded.

In recent years, there has been an increasing focus on the risk that terrorists might detonate a so-called dirty bomb radioactivity dispersion device (see Section 7 of Chapter 1). It is considered likely that a 'dirty bomb' attack would have maximum societal impact if carried out in a highly populated area (Andersson, 2005; Sohier and Hardeman, 2006). Due to their widespread use over decades (e.g. in now possibly 'orphaned' sources), comparatively high radiotoxicity, and relatively long half-lives, some of the radionuclides of primary importance would in this context be ^{90}Sr, ^{137}Cs, ^{134}Cs, ^{60}Co and ^{192}Ir (Sohier and Hardeman, 2006; Ferguson, 2003). Unless the explosion leads to evaporation of the contaminants, most of the contaminant particles generated by this type of explosions would be very large, but there would also likely be a significant release of particles in the size range of only a few microns (Andersson, 2005). These latter particles could disperse over a large city area. Again, the physical (e.g. metal, powdered, solution, ceramic) and chemical form of the applied source can greatly influence the post-deposition migration in the environment. Radionuclide compounds that might plausibly be of concern range from the virtually insoluble to the readily soluble, for example depending on their previous application. The explosion process may also, according to type, lead to phase transition and radionuclide volatilisation, thus influencing contaminant solubility in the environment (Harper et al., 2007).

3. Migration of Radionuclides in Areas of Soil in an Inhabited Environment

In assessing the consequences of airborne releases affecting inhabited areas, it is often estimated that the external dose from deposited contaminants will be dominated by the contribution from areas of soil (see Chapter 6). One of the main reasons for this is that the deposited contamination in soil is not removed over time from the surface by wind and weather, as it will be on many other types of surface in the inhabited environment. For many radionuclides the downward migration in undisturbed soil will be very slow, which implies that through natural processes, the external dose rate may only be reduced slowly (Andersson et al., 2002b; Brown et al., 2006).

3.1. Selective fixation of caesium in soil minerals

Two radionuclides of particular interest in connection with large reactor accidents (and other types of incidents including explosions of nuclear bombs) are ^{134}Cs and ^{137}Cs. The reasons for this are manifold, including high fission yields, comparatively long physical half-lives, high solubility of condensation particles and the strong gamma and beta radiation emitted (Evans et al., 1983). As mentioned above, ^{134}Cs and ^{137}Cs would also be among the primary candidates for dispersion from a 'dirty bomb' (many strong existing industrial sources, some of which have been 'orphaned' over the years, are in chloride form – i.e. readily soluble). The initial vertical distribution of radiocaesium in soil will greatly depend on the deposition process and the environmental conditions during a short subsequent period. The Chernobyl accident showed that, presumably due to infiltration of contaminated rainwater over the first days after deposition, the majority of the contamination may locally penetrate some centimetres into the soil. Soil type and structure (e.g. open soil porosity), season, rain amounts and rain intensity are among the key factors determining this initial penetration of radionuclides. This penetration can be particularly pronounced for soils with a thick organic horizon. Schimmack et al. (1994) showed that, at low rain intensities, the organic surface layer of a forest soil (0–4 cm) was an effective sink for deposited caesium and cobalt contaminant tracers, but at high rain intensity (30 mm h^{-1}), only 70% was retained in the organic layer, while the rest of the caesium and cobalt rapidly became distributed over an underlying ca. 15 cm deep layer. Contamination that was initially deposited on a grass cover has typically been found to migrate to the soil surface with a natural weathering half-life of about 1–2 weeks (Martin, 1964; Milbourn and Taylor, 1965; Heinemann and Vogt, 1980; Mück and Gerzabek, 1995), largely depending on the occurrence of rain showers and their intensity.

Later on, the contamination will much more slowly migrate vertically down into the soil. Here, both the organic and inorganic soil components will show an affinity for different types of contaminants, thereby limiting their downward migration (Coughtrey et al., 1986). It has been known for many decades that the inorganic components of the soil in particular have a remarkable capacity for adsorbing and strongly retaining caesium cations (e.g. Schultz et al., 1959; Sawhney, 1967). This capacity of certain clay minerals to extract caesium from aqueous solutions, even in the presence of excessive amounts of other cations, is in fact exploited in the treatment of radioactive waste from nuclear fuel processing plants and in underground repositories for radioactive waste.

The reason for the strong fixation of the caesium cation by certain clays lies in the layered structure of these minerals. Figure 2 shows the typical unit cell structure of a three-layer clay mineral. This unit cell consists of an octahedral Al–O–OH layer, which is tied in between two tetrahedral silicon oxide layers. As shown, the aluminium, silicon or other cations are sited in 'cavities' of the lattice formed by the negatively charged ions. During crystallisation, some of the silicon and aluminium ions are substituted in the structure with ions of lower valency, giving rise to a negative net charge on the crystal. This negative charge is balanced by the presence of other positively charged ions in the 'interlayer' gap between the crystal layers. The strength of fixation of these ions in the interlayer

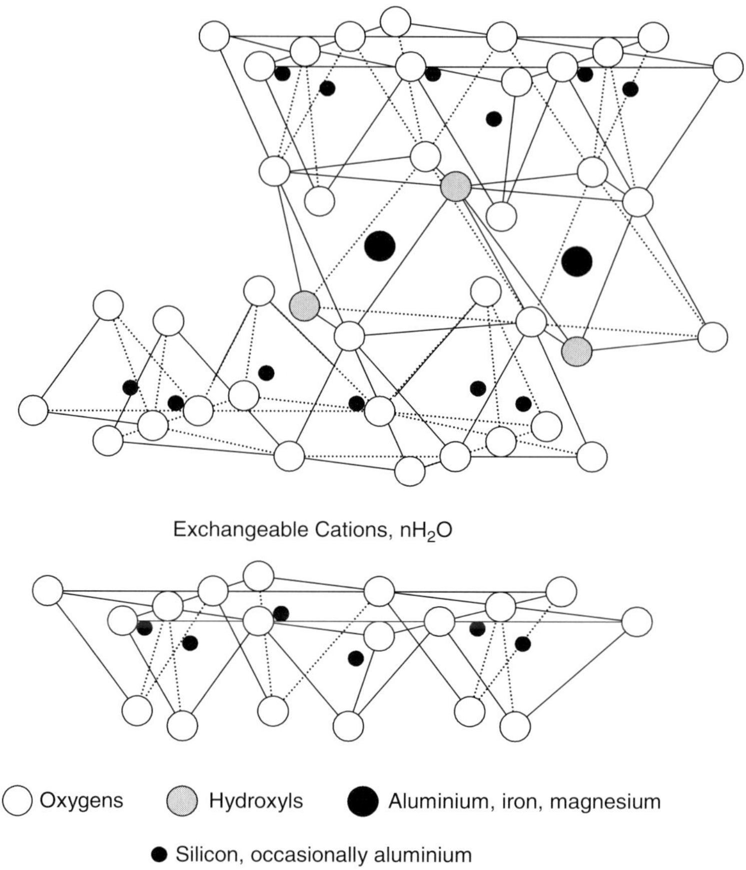

Figure 2 Diagrammatic sketch showing the lattice structure of three-layer clay minerals (here shown for montmorillonite; after Grim, 1965).

positions will depend on the negative charge density of the crystal. Amongst common three-layer clays this increases in the order:

montmorillonite < vermiculite < illite

In illite, the content of aluminium is high compared with silicon. This results in a highly negative lattice surface charge. As a consequence of this high charge, the illite unit cell layers are tightly held together with a non-expandable *c*-axis of 10 Å (the unit cells are stacked on top of each other in the *c*-axis direction) (Grim, 1965; Tamura and Jacobs, 1961). The end-surfaces of the crystal, as shown in Figure 2, consist of hexagonally arranged oxygen ions. The hexagonal-shaped opening thus formed on the surface has a diameter of about 2.8 Å, which is similar to that of the oxygen ion

(Nommik and Vahtras, 1982). In nature, this is where the balancing (typically potassium) cations are located. The attraction between these cations and the negatively charged crystal surfaces is sufficiently strong in illite to facilitate the collapse of the interlayer spacing. The fixation of the interlayer cations is therefore virtually irreversible, and the *c*-axis is fixed. Normally, interlayer positions of this type cannot be reached by cations from the soil solution. However, due to natural weathering of clay minerals in the soil, the edges of the crystals will often 'fray' from the edges inwards (Boyle et al., 1967; Sawhney and Voigt, 1969), so that they are partially open to some distance into the mineral. This is illustrated in Figure 3. To the left, potassium ions are shown, which are virtually irreversibly fixed in the hexagonal crystal end-surface openings of illite. Next, a caesium ion is shown, which can enter an only slightly opened site at the frayed edge through exchange with a potassium ion. To the right, large, hydrated calcium cations are shown in the matrix. These are too large to enter the strong fixation sites and have too little surface charge density to cause collapse of the interlayer. They will thus remain exchangeable. Cation fixation in clays will generally increase with decreasing tendency to hydration and increasing polarisability of the cation (Shainberg and Kemper, 1967). Hydration of the caesium ion is relatively low, and the

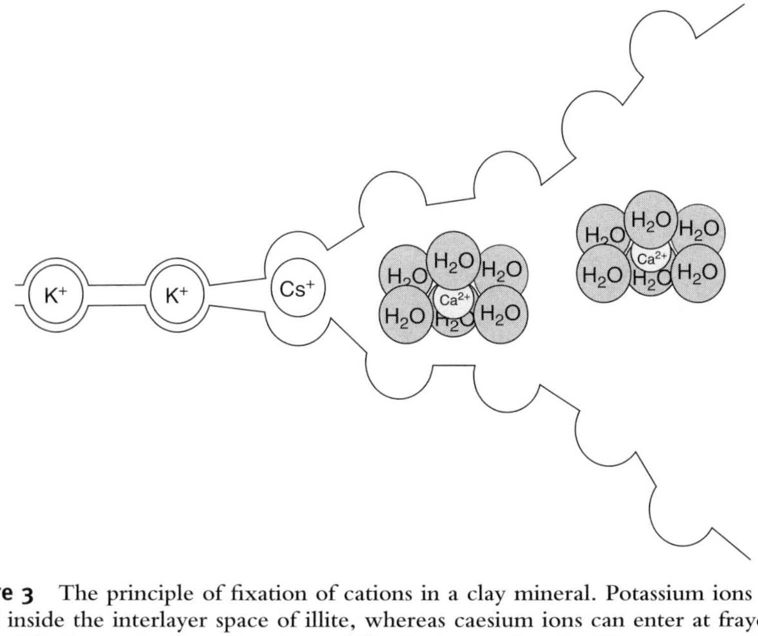

Figure 3 The principle of fixation of cations in a clay mineral. Potassium ions are fixed inside the interlayer space of illite, whereas caesium ions can enter at frayed edges. The larger, hydrated ions have difficulty in entering the illite structure, but can enter similarly structured minerals with considerably greater interlayer spacing (e.g. montmorillonites) (after Kinniburgh and Jackson, 1981, and Sawhney, 1972).

polarisability is high compared with that of smaller cations. Consequently, selectivity of caesium ions in 'high energy' fixation sites in clays is great for caesium.

Other types of similar clay minerals with essentially the same crystal composition as shown in Figure 2, but a different substitution formula for the cations inside the crystal lattice, will have significantly smaller surface charge density. For instance, montmorillonite has substantially less replacement of aluminium for silicon, and its unit cell is thus much more loosely held together than that of illite. This means that water and highly hydrated molecules of various dimensions can enter the interlayer space in montmorillonite. Such large molecules with low surface density can act as 'pillars' between the unit layers of the crystal, preventing crystal interlayer collapse. Therefore, interlayer cations like potassium and ammonium, which would cause a collapse in an illite crystal, will in the montmorillonite interlayer spacing often remain exchangeable with other cations. However, sorption of progressively larger amounts of such cations by montmorillonites and vermiculites can produce a collapse of their interlayers. Montmorillonites collapse to give a 12 Å *c*-axis spacing on saturation with potassium or caesium, whereas vermiculites collapse in the *c*-axis to 10 and 10.8 Å on saturation with, respectively, potassium and caesium (Sawhney, 1972). It has been demonstrated by Sawhney (1970) that at relatively moderate concentrations of added cations, the selectivity for potassium and caesium sorption in frayed edge positions of illite is significantly greater than that in interlayer positions of vermiculite. At high concentrations, the frayed edge sites of illite will become saturated before the interlayer sites of vermiculite.

The cation exchange sites in clays can conveniently be distinguished into three categories, according to the strength of fixation (Cremers et al., 1988; de Preter, 1990; Tamura and Jacobs, 1961; Jackson, 1963):

- 'Regular exchange sites' (RES), which are located on the outer surfaces of the clay minerals.
- 'Frayed edge sites' (FES), which are cation exchange sites of high selectivity, which can in illite only be entered by some cations. Low-hydration cations are selectively sorbed here, causing a collapse of the frayed edges of the crystal.
- 'High-affinity sites' (HAS), which are located inside the interlayer space of the clay minerals. Exchange of cations located here is only possible if the crystal lattice is capable of expanding or if it is dissolved.

Cremers et al. (1988) carried out experiments to investigate the selective 'frayed edge' fixation capacity of clays. The theory is as follows: the large bulky complex silver thiourea (AgTU) has been shown to block the non-specific cation exchange sites. On these sites, experiments have revealed that the AgTU ions have a higher preference than for instance caesium. However, due to steric or selectivity reasons, there are some specific sites that the AgTU ions cannot reach. The subsequent caesium sorption should therefore be restricted to these selective adsorption sites. However, the presence of the AgTU complex may possibly prevent the entrance of caesium ions to the highly caesium-specific interlayer frayed edge positions. The number of highly selective sites found by this method was remarkably small: of the order of 2% of the total mineral cation exchange capacity (Cremers et al., 1988). Although only constituting a small proportion of the total cation exchange capacity of the soil, however, the FES play a very important role in the fixation of monovalent ions with little tendency to take up water and high polarisability (Andersson, 1991). It has been estimated that the natural relationship in mineral soils between caesium located in specific FES and RES is of the order of 50 (Wauters et al., 1994). In experiments with pure clay minerals, it has been demonstrated that whereas the potassium selectivity in RES in illite was only about twice as great as that of calcium, the selectivity in frayed edge positions was about 500 times greater (Bolt et al., 1963). The selectivity of sorption of monovalent cations in clays generally decreases in the order (Krouglov et al., 1998):

$$Cs^+ > Rb^+ > NH_4^+ > K^+ > Na^+$$

All the larger, hydrated cations, for example magnesium and calcium and its metabolically similar element, strontium, will be held significantly less strongly on clays.

Lamm and Nadafy (1973) describe the processes of fixation and remobilisation of cations in clays in the following way:

$$\begin{matrix}\text{Soluble} \\ \text{cations}\end{matrix} \overset{\text{Fast}}{\rightleftarrows} \begin{matrix}\text{Exchangeable} \\ \text{cations}\end{matrix} \overset{\text{Slow}}{\rightleftarrows} \begin{matrix}\text{`Intermediate'} \\ \text{cations}\end{matrix} \overset{\text{Very slow}}{\rightleftarrows} \begin{matrix}\text{Fixed} \\ \text{cations}\end{matrix}$$

Caesium cations will rapidly settle in 'caves' of a size that allows the ions to diffuse in. In the 'intermediate' phase, the cations can be considered to occupy positions on frayed interlayer edges. Here, the presence of positively charged cations will very slowly cause a collapse of the frayed edge, leading to fixation. The rate at which this process takes place is exceedingly slow. Squire and Middleton (1966) described an experiment where 3 years elapsed before the caesium ions were virtually completely fixed in a clay soil. Results obtained after the Chernobyl accident (Roed et al., 1996, 1998; Krouglov et al., 1998) suggest that the caesium fixation process was complete after some 3–7 years. The duration of the fixation process depends on the type and conditions (e.g. with respect to contents of competing cations) of the soil.

It should be noted that large, hydrated ions, such as magnesium and calcium, may be able to expand clay layers held together by potassium or ammonium ions, but not layers held together by caesium, which has particularly low hydration energy (Sawhney, 1972).

Ammonium and potassium ions can to some extent release caesium by ion exchange from FES, where collapse has not yet been induced. The selectivity in this context of the ammonium ion over potassium can be illustrated by the results of application of countermeasures to reduce plant uptake of radiocaesium. Here fertilisation with potassium is a method that is often recommended, as the added potassium in soil solution will reduce the caesium uptake by dilution (Nisbet et al., 2004; Andersson et al., 2000). However, the fertiliser should not contain ammonium, as this has been found to significantly increase the caesium content of the soil solution (Renaud and Maubert, 1997). The addition of potassium fertiliser will also increase potassium concentrations in frayed edge positions, thus inducing lattice interlayer collapse (Hird et al., 1996). Thereby, the fixation of caesium ions already present between the layers will become even stronger. Conversely, an interlayer collapse induced by potassium can slow down the process of strong fixation of caesium from soil solution. Large cations, like calcium and magnesium, may keep the frayed edge gaps open and readily accessible to caesium ions.

3.2. Retention in soil of contaminant ions by different mechanisms

Organic material often offers a very high number of cation exchange sites. The total cation exchange capacity of the organic part of a soil may be hundreds of times greater than that of the clay part (Schnitzer and Kahn, 1978). The organic cation exchange capacity of the soil decreases with decreasing soil pH (Brady, 1970).

It may be a matter of definition whether cations are prevented from migrating by the mineral fraction or the organic fraction of the soil, since they can form a complex of organic material bound to the surfaces and edges of a mixture of minerals by inter-ion Coulomb forces and van der Waals attractions between molecules. In practice, the fraction of organic matter combined with inorganic matter is less than 50% only in extremely sandy soils (Schnitzer and Kahn, 1978). In fact, humic molecules sorbed at edges of clay minerals may prevent the mineral layers from collapsing, thus maintaining a high number of frayed edge or interlayer exchange sites for, for example the caesium cation to diffuse into. However, since interlayer collapse cannot be induced, these cations are not irreversibly fixed in the soil (Krouglov et al., 1998).

Two possible mechanisms are envisaged for fixation of ions by organic matter in the soil. One is by simple cationic exchange with more or less loosely bound ions and the other is by chelation. Chelation generally produces a complex, which is not easily disturbed by water flow or microbial attacks. A chelate is a complex which is formed when an ion and an organic molecule are bound to each other by more than one bond, thereby forming a 'cyclic' structure. Although chelate formations can have different degrees of stability (Allison, 1973), they are usually very stable. The stability increases with the number of atoms which form a 'ring' with the captured cation and with the number of 'rings' participating (Chaberek and Martell, 1980).

Monovalent ions like caesium will be weakly bound to organic material following simple cation exchange. However, the retention in organic-rich surface layers, for example in some forest soils contaminated by the Chernobyl accident, has been found to be relatively high, and the downward migration is thus slow (Junker et al., 1998). This may possibly reflect a slow decomposition of contaminated litter or the role of organic molecules in maintaining frayed edge or interlayer exchange sites in a very small mineral fraction of the soil. However, the mobility of Chernobyl caesium has, as expected, generally been found to be considerably greater in peaty soils than in mineral soils (Askbrant et al., 1996). In particular, the retention of radioactive contaminants by organic matter may be sensitive to changes in pH, for example brought about by the presence in the soil of other pollutants. Studies conducted in the vicinity of a large copper–nickel smelter on the Kola Peninsula have indicated that residence half-times of ^{137}Cs and ^{90}Sr in organic top layers of the soil decrease with decreasing distance from the smelter. This is believed to be connected to the inhibited conversion of litter to humic substances near the smelter (Bunzl et al., 2001).

It has been reported (Duff et al., 1963; Sørensen, 1990, personal communication) that the activity of certain acid-producing bacteria may cause solubilisation and subsequent binding of cations from various humic substances, and in certain cases (2-ketogluconic acid) even from minerals which are normally considered insoluble. It has also been found that smaller chelating molecules like fulvic acid may have an important influence on the downward migration of cations in the soil.

Strontium ions are loosely held on the outer surfaces of minerals, and are generally more strongly bound in peat soils than in mineral soils (van Bergeijk et al., 1992). This probably reflects the chelation of the strontium ion with humic substances, such as humic acid. The most important naturally occurring ions competing for the same ion exchange sites as strontium are calcium and magnesium. It has for instance been demonstrated in field experiments that the amounts of soluble calcium present in the soil very strongly influences the content of strontium in soil solution (Lönsjö and Haak, 1975). Organic soils preferentially sorb calcium over strontium (Juo and Barber, 1969). As mentioned above, much of the strontium deposited after the Chernobyl accident was in complex physico-chemical forms, and often ionic binding of Chernobyl strontium is delayed for a period of some years until weathering of the deposited particles leads to the release of the strontium ion to the soil matrix. Strontium particles originating from a radioisotope thermoelectric generator (RTG) would often be in ceramic form (as strontium titanate), and can be powdered, but is insoluble in water, and thus relatively inert in the environment if spread with a malicious dispersion device, such as a 'dirty bomb'. However, existing strong strontium sources that could fall into the wrong hands can have various chemical forms, including soluble compounds (Ferguson et al., 2003; Argonne, 2005).

Ruthenium ions may assume various chemical forms (including complexes) in soils. In general, the cationic forms will be fixed, whereas the anionic forms will remain mobile in the soil (Coughtrey et al., 1986). Under acid conditions the +3 and +4 oxidation states will dominate (Essington et al., 1963). Ruthenium in nitrosyl complexes in soil will be mobile (Schultz, 1965). However, Jones (1960) found that nitrosyl ruthenium forms insoluble complexes with the surface layer of ferric hydroxide on sediment sand and silt particles. Ruthenium ions have been found to be significantly less efficiently retained by clays than caesium ions (Livens and Baxter, 1988), presumably due to hydration and polarisability, as described above. However, the effect of fixation by organic ligands would be more marked for multivalent ions like ruthenium. Consequently, organic material plays an important role in the binding of ruthenium in soils. Investigations have in general shown that ruthenium is less mobile in soil than is strontium (Nishita et al., 1956; Bunzl et al., 1984).

Plutonium behaviour in soil is relatively complex, owing to its many different natural oxidation stages. It has been reported that plutonium is relatively strongly retained by both clays and organic matter, as well as in hydrous oxides, and its vertical migration in soil is consequently slow (Coughtrey et al., 1986; Bunzl et al., 1998; Livens and Baxter, 1988). Cook et al. (1984) found that in the immediate environment of the Dounreay Nuclear Power Development Establishment, the major part of the plutonium in soil was held in insoluble organic chelated complexes. The plutonium radionuclides that might occur in the environment after a contaminating incident are, with the exception of the very weak beta emitter ^{241}Pu, alpha emitters, and would thus only constitute a hazard to humans if they enter the body (e.g. through inhalation or consumption).

Iodine is readily soluble in water, and has been reported to bind strongly in soils. Iodine in soil has been found to be associated with clays as well as organic materials and metal oxides (Koch-Steindl and Pröhl, 2001; Kocher, 1991). Analyses indicate that the residence time of iodine in the top 30 cm soil layer is of the order of 40 years. Hou et al. (2003) found that in areas of Russia contaminated by the Chernobyl accident, about 70% of the radioiodine in soil samples is bound to oxides and organic matter. About three-fourths of the radioiodine contamination was so strongly bound here that it was not extractable by 30% H_2O_2–HNO_3 (at pH 2).

Cobalt in soil is to a great extent sorbed on the surfaces of clay minerals, and not in selective mineral fixation sites (Norrish, 1975; Solovitch-Vella and Garnier, 2006). Only a small fraction appears to be associated with organic substances. It has nevertheless in general not been reported to be highly mobile in soils (Willard, 1975). Rahman et al. (2008) reported comparatively high cobalt mobility in soils with low pH, exchangeable K^+ and clay content. Strong ^{60}Co sources that might be applied in a malicious attack, if they come into the wrong hands, include teletherapy sources, which are mostly in metallic form (Ferguson et al., 2003). However, metallic cobalt can be dissolved in acid to produce liquid sources that can be easier to disperse and would have enhanced environmental mobility.

It should be noted that, during periods of frost, deposited contaminants can also be retained in snow cover on the ground, until the first thaw. In an experiment on a soil area with a slope of about 3°, it was found that about half of a ^{90}Sr contamination was carried away from the area with run-off water during snow melt (Qvenild and Tveten, 1986). In this case, an underlying ice surface provided a shield against deep soil contamination. The corresponding removal of a ^{137}Cs tracer in another experiment carried out by the same workers was much less (Qvenild and Tveten, 1984).

3.3. Binding strength and migration of contaminants in areas of soil

Looking at Figure 4, which shows the vertical distribution of ^{134}Cs, ^{137}Cs and ^{106}Ru in 1990 in a mineral soil in Russia, which had been contaminated 4 years earlier by the Chernobyl accident, it is remarkable that about 95% of the radiocaesium still lay in the top ca. 2 cm layer. This is a generic observation for many mineral soils (e.g. Andersson and Roed, 1994; Salbu et al., 1994; Ivanov et al., 1997). It is also evident that the ruthenium is, as would be expected from the above considerations, more mobile than the caesium. However, a simple vertical profile merely gives information on the extent of contaminant migration at a given point in time for the particular soil. In order to investigate the strength of fixation of the contaminants in the soil and thus enable estimation of the future behaviour of the contaminants, it is valuable to subject samples to a sequence of chemical extraction procedures, which will yield information on the type and strength of the binding in the soil.

The reagents used in sequential chemical extractions are selected to distinguish between different binding mechanisms of contaminants in soil or other materials. Many commonly used procedures are essentially further developments and variants of the method devised by Tessier et al. (1979) for speciation of particulate trace metals in solid materials, although targeted for specific tasks. Typically, the first extraction may be performed with an inert solution, for example 1 M ammonium acetate at pH 7, to extract

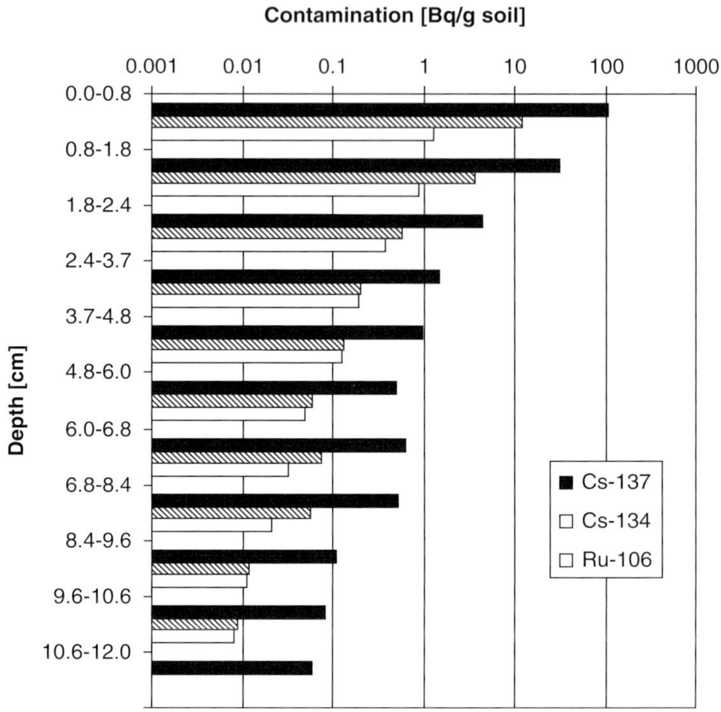

Figure 4 Vertical soil profiles of ^{134}Cs, ^{137}Cs and ^{106}Ru in the summer of 1990, measured in a mineral soil in Russia (data from Andersson and Roed, 1994).

contaminants that are weakly held and can be readily exchanged (Riise et al, 1990; Andersson and Roed, 1994). The next extractant could then be a solution of hydroxylamine hydrochloride, mildly reducing the sample, followed by hydrogen peroxide (adjusted to pH of about 2 with nitric acid), which oxidises the organic fraction of the soil and releases the radionuclides bound to this fraction. The latter would also oxidise fuel particles, leading to a release of radionuclides bound in particle matrix (Salbu et al., 1998). A final extraction with a strong solution of nitric acid would lead to a release of part of the strongly bound contamination. The residue of contamination after the five sequential extractions would represent very strongly (mineral) bound contamination (Andersson et al., 2002a).

Figure 5 shows an example of the results of sequential extractions on a soil sample taken in 1992 near the Chernobyl nuclear power plant. In line with the above descriptions of different sorption and fixation mechanisms, it is clearly seen that a large part of the strontium contamination here is bound to oxidisable particles, whereas the caesium contamination is strongly bound, due to the specific clay mineral fixation mechanisms. In other contaminated areas, where the strontium contamination is not

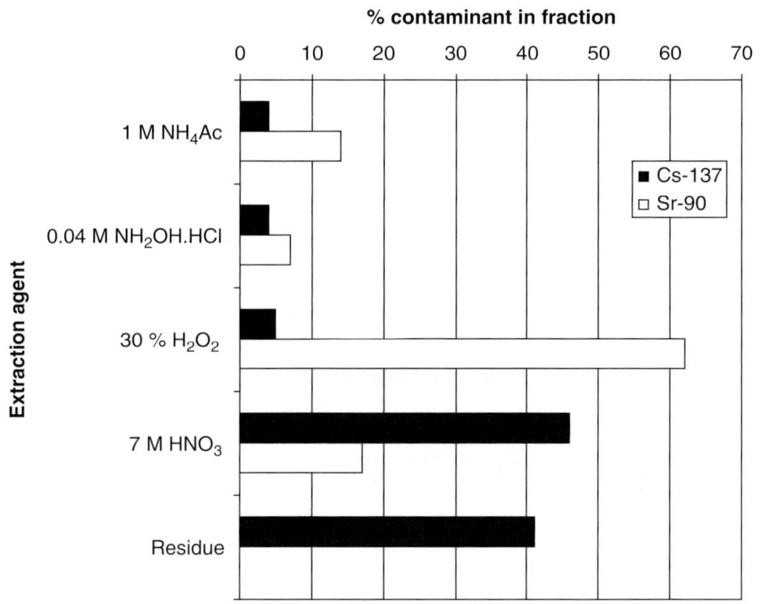

Figure 5 Radiostrontium and radiocaesium released by sequential extractions from a soil sample taken in 1992 near the Chernobyl Nuclear Power Plant (data from Salbu et al., 1998).

associated with particles, most of it has been reported to be released by the two mildest extractions (Salbu et al., 1998). Figure 6 shows an example of how the caesium from Chernobyl became slightly stronger associated with the soil over the period of 4–10 years after the contamination took place.

Figure 7 shows how the fixation of ^{137}Cs contamination in a soil (identical to that described in Figure 4) can vary between soil layers. The fact that soil fixation of caesium increases with time may be reflected in the smaller fraction of the most strongly bound residue in the deeper soil layers. Figure 8 demonstrates that ruthenium in the same soil is bound completely differently (the first, second and fourth extraction resulted in releases that were below the detection limit – i.e. each less than 3% of the total). Here, much of the contamination is in the oxidisable fraction, and although the soil profiles in Figure 4 show that in 1990 a very large part of both the caesium and ruthenium was present in the upper ca. 2 cm of the soil, it is clear that the ruthenium can much more easily be leached than the caesium. Therefore, it should be expected that the vertical ruthenium profile in the soil would subsequently become significantly deeper. However, due to the comparatively short physical half-life of ^{106}Ru, it has not been possible to investigate this over long time periods.

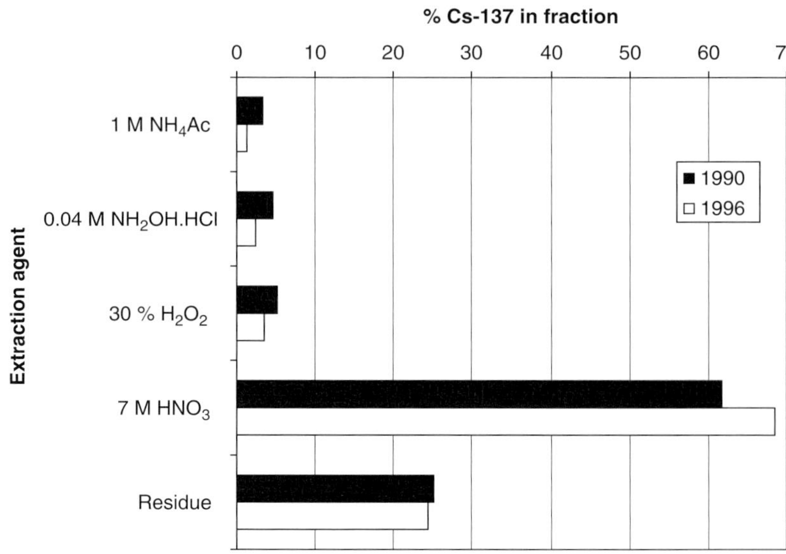

Figure 6 Changes with time in fixation of radiocaesium in a soil sample taken near the Chernobyl Nuclear Power Plant (data from Salbu et al., 1998).

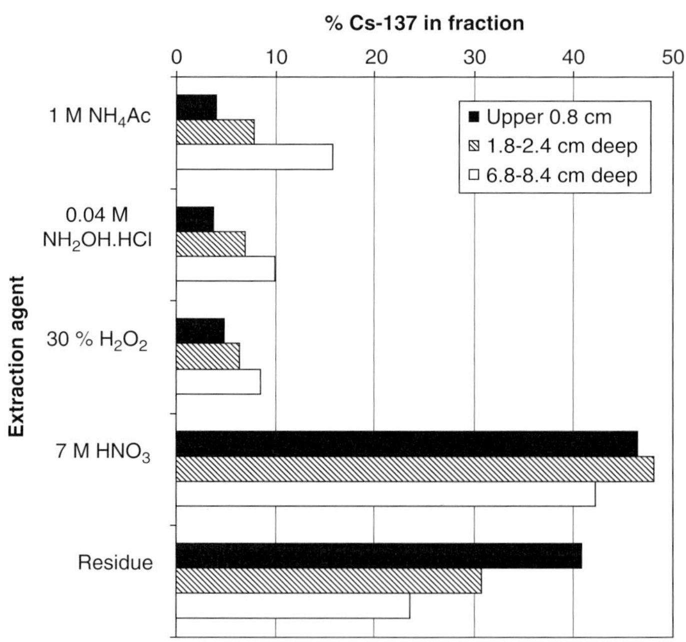

Figure 7 Radiocaesium (^{137}Cs) released by sequential extractions from three layers of a soil sample taken in 1990 near the Chernobyl Nuclear Power Plant (data from Andersson and Roed, 1994).

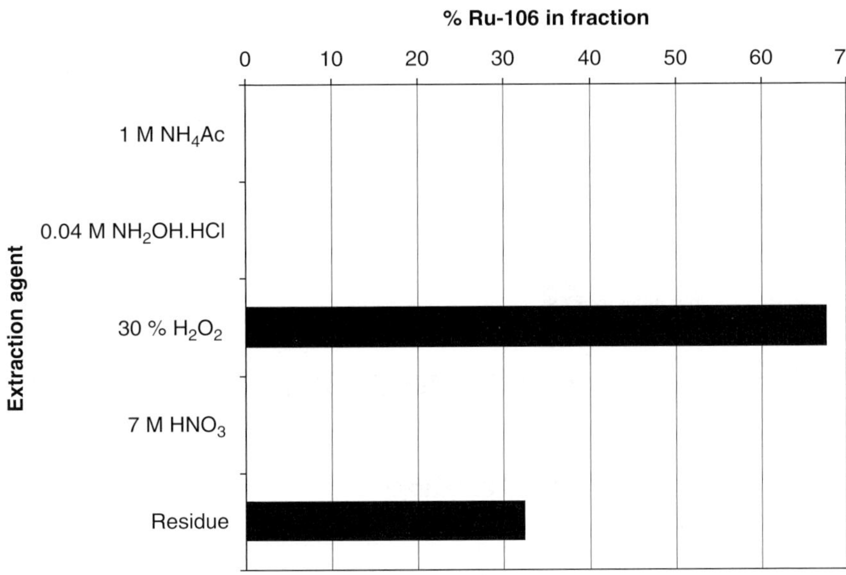

Figure 8 Radioruthenium (^{106}Ru) released by sequential extractions from a soil sample taken in 1990 near the Chernobyl Nuclear Power Plant (data from Andersson and Roed, 1994).

Figure 9 shows an example of the influences of soil composition (clay content) on contaminant fixation/migration. Both illustrations show the vertical radiocaesium distribution in an undisturbed soil in the Bryansk Region (Russia), 10 years after the Chernobyl accident (Roed et al., 1998). The conditions of contaminant deposition are reported to be identical. However, the profile shown to the left in the figure is for a very shallow, sandy soil, whereas the one to the right had clay content comparable to that typically found in Scandinavian soils.

Different types of models exist for prediction of future depth profiles of contaminants in soil, which may for instance be applied for dose estimation in decision support systems. The most commonly used of these can broadly be categorised as follows:

- Convective–dispersive models, describing the activity at a given time and depth in the soil by a formula, where decisive parameters can be the dispersion coefficient, the pore water velocity and distribution coefficient in the soil, as well as the bulk density and porosity of the soil. Such physically based models often provide a high degree of flexibility (Konshin, 1992; Schuller et al., 1997; Szerbin et al., 1999; Bunzl et al., 2000; Krstic et al., 2004; Jones et al., 2006; Kirchner et al., 2008).
- Compartment models, where the soil is divided into a number of layers of arbitrary thickness. Such models generally assume that the transfer rate

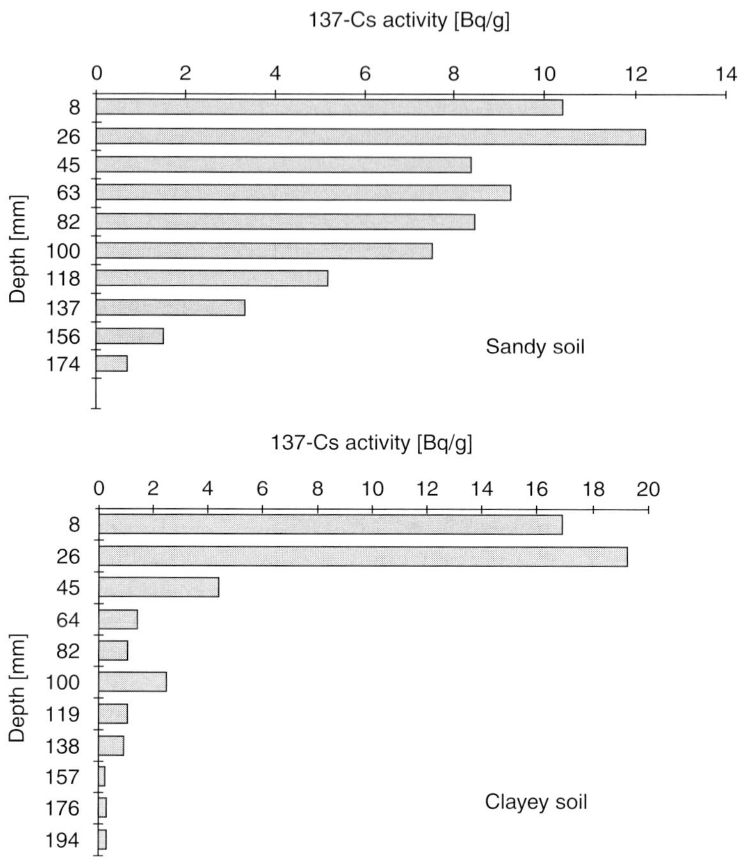

Figure 9 Vertical radiocaesium profiles in two soils in the Bryansk region. Samples taken 10 years after the Chernobyl accident contaminated the area.

for activity in a given layer, m, can be written as $\mathrm{d}X_m/\mathrm{d}t = \sum_{n=1}^{P} S_{nm}X_n - \sum_{n=1}^{P} S_{mn}X_m - L_mX_m$, where X_m and X_n represent the activity in compartments m and n , respectively, at a time t. S_{nm} is a transfer coefficient from layer n to m, and L_m is a transfer coefficient representing loss from the system, for example by radioactive decay (Crick et al., 1987; Andersson, 1990; Bunzl et al., 2000).

- Models fitting measured data to an exponential or other function dynamically describing the vertical distribution of contaminants (Chamard et al., 1993; Velasco et al., 1997).

Common for most applications of such models is the assumption that migration rates will not change significantly with time. As discussed above, however, that assumption cannot be assumed to be generally valid.

Naturally, many of the areas of soil in an inhabited area are covered by vegetation, and this will intercept part of the contamination falling on the area, depending on the mode of deposition (wet or dry). As mentioned in Section 3.1, the migration of contaminants from a grass cover to the upper soil layer will typically take place over days or weeks. Contamination on trees and bushes can be very important to take into account when estimating external doses in an inhabited area. This is partly because of the high dry deposition velocity to such vegetation, particularly if in leaf, and partly because much of the radiation passing into a dwelling after being emitted from contamination on trees will go through windows, which will offer very little protective attenuation. A significantly larger proportion of the radiation entering a dwelling after emission from contamination on horizontal surfaces will have passed through thick walls and lost energy.

It is important to distinguish between coniferous and deciduous vegetation. According to the findings of Roed (1988), after the Chernobyl accident, the needles or leaves received some 98% of the bulk aerosol deposition on a tree. For elemental iodine, this value was found to be 75%. Deposition to bare trees is thus in general of comparatively low importance – especially for aerosols. Whereas the foliage of deciduous trees or bushes will normally be lost over a single autumn, taking with it by far the most of the contamination, conifers will retain contaminated foliage over a time span of typically 2–7 years, depending on the species (Jones et al., 2006). According to the findings of Mamikhin and Klyashtorin (2000), the retention half-life of contamination deposited on the stem and branches of a tree is of the order of 2 years.

Another type of vegetation that may be found in inhabited areas is edible crops grown in kitchen gardens. Kitchen garden products grown in inhabited areas constitute a significant part of the diet in many European countries. For instance, a common Swedish household annually consumes 57 kg of home-grown vegetables (Rosén, 2006). Almost half of this is potatoes, but other common kitchen garden products include apples, berries, carrots, onions and lettuce. The value of this home-grown food corresponds to about 11% of the total value of the vegetables consumed in Sweden. Kitchen gardening may in other regions of the world be even more common (Kaduka, 2003, personal communication). The transfer of radioactive matter to edible crops is very extensively described elsewhere (e.g. Eriksson, 1977; IAEA, 1994; Nisbet et al., 1999; Kostiainen et al., 2002). However, the investigations made so far have focused on the growth of crops in large agricultural areas. Kitchen gardens in inhabited areas have much smaller dimensions, and the obviously different culturing practice in such ground lots within an inhabited community would be expected to result in very different crop growth conditions, and therefore different contaminant transfer factors.

4. Migration of Radionuclides on Anthropogeneous Surfaces in an Inhabited Environment

As mentioned above, contamination deposited on grass/soil would generally be expected to contribute considerably to the long-term external dose received in suburban or other types of environments characterised by large open, unpaved areas. However, in city areas, where most people live or work, open, unpaved areas, if any, would often be small. Here, other types of contaminated areas (e.g. paved horizontal areas and building surfaces) can give high time-integrated contributions to external dose. Some anthropogeneous surfaces in the inhabited environment retain very little of the deposited contamination against wind and weather, whereas other types of surfaces may retain much of the deposited contamination, even after having been exposed to the weather for many years (e.g. Roed et al., 1998). The inhabited environment exhibits a complexity of surfaces of different material characteristics and orientation, on which deposited contaminants will behave differently. Prior to the Chernobyl accident, it was in general not considered likely that any plausible incident leading to airborne dispersion of radioactive contaminants would significantly affect inhabited areas. Therefore, the detailed study of the behaviour of radionuclides on anthropogeneous outdoor surfaces was practically initiated by that single event. However, it also means that a very large part of the investigations that have been made of contaminant behaviour in inhabited areas relates to radiocaesium, which was the contaminant of concern in the Chernobyl case. In general, not only soil, but also most construction materials retain caesium stronger than other contaminant ions, which are more easily removed from surfaces by natural processes. Generic external dose estimates for inhabited areas based on data for the environmental migration of caesium may thus be highly conservative, and quite possibly misguiding.

4.1. Migration of contamination deposited on roofs

Already soon after the Chernobyl accident, it was clear that various radionuclides behaved very differently on different types of roofing material. When an area receives rainfall, part of the rainwater will not be retained by the surface to which it is deposited, but runs off. This run-off water will carry with it a part of the contamination in the deposited rain, depending on the amount of rainfall and characteristics of the contaminants, as well as the type and slope of the surface from which the run-off occurs. At Risø National Laboratory in Denmark the deposition and run-off of contaminants was measured in connection with the first rainfall after the

Chernobyl accident, which occurred on 8 May 1986. The dry deposition that had until then occurred to the roof pavings was reported to be small in comparison (Roed, 1987). Table 1 shows the amounts of run-off water measured per unit of horizontally projected surface area on each of six different types of roofs. The total amount of rainfall was 9.2 mm. Very nearly all the deposited rainwater ran off the smooth impermeable silicon-treated Eternit roof with 45° slope, whereas for instance only slightly more than half of the rainwater ran off the red clay tiles, which clearly had much higher open porosity. However, as can be seen from Table 1, even the rainwater which virtually completely ran off the silicon-treated Eternit left about one-fourth of the deposited caesium and larger fractions of all other radionuclides, except iodine, on the roof. The results in Table 1 are in reasonable agreement with findings from earlier experiments employing artificial contamination in water (Qvenild and Tveten, 1983).

After the following 22 days, it was found that the rain had washed off different fractions of the remaining contaminants on the roofs. The total rainfall over this period was measured to be 320 mm. Table 2 shows the amounts of rainwater running off the roof measured per unit of horizontally projected surface area on each of six different types of roofs. The original data (Roed, 1987) show that the figures in Table 2 are the result over the period investigated of a combined process of wash-off removal by rain and further deposition/run-off with rain of contaminants from increasingly depleted air concentrations.

Table 1 Amounts of rainwater at Risø running off different types of roofs and fractions of different contaminant radionuclides removed with the run-off water, during the first rain shower after the Chernobyl accident.

Roof type	Run-off (mm water)	Removed %						
		^{137}Cs	^{134}Cs	^{131}I	^{103}Ru	^{106}Ru	^{140}La	^{140}Ba
Concrete tile, 45° slope	7.4	42	38	100	73	58	32	31
Red clay tile, 45° slope	5.0	32	29	57	47	40	31	32
Corrugated Eternit, 45° slope	7.8	13	12	100	36	35	32	32
Silicon-treated Eternit, 45° slope	9.2	78	71	100	59	37	67	68
Corrugated Eternit, 30° slope	5.6	20	19	100	45	41	37	32
Silicon-treated Eternit, 30° slope	6.3	82	75	100	53	51	78	78

Source: Data from Roed (1987).

Table 2 Amounts of rainwater at Risø running off different types of roofs and fractions of different contaminant radionuclides removed with the run-off/wash-off water, during the first 22 days after the first rain shower following the Chernobyl accident.

Roof type	Run-off/ wash-off (mm water)	Removed %						
		^{137}Cs	^{134}Cs	^{131}I	^{103}Ru	^{106}Ru	^{140}La	^{140}Ba
Concrete tile, 45° slope	29.7	10	10	100	93	86	22	22
Red clay tile, 45° slope	26.7	19	19	47	53	49	20	17
Corrugated Eternit, 45° slope	29.4	6	6	100	88	81	14	15
Silicon-treated Eternit, 45° slope	30.4	70	70	100	95	90	69	56
Corrugated Eternit, 30° slope	29.5	4	4	100	60	56	18	17
Silicon-treated Eternit, 30° slope	26.5	48	48	100	75	68	58	58

Source: Data from Roed (1987).

It is quite clear from the data in Table 2 that the caesium radionuclides are in general much more strongly retained on the surfaces than are other contaminants (except perhaps on the silicon-treated surfaces, which are very smooth). A wide range of urban construction materials contain the micaceous mineral illite, which is believed to be the cause of the strong fixation of the caesium ion in soil (see Section 3). For instance, illite occurs naturally in limestone used for producing cement (Sprung and Siebel, 1991), and it is sometimes added to cement-based materials to make stronger and more ductile products (Moukwa et al., 1993). Also the mineral tricalcium disilicate hydrate (tobermorite) is sometimes applied for this purpose (Sato et al., 1997), and this mineral, which has a layer structure similar to that of illite, has also been reported to exhibit some selectivity for caesium in solutions containing competing cations (Shrivastava and Shrivastava, 1998; Ma et al., 1996). de Preter (1990) found that slate products like Eternit also contain minerals capable of strongly and highly selectively binding the caesium ion.

Illite is usually present in the materials used to make clay tiles and bricks (Andersson et al., 1995; Strazzera et al., 1997). Although such building materials are fired at high temperatures, it might thus still be that any strong caesium fixation in these could be related to the presence of detrital micaceous structures. As an example, consider the changes which illite undergoes by firing, for instance in brick and tile production. According to

Wahl (1965), the first change is a partial loss of loosely bound water at 100°C, but the final dehydration will take place rapidly between 450 and 550°C. For all clays of the three-layer type (see Figure 2) this process will bring about a reduction of the interlayer spacing as the water contributes to the retention of this spacing. Apart from that, no marked change in structure will occur until a temperature of 850–900°C is reached. Here the initial structure will begin to vanish. On further heating, the material will persist until about 1,100°C (Cole and Segnit, 1963). Amorphous vitrification of the clay will set in at approximately 1,200°C. As clay roof tiles are usually fired at 900–1,200°C (Frederiksholm Tile Factory, 1989, personal communication), some of them are likely to contain considerable amounts of intact mica. In fact, caesium selectivity could be improved in some clays of the montmorillonite type by firing at 600–700°C, whereby the interlayer spacing is reduced (Tamura and Jacobs, 1961). Andersson (1991) demonstrated the effect of clay tile firing temperature on caesium fixation by soaking pieces of common British clay tiles, which were identical except that one had been fired at 700°C and the other at 1,100°C, in a solution of ^{137}Cs. After the ^{137}Cs had adsorbed on the tile pieces they were left to soak for some time in solutions of various cation exchangers (see Figure 10). This displaced about five-sixths of the caesium on the tile fired at 1,100°C, but less than one-fifth of the caesium on the tile fired at 700°C. X-ray diffraction analysis of the clay material prior to firing showed a high content of illite, and whereas the densities of the two tile pieces fired at different temperatures were similar, the BET (Brunauer–Emmett–Teller) surfaces of the tiles fired at 700 and 1,100°C were found to be, respectively, 24.4 and $0.3\,m^2\,g^{-1}$, indicating that the high microporosity of the original layered mineral material was lost at the higher firing temperature and only macropores (cracks) were left.

Andersson (1991) also demonstrated that the number of strong (virtually irreversible) caesium fixation sites on an ordinary clay tile from a roof in Oxfordshire was great compared with the typical amounts of naturally occurring caesium depositing in British rainwater. In fact, it would take thousands of years just to fill up these sites, which are the ultimate destinations of adsorbed caesium on clay tiles. This means that even tiles that have been exposed to the weather for centuries can still be expected to have a marked capacity for absorbing and retaining fallout caesium. As expected, Real et al. (2002) showed that strontium ions are much more easily removed from clay- and concrete-based construction material samples by soaking in water than are caesium ions, and Andersson (1991) demonstrated that ruthenium is less strongly retained on clay tiles than caesium.

Table 3 shows the long-term (over a period of 14.5 years) decline in Chernobyl radiocaesium contamination level on the different types of roofs mentioned above. Each of the figures shown represents the average value of

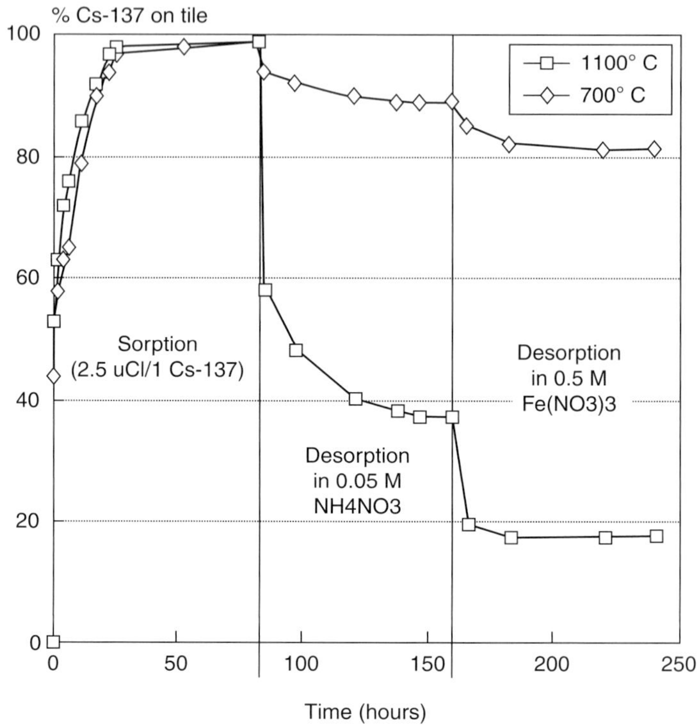

Figure 10 Sorption and desorption of ^{137}Cs on clay roof tile pieces fired at different temperatures. Desorption was carried out in solutions of, respectively, ammonium and ferric nitrate (data from Andersson, 1991).

a large series of measurements on individual sheets from the same roof, which showed very little variation (less than 15%). The results are consistent with measurement data obtained in Germany over the first 2 years after the Chernobyl accident (Jacob et al., 1987; Roed and Jacob, 1990), and the generally high retention of ^{137}Cs by clay tiles and slates over the earliest months agrees with the findings of Sandalls and Gaudern (1988) on wet-contaminated roofs in West Cumbria, UK. The results are also in reasonably good agreement with those of another series of measurements carried out at Risø, Denmark. Here clay roof tiles that had been contaminated by Chernobyl fallout in Gävle, Sweden, were brought to Risø after 4 years of exposure to wind and weather (Andersson et al., 1995). The contamination level at this stage was assessed and the tiles were then exposed to Danish weather after having been placed on a specially constructed scaffold. A decrease in the radiocaesium contamination level by between 28% and 35% was recorded over the following 19 months. In a pre-Chernobyl experiment, Wilkins (1987) found that the removal of caesium from roof tiles appeared only to occur during periods of rainfall.

Table 3 Measured ^{137}Cs contamination levels on different types of roofs at Risø, Denmark (given as percentages of the initial contamination, 760 Bq m^{-2}, as measured from a 10 m^2 rain sampler).

Roof type	11 days	100 days	7 months	13 months	19 months	25 months	14.5 years
Concrete tile, 45° slope	59	49	45	40	35	30	24
Red clay tile, 45° slope	68	62	59	55	48	47	25
Corrugated Eternit, 45° slope	87	81	79	76	73	68	41
Silicon-treated Eternit, 45° slope	21	15	13	12	12	11	7
Corrugated Eternit, 30° slope	80	75	72	69	67	64	35
Silicon-treated Eternit, 30° slope	19	14	11	10	9	9	5

Source: Data from Andersson et al. (2002b).

Measurements made in the Novozybkov area of Russia in 1995 and 1997 (Fogh et al., 1999; Roed et al., 1998) showed that, under special conditions where slate roofs are overgrown with algae and fungi and nothing is done to clean the roof, a considerably greater part of the initially deposited caesium may be retained. This phenomenon may be ascribed to the large and rough surface of the algae that covered the asbestos roof, acting as a contaminant sink and offering a great number of ion exchange sites. It is known that metal ions may become strongly bound in various types of algae (Lujan et al., 1994).

On the basis of the above assessments, it would appear that the natural reduction (by weathering and radioactive decay) in caesium contamination level on roofs can generally be described by a function of the type:

$$C(t) = 0.5C_0 \exp\left(\frac{-t \times \ln(2)}{T(1/2)}\right)(\exp(-\lambda_1 t) + \exp(-\lambda_2 t)) \qquad (1)$$

where $C(t)$ is the contamination level at time t, C_0 is the contamination level at time 0, $T(1/2)$ is the radiological half-life of ^{137}Cs (30.07 years) and λ_1 and λ_2 are rate constants corresponding to half-lives of, respectively, some 1–4 and 25–50 years. If contamination takes place in snow or to a snow cover, both the 'run-off' and 'weathering' processes may be delayed until the first thaw (Qvenild and Tveten, 1984, 1986).

It should be stressed that if the deposited contamination is in the form of large, insoluble particles, as might for instance result from physical fractionation in connection with some types of 'dirty bomb' explosions (Harper et al., 2007), it would be expected to be removed much more quickly from impermeable surfaces by the weather (Andersson, 2005). Also, other contaminants than caesium would, as indicated by the values in Table 2, be likely to be weathered off the surface comparatively more quickly.

It should also be noted that rainwater which runs over a contaminated roof and is led to the roof gutter may carry with it much contamination. This has in some cases over time been reported to result in rather strong and deep vertical contaminant profiles over small areas of soil around the down pipe (Roed et al., 1996). Contaminated wash-off water led to sewers might also give rise to problems with sludge management in sewage treatment plants (Gallego, 2006).

4.2. Migration of contamination deposited on walls

As clay bricks are usually fired at very high temperatures (of the order of 1,500°C) the mica fixation theory is irrelevant in this case. Hill (1960) demonstrated that, on firing above about 1,100°C, illite undergoes shrinkage and there is a marked decrease in porosity along with a marked coarsening of the pore structure (as also discussed under Section 4.1). This difference between clay brick and clay tile characteristics was also detected by Sandalls (1987) in the measurements of open porosity and surface area of samples of different construction materials. Laboratory experiments by Sandalls (1987) and Andersson (1991) showed that caesium on clay bricks was considerably less strongly bound than caesium on clay tiles. During measurement campaigns to the Gävle area of Sweden, which is one of the areas outside the former Soviet Union that were most strongly contaminated by the Chernobyl accident, *in situ* measurements were made of the ^{137}Cs contamination levels on walls. The results of the measurements on different types of walls in Gävle over the first decade after the Chernobyl accident are shown in Table 4.

It is clearly seen from this data set that although there are no specific caesium fixation mechanisms in the clay brick walls, the natural decrease in contamination level through weathering processes is exceedingly slow. This is in line with findings made on sandstone and clay brick walls in the town of Pripyat, very near the Chernobyl power plant (Roed and Andersson, 1996). The reason for this is probably that the main mechanism responsible for the contaminant migration on surfaces in an inhabited area is wash-off with rainwater. However, due to their vertical orientation, this process has very little influence on the contamination on walls.

Some samples taken in 1991 from a plastered wall in Gävle at a height of 40 cm showed several times higher levels of both ^{137}Cs and ^{40}K than did samples taken at heights of 80 or 120 cm. This indicates that soil (or street

Table 4 Measured ^{137}Cs contamination levels ($kBq\,m^{-2}$) on various types of walls in Gävle, Sweden, where it rained as the contaminated cloud from Chernobyl passed.

Wall type	Orientation	1987	1988	1991	1993	1994	1996
Plastered	South-east		0.78 ± 11%			0.53 ± 12%	
Plastered	South-west		0.55 ± 16%			0.21 ± 27%	
Red clay	South	1.93 ± 9%	1.65 ± 9%				
Red clay	North		3.93 ± 7%	1.85 ± 7%	2.13 ± 15%	1.89 ± 7%	1.36 ± 10%
Yellow clay	North		0.42 ± 15%			0.14 ± 36%	
Yellow clay	South		1.06 ± 11%	0.80 ± 10%	0.76 ± 10%		
Plastered	North	0.99 ± 10%		0.98 ± 10%			
Plastered	East		1.14 ± 10%			0.59 ± 12%	

Source: Data from Andersson et al. (2002b).

Note: The contamination level on grassed reference surfaces in the area was in the range of 150–200 $kBq\,m^{-2}$.

dust) splash may significantly influence the contamination level on the lower part of the wall.

Generally, the half-life of the weathering process appears to be as long as 5–10 years. Therefore, contamination of walls is in general not quite as unimportant for the long-term dose (Andersson and Roed, 2006) as one might imagine from the low deposition velocities to these surfaces quoted in Section 4.3 of Chapter 2. The influence of material differences on this time factor is indistinguishable from the available data sets. Also some contaminant migration is likely to occur *into* the structure of porous construction materials (Sandalls, 1987; Roed and Andersson, 1996). Even the slightest penetration could have great bearing on the beta dose rate from the contaminants.

Naturally, dwellings also comprise window surfaces, which are much smoother than walls, and which are likely to be polished at regular time intervals. Also the deposition velocity of contaminants on such surfaces has in general been found to be lower than that to other parts of the wall (Roed, 1990). However, this by no means implies that houses with large window areas are safer to live in if an airborne contamination occurs. The thin window material will not provide much shielding against radiation from the usually comparatively strongly contaminated outdoor surfaces.

4.3. Migration of contamination deposited on horizontal paved surfaces

Experiments made by Andersson (1991) showed that radiocaesium cations are not intercepted or retained well by bitumen surfaces. In the same publication, it was demonstrated that the downward penetration of caesium into an asphalted road surface subjected to weathering and traffic for 2 years after having been contaminated by the Chernobyl accident was less than 1 mm. However, de Preter (1990) showed that street dust, which will normally cover much of a street surface in a thin layer and which to some extent originates from erosion and weathering of urban surfaces, contains high numbers of selective and virtually irreversible caesium fixation sites, indicating contents of minerals of similar characteristics to those responsible for the selective caesium fixation on other surfaces in the inhabited area, including soil. Any contamination bound to street dust would inevitably be removed over a relatively short period of time by the weather and mechanical impact (e.g. traffic and routine street cleaning). This ties in with the results of a series of measurements made on horizontal asphalt and concrete-paved areas in Gävle, Sweden (see Figure 11). On these measurements, statistical uncertainties ranged between 5% and 20%. The variation between data recorded during the same measurement campaign (i.e. at virtually the same time) is, however, greatly dominated by variations in traffic in different areas of the city. As can be seen, less than 2% of the

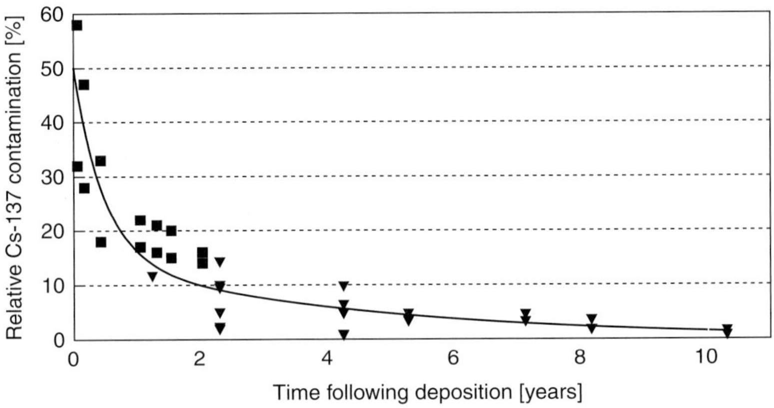

Figure 11 Weathering of Chernobyl ^{137}Cs on streets and pavements in Gävle. The relative contamination levels are given in percentages of the level on a corresponding reference surface (a cut lawn). Measurements marked with squares were carried out by Karlberg (1988), whereas measurements carried out by Andersson et al. (2002b) are marked with triangles.

caesium initially deposited on these surfaces after the Chernobyl accident was left after a decade had passed. Also shown in this figure is a modelling fit according to the expression:

$$C_r(t) = 50\% \ (A_1 \exp(-\lambda_1 t) + A_2 \exp(-\lambda_2 t)) \tag{2}$$

in which $C_r(t)$ is the street contamination level at time t, relative to the contamination level on a grassed reference surface in the area, and A_1 and A_2 are equal to, respectively, 0.7 and 0.3, whereas λ_1 and λ_2 are rate constants corresponding to half-lives of, respectively, 120 days and 3 years. These results are in good agreement with a series of measurements made in Bavaria, Germany, over the same period (Roed and Jacob, 1990; Roed, 1990). The physical reason for the slow component in the above expression might be that the contamination level is so low on these surfaces after a few years compared with, for example those of adjacent soil areas that the transfer of contaminated soil particles to the streets compensates for much of the contaminant loss from the streets through weathering. Since no significant element-specific fixation in this case seems to occur to the surface itself, other contaminant ions may be expected to be removed equally fast from streets by natural processes. Warming (1982, 1984) showed that most of the ^{86}Rb, ^{103}Ru and ^{140}Ba sprayed onto asphalt and concrete pavements in a solution could be removed by simple firehosing a few days later.

In special cases, where there is very little traffic, street surfaces may, however, remain highly contaminated over longer periods. For instance, in the deserted, heavily contaminated town of Pripyat, only about 3 km away from the Chernobyl power plant, it has been estimated that one-fifth of the initial radiocaesium deposition to the road surface remained after 7 years (Andersson et al., 2002b). If the depletion processes in Gävle and Pripyat had been the same, the remaining contaminant fraction in Pripyat would in the long run be expected to be the smallest, since the contamination seemed easier to remove by forced decontamination in Pripyat. The explanation may be that the contamination in Pripyat, very near the Chernobyl power plant, took place with large (insoluble) core fragment particles. In this respect, it is worth noting that Clark and Cobbin (1964) recorded the extremely high decontamination factor of 50 when hosing water on a street contaminated by particles in the 44–100 μm range. Such large particles would be very easy to remove, and routine street cleaning together with natural weathering processes would therefore be expected to rapidly result in very low street contamination levels. This could have relevance for large, not readily soluble particles generated by physical fractionation in connection with 'dirty bomb' explosions.

REFERENCES

Aarkrog, A., H. Dahlgaard, M. Frissel, L. Foulquier, N. V. Kulikov, I. V. Molchanova, C. Myttenaere, S. P. Nielsen, G. G. Polikarpov, and P. I. Yushkov. (1992). Sources of anthropogenic radionuclides in the Southern Urals. *Journal of Environmental Radioactivity*, 15, 69–80.

Allison, F. E. (1973). *Soil Organic Matter and Its Role in Crop Production.* Elsevier, USA, New York.

Andersson, K. G. (1990). TACTUS: A code for simulation of the flow of caesium-137 in urban surroundings. *Proceedings of the Symposium on Recovery Operations in the Event of a Nuclear Accident or Radiological Emergency*, IAEA-SM-316/48, ISBN 92-0-020290-x, IAEA, Vienna, pp. 217–228.

Andersson, K. G. (1991). Contamination and decontamination of urban areas. PhD Thesis, Risø National Laboratory, Roskilde, Denmark.

Andersson, K. G. (2005). Dose contributions from malicious radionuclide dispersion in an urban area. In: *Radiological Protection in Transition. Proceedings of the XIV Regular Meeting of the Nordic Society for Radiation Protection*, Rättvik, Sweden, 27–31 August 2005, SSI Report 2005:15, Swedish Radiation Protection Authority, ISSN 0282-4434, pp. 59–62.

Andersson, K. G., C. L. Fogh, and J. Roed. (2002a). Mobility of Chernobyl ^{137}Cs in undisturbed soil: Analysis methods and techniques. *Radioprotection – Colloques*, 37(C1), 231–236.

Andersson, K. G., A. Rantavaara, J. Roed, K. Rosén, B. Salbu, and L. Skipperud. (2000). *A Guide to Countermeasures for Implementation in the Event of a Nuclear Accident Affecting Nordic Food-Producing Areas.* NKS/BOK1.4 Project Report NKS-16. ISBN 87-7893-066-9.

Andersson, K. G., and J. Roed. (1994). The behaviour of Chernobyl ^{137}Cs, ^{134}Cs and ^{106}Ru in undisturbed soil: Implications for external radiation. *Journal of Environmental Radioactivity*, 22, 183–196.

Andersson, K. G., and J. Roed. (2006). Estimation of doses received in a dry-contaminated living area in the Bryansk Region, Russia, since the Chernobyl accident. *Journal of Environmental Radioactivity*, 85(2-3), 228–240.

Andersson, K. G., J. Roed, and C. L. Fogh. (2002b). Weathering of radiocaesium contamination on urban streets, walls and roofs. *Journal of Environmental Radioactivity*, 62, 49–60.

Andersson, K. G., J. Roed, H. G. Paretzke, and J. Tschiersch. (1995). Modelling of the radiological impact of a deposit of artificial radionuclides in inhabited areas. In: *Deposition of radionuclides, their subsequent relocation in the environment and resulting implications* (Eds J. Tschiersch). Report EUR 16604 EN. European Commission, ISBN 92-827-4903-7, pp. 83–94.

Argonne. (2005). *Radiological Dispersal Device (RDD), Human Health Fact Sheet.* Argonne National Laboratory, EVS, Argonne, IL, USA.

Askbrant, S., J. Melin, J. Sandalls, G. Rauret, R. Vallejo, T. Hinton, A. Cremers, C. Vandecastelle, N. Lewyckyj, Y. A. Ivanov, S. K. Firsakova, N. P. Arkhipov, and R. M. Alexakhin. (1996). Mobility of radionuclides in undisturbed and cultivated soils in Ukraine, Belarus and Russia six years after the Chernobyl fallout. *Journal of Environmental Radioactivity*, 31(3), 287–312.

Baltensperger, U., H. W. Gäggeler, D. T. Jost, B. Zindler, and P. Haller. (1987). Chernobyl radioactivity in size-fractionated aerosol. *Journal of Aerosol Science*, 18(6), 685–688.

van Bergeijk, K. E., H. Noordijk, J. Lembrechts, and M. J. Frissel. (1992). Influence of pH, soil type and organic matter content on soil-to-plant transfer of radioacesium and strontium as analysed by a nonparametric method. *Journal of Environmental Radioactivity*, 15, 265–276.

Bolt, G. H., M. E. Sumner, and A. Kamphorst. (1963). A study of the equilibriums between three categories of potassium in an illitic soil. *Soil Science Society of America Proceedings*, 27(3), 294–299.

Bondietti, E. A., J. N. Brantley, and C. Rangarajan. (1988). Size distributions and growth of natural and Chernobyl-derived submicron aerosols in Tennessee. *Journal of Environmental Radioactivity*, 6, 99–120.

Boyle, J. R., G. K. Voigt, and B. L. Sawhney. (1967). Biotite flakes: Alteration by chemical and biological treatment. *Science*, 155, 193–195.

Brady, N. S. (1970). *The Nature and Properties of Soils.* 8th ed., Macmillan Publishing Co. Inc., New York.

Brown, J., K. G. Andersson, J. A. Jones, R. Meckbach, H. Müller, and J. Roed. (2006). Requirements of future models for inhabited areas. *Journal of Environmental Radioactivity*, 85, 344–360.

Bunzl, K., H. Bachhuber, and W. Schimmack. (1984). Distribution co-efficients of ^{137}Cs, ^{85}Sr, ^{141}Ce, ^{103}Ru, ^{131}I and ^{95m}Tc in the various horizons of cultivated soils in Germany. In: *Proceedings of the International Symposium on Recent Investigations in the Zone of Aeration* (Eds P. Udulft, B. Mekel, and K. M. Prosl), pp. 567–577.

Bunzl, K., W. Kracke, W. Schimmack, and L. Zelles. (1998). Forms of fallout ^{137}Cs and $^{239+240}$Pu in successive horizons of a forest soil. *Journal of Environmental Radioactivity*, 39, 55–68.

Bunzl, K., M. Puhakainen, I. Riekkinnen, P. Karhu, W. Schimmack, T. Heikkinen, T. Jaakola, V. Nikonov, V. Pavlov, T. Rahola, K. Rissanen, M. Suomela, M. Tillander, and M. Äyräs. (2001). Fallout ^{137}Cs, ^{90}Sr and $^{239+240}$Pu in soils polluted by heavy metals: Vertical distribution, residence half-times, and external gamma-dose rates. *Journal of Radioanalytical and Nuclear Chemistry*, 247(1), 15–24.

Bunzl, K., W. Schimmack, K. Kreutzer, and R. Schierl. (1989). The migration of fallout ^{134}Cs, ^{137}Cs and ^{106}Ru from Chernobyl and ^{137}Cs from weapons testing in forest soil. *Zeitschrift für Pflanzenernährung und Bodenkunde*, 152, 39–44.

Bunzl, K., W. Schimmack, L. Zelles, and B. P. Albers. (2000). Spatial variability of the vertical migration of fallout ^{137}Cs in the soil of a pasture, and consequences for long-term predictions. *Radiation and Environmental Biophysics*, 39, 197–205.

Chaberek, S., and A. E. Martell. (1980). *Organic Sequestering Agents, a Discussion of the Chemical Behaviour and Applications of Metal Chelate Compounds in Aqueous Systems.* Wiley, New York.

Chamard, P., R. H. Velasco, M. Belli, G. Di Silvestro, G. Ingrao, and U. Sansone. (1993). Caesium-137 and strontium-90 distribution in a soil profile. *The Science of the Total Environment*, 136(3), 251–258.

Clark, D. E., and W. C. Cobbin. (1964). *Removal of Simulated Fallout from Pavements by Conventional Street Flushers.* USNRDL-TR-797. US Naval Radiological Defence Laboratory.

Cole, W. F., and E. R. Segnit. (1963). High temperature phases developed in some kaolinite–mica–quartz clays. *Transactions and Journal of the British Ceramic Society*, 62, 375–392.

Cook, G. T., M. S. Baxter, H. J. Duncan, J. Toole, and R. Malcolmson. (1984). Geochemical association of plutonium in the Caithness environment. *Nuclear Instruments and Methods in Physics Research*, 223, 517–522.

Coughtrey, P. J., M. C. Thorne, and D. Jackson. (1986). *Radionuclide Distribution and Transport in Terrestrial and Aquatic Ecosystems.* A.A. Balkema, Rotterdam.

Cremers, A., A. Elsen, P. de Preter, and A. Maes. (1988). Quantitative analysis of radiocaesium retention in soils. *Nature*, 335, 247–249.

Crick, M. J., Z. Hussain, and A. Walmsley. (1987). *Identification of Important Parameters in Urban Dose Assessment.* National Radiological Protection Board, UK.

Devell, L. (1987). *Parametervärden vid radiologisk konsekvensanalys av reaktorhaveri. Några erfarenheter från Tjernobylhaveriet (in Swedish).* Studsvik Report NP-87/89.

Dorrian, M. D., and M. R. Bailey. (1995). Particle size distributions of radioactive aerosols measured in workplaces. *Radiation Protection Dosimetry*, 60(2), 119–133.

Duff, R. B., D. M. Webley, and R. O. Scott. (1963). Solubilisation of minerals and related materials by 2-ketogluconic acid-producing bacteria. *Soil Science*, 95, 105–114.

Eriksson, Å. (1977). *Direct Uptake of Vegetation of Deposited Materials. I: Retention of Nuclides and Simulated Fallout Particles in Pasture Grass.* Report SLU-IRB-42. Department of Radiobiology, Agricultural College of Sweden, Uppsala.

Eriksson, M. (2002). *On Weapons Plutonium in the Arctic Environment (Thule, Greenland).* Risø Report Riso-R-1321. ISBN 87-550-3006-8, Risø National Laboratory, Denmark.

Essington, E., H. Nishita, and A. Wallace. (1963). Effect of chelating agents on the uptake of Y-91, Ce-144, Ru-106, and Pm-147 by beans grown in calcareous soil. *Soil Science*, 95, 331–337.

Evans, D. W., J. J. Alberts, and R. A. Clark. (1983). Reversible ion-exchange fixation of caesium-137 leading to mobilisation from reservoir sediments. *Geochimica et Cosmochimica*, 47, 1041–1049.

Ferguson, C. D. (2003). *Reducing the Treats of RDDs.* IAEA Bulletin 45/1.

Ferguson, C. D., T. Kazi, and J. Perera. (2003). *Commercial Radioactive Sources.* Occasional Paper No. 11. Center for Nonproliferation Studies, Monterey Institute of International Studies, California 93940, USA, ISBN 1-885350-06-6.

Fogh, C. L., K. G. Andersson, A. N. Barkovsky, A. S. Mishine, A. V. Ponomarjov, V. P. Ramzaev, and J. Roed. (1999). Decontamination in a Russian settlement. *Health Physics*, 76(4), 421–430.

Gallego, E. (2006). MUD: A model to investigate the migration of Cs-137 in the urban environment and drainage and sewage treatment systems. *Journal of Environmental Radioactivity*, 85(2-3), 247–264.

Grim, R. E. (1965). The clay mineral concept. *American Ceramic Society*, 44, 687–692.

Harper, F. T., S. V. Musolino, and W. B. Wente. (2007). Realistic radiological dispersal device hazard boundaries and ramifications for early consequence management decisions. *Health Physics*, 93(1), 1–16.

Heinemann, K., and K. J. Vogt. (1980). Measurements of the deposition of iodine onto vegetation and the biological half-life of iodine on vegetation. *Health Physics*, 39, 463–474.

Hill, R. D. (1960). A study of pore-size distribution of fired clay bodies. *Transactions and Journal of the British Ceramic Society*, 59, 189–197.

Hilton, J., R. S. Cambray, and N. Green. (1992). Chemical fractionation of radioactive caesium in airborne particles containing bomb fallout, Chernobyl fallout and atmospheric material from the Sellafield site. *Journal of Environmental Radioactivity*, 15, 103–111.

Hird, A. B., D. L. Rimmer, and F. R. Livens. (1996). Factors affecting the sorption and fixation of caesium in acid organic soil. *European Journal of Soil Science*, 47(1), 97–104.

Horn, H.-G., H. Bonka, and M. Maqua. (1987). Measured particle bound activity size distribution, deposition velocity, and activity concentration in rainwater after the Chernobyl accident. *Journal of Aerosol Science*, 18(6), 681–684.

Hou, X. L., C. L. Fogh, J. Kucera, K. G. Andersson, H. Dahlgaard, and S. P. Nielsen. (2003). Iodine-129 and caesium-137 in Chernobyl contaminated soil and their chemical fractionation. *The Science of the Total Environment*, 308, 97–109.

Hunt, C. E. L., D. S. Cox, and F. C. Iglesias. (1994). Fission-product release during accidents – An accident management perspective. *Nuclear Engineering and Design*, 148(2-3), 205–216.

IAEA. (1994). *Handbook of Parameter Values for the Prediction of Radionuclide Transfer in Temperate Environments*. ISBN 92-0-101094-X IAEA-Technical Reports Series No. 364, STI-DOC-10-364. IAEA, Vienna, Austria.

Ivanov, Y. A., N. Lewyckyj, S. E. Levchuk, B. S. Prister, S. K. Firsakova, N. P. Arkhipov, A. N. Arkhipov, S. V. Kruglov, R. M. Alexakhin, J. Sandalls, and S. Askbrant. (1997). Migration of Cs-137 and Sr-90 from Chernobyl fallout in Ukrainian, Belarussian and Russian soils. *Journal of Environmental Radioactivity*, 35(1), 1–21.

Jackson, M. L. (1963). Interlayering of expansible layer silicates in soils by chemical weathering. *Clays and Clay Minerals*, 11, 29–46.

Jacob, P., R. Meckbach, and H. M. Müller. (1987). Reduction of external exposure from deposited Chernobyl activity by run-off, weathering, street cleaning and migration in the soil. *Radiation Protection Dosimetry*, 21, 51–57.

Jones, A., T. Charnock, L. Singer, J. Roed, K. G. Andersson, S. Thykier-Nielsen, T. Mikkelsen, P. Astrup, J. C. Kaiser, H. Müller, G. Pröhl, W. Raskob, S. Hoe, L. H. Jacobsen, L. Schou-Jensen, and F. Gering. (2006). *Description of the Modelling of Transfer and Dose Calculations within ERMIN*. Draft of deliverable D5C2R2 of the CEC-EURANOS project, EURANOS(CAT2)-TN(05)-04.

Jones, R. F. (1960). The accumulation of nitrosyl ruthenium by fine particles and marine organisms. *Limnology and Oceanology*, 5, 312–325.

Jost, D. T., H. W. Gäggeler, U. Baltensperger, B. Zinder, and P. Haller. (1986). Chernobyl fallout in size-fractionated aerosol. *Nature*, 324, 22–23.

Junker, H., J. M. Jensen, H. B. Jørgensen, J. Roed, K. G. Andersson, P. D. Kofman, E. Bøllehuus, L. Baxter, and A. Grebenkov (1998). *Chernobyl Bioenergy Project, Final Report*, Phase 1. Elsamprojekt Report No. EP 8204.

Juo, A. S., and S. A. Barber. (1969). An explanation for the variability in Sr–Ca exchange selectivity of soils, clays and humic acid. *Soil Science Society of America Proceedings*, 33, 360–364.

Karlberg, O. (1988). *Weathering and Migration of Chernobyl Fallout in Sweden*. Studsvik Report NP 216 EA, Studsvik, Sweden.

Kashparov, V. A., N. Ahamdach, S. I. Zvarich, V. I. Yoschenko, I. M. Maloshtan, and L. Dewiere. (2004). Kinetics of dissolution of Chernobyl fuel particles in soil in natural conditions. *Journal of Environmental Radioactivity*, 72, 335–353.

Kashparov, V. A., Y. A. Ivanov, S. I. Zvarisch, V. P. Protsak, Y. V. Khomutinin, A. D. Kurepin, and E. M. Pazukhin. (1996). Formation of hot particles during the Chernobyl nuclear power plant accident. *Nuclear Technology*, 114(2), 246–253.

Kinniburgh, D. G., and M. L. Jackson. (1981). Cation adsorption by hydrous metal oxides and clay. In: *Adsorption of Inorganics at Solid-Liquid Interfaces* (Eds M. A. Anderson and A. J. Rubin). Ann Arbor Science Publishers, Michigan, ISBN: 0250402262, pp. 91–160.

Kirchner, G., F. Strebl, P. Bossew, S. Ehlken, and M. H. Gerzabek. (2008). Vertical migration of radionuclides in undisturbed grassland soils. *Journal of Environmental Radioactivity* (in press).

Kocher, D. C. (1991). A validation test of a model for long-term retention of ^{129}I in surface soils. *Health Physics*, 60, 523–531.

Koch-Steindl, H., and G. Pröhl. (2001). Considerations on the behaviour of long-lived radionuclides in the soil. *Radiation and Environmental Biophysics*, 40, 93–104.

Konoplev, A. V., A. A. Bulgakov, V. E. Popov, and T. I. Bobovnikova. (1992). Behaviour of long-lived Chernobyl radionuclides in a soil–water system. *Analyst*, 117, 1041–1047.

Konshin, O. V. (1992). Applicability of the convection-diffusion mechanism for modeling migration of ^{137}Cs and ^{90}Sr in the soil. *Health Physics*, 63(3), 291–300.

Kostiainen, E., R. Hänninen, K. Rosén, E. Haak, Å. Eriksson, S. P. Nielsen, M. Keith-Roach, and B. Salbu. (2002). *Transfer Factors for Nuclear Emergency Preparedness, NKS-78*, Nordic Nuclear Safety Research, Roskilde, Denmark, ISBN 87-7893-134-7.

Krouglov, S. V., A. D. Kurinov, and R. M. Alexakhin. (1998). Chemical fractionation of ^{90}Sr, ^{106}Ru, ^{137}Cs and ^{144}Ce in Chernobyl contaminated soils: An evolution in a course of time. *Journal of Environmental Radioactivity*, 38(1), 59–76.

Krstic, D., D. Nikezic, N. Stevanovic, and M. Jelic. (2004). Vertical profile of ^{137}Cs in soil. *Applied Radiation and Isotopes*, 61, 1487–1492.

Kryshev, I. I. (1996). Dose reconstruction for the areas of Russia affected by ^{131}I contamination. *Radiation Protection Dosimetry*, 64(1/2), 93–96.

Lamm, C. G., and M. H. Nadafy. (1973). Plant nutrient availability in soils, III studies on potassium in Danish soils, 6. The rate of release and availability of fixed potassium. *Agrochimica*, 17, 435–444.

Lange, C. (1995). Indoor deposition and the protective effect of houses against airborne pollution. PhD Thesis, Risø National Laboratory, Roskilde, Denmark.

Livens, F. R., and M. S. Baxter. (1988). Chemical associations of artificial radionuclides in Cumbrian soil. *Journal of Environmental Radioactivity*, 7, 75–86.

Lönsjö, H., and E. Haak. (1975). *Uptake of ^{90}Sr by Agricultural Crops as Influenced by Soil Type, Liming Rate and PK-Fertilization in a Long Term Microplot Experiment*. Report 31. Radiobiology, Uppsala.

Lujan, J. R., D. W. Darnall, P. C. Stark, G. D. Rayson, and J. L. Gardeatorresdey. (1994). Metal ion binding by algae and higher plant tissues – A phenomenological study of solution pH dependence. *Solvent Extraction and Ion Exchange*, 12(4), 803–816.

Ma, W., P. W. Brown, and S. Komarneni. (1996). Sequestration of cesium and strontium by tobermorite synthesized from fly ashes. *Journal of the American Ceramic Society*, 79(6), 1707–1710.

Mamikhin, S. V., and A. L. Klyashtorin. (2000). Mathematical model of ^{137}Cs dynamics in the deciduous forest. *Journal of Environmental Radioactivity*, 47, 101–114.

Martin, W. E. (1964). Losses of Sr-90, Sr-89 and I-131 from fallout-contaminated plants. *Radiation Botany*, 4, 275–284.

Milbourn, G. M., and R. Taylor. (1965). The contamination of grassland with radioactive strontium – I. Initial retention and loss. *Radiation Botany*, 5, 337–347.

Moukwa, M., B. G. Lewis, S. P. Shah, and C. Ouyang. (1993). Effect of clays on fracture properties of cement-based materials. *Cement and concrete Research*, 23(3), 711–723.

Mück, K., and M. H. Gerzabek. (1995). Trends in caesium activity concentrations in milk from agricultural and semi-natural environments after nuclear fallout. *Bodenkultur*, 46(4), 337–353.

Mück, K., G. Pröhl, I. Likhtarev, L. Kovgan, R. Meckbach, and V. Golikov. (2002). A consistent radionuclide vector after the Chernobyl accident. *Health Physics*, 82(2), 141–156.

Nisbet, A. F., J. A. Mercer, N. Hesketh, A. Liland, H. Thørring, T. Bergan, N. A. Beresford, B. J. Howard, J. Hunt, and D. H. Oughton. (2004). *Datasheets on Countermeasures and Waste Disposal Options for the Management of Food Production Systems Contaminated Following a Nuclear Accident*. NRPB-W58, ISBN 0 85951 539 7. National Radiological Protection Board, Oxon, UK.

Nisbet, A. F., R. F. M. Woodman, and R. G. E. Haylock. (1999). *Recommended Soil-to-Plant Transfer Factors for Radiocaesium and Radiostrontium for Use in Arable Systems*. NRPB-R304. National Radiological Protection Board, Chilton, Didcot.

Nishita, H., B. W. Kowalewsky, A. J. Steen, and K. H. Larson. (1956). Fixation and extractability of fission products contaminating various soils and clays: I ^{90}Sr, ^{91}Y, ^{106}Ru, ^{137}Cs, ^{144}Ce. *Soil Science*, 81, 317–326.

Nommik, H., and K. Vahtras. (1982). Retention and fixation of ammonium and ammonia in soils. In: *Nitrogen in Agricultural Soils*, No. 22 in the Series Agronomy (Ed. F. J. Stevenson). Madison, WI, pp. 123–133.

Norrish, K. (1975). Geochemistry and mineralogy of trace elements. In: *Trace Elements in Soil–Plant–Animal Systems* (Eds D. J. D. Nicholas and A. R. Egan). Academic Press, New York, pp. 55–81.

Petryaev, E. P., S. V. Ovsyannikova, S. Y. Rubinchik, I. J. Lubinka, and G. A. Sokolik. (1991). State of radionuclides of the Chernobyl fallout in soils of Belarus. *Proceeding of AS of BSSR, Physico-Energetical Sciences*, 4, 48–55.

de Preter, P. (1990). Radiocaesium retention in the aquatic, terrestrial and urban environment: A quantitative and unifying analysis. Doctoraatsproefschrift Nr. 190 aan de Fakulteit der Landbouwwetenschappen van de Katholieke Universiteit te Leuven.

Qvenild, C., and U. Tveten. (1983). *Adsorption of Cs-134 onto two Different Types of Roof Materials*. IFE Report IFE/KR/E-83/011 +V. Institute for Energy Technology, Kjeller, Norway, ISBN 82-7017-051-8.

Qvenild, C., and U. Tveten. (1984). *Adsorption of Cs-134 onto two Different Types of Roof Materials during Summer and Winter Conditions*. IFE Report IFE/KR/E-84/013 +V. Institute for Energy Technology, Kjeller, Norway, ISBN 82-7017-066-6.

Qvenild, C., and U. Tveten. (1986). *Run-off of Strontium with Melting Snow in Spring*. IFE Report IFE/KR/E-86/003. Institute for Energy Technology, Kjeller, Norway, ISBN 82-7017-082-8.

Rahman, M. M., M. M. Chand, A. Koddus, M. M. Rahman, M. A. Zaman, and G. Voiggt. (2008). Transfer of radiocobalt from soil to selected plant species in tropical environments. *Journal of Environmental Radioactivity*, 99, 658–664.

Real, J., F. Persin, and C. Camarasa- Claret. (2002). Mechanisms of desorption of ^{134}Cs and ^{85}Sr aerosols deposited on urban surfaces. *Journal of Environmental Radioactivity*, 62, 1–15.

Renaud, Ph., and H. Maubert. (1997). Agricultural countermeasures in the management of a post-accidental situation. *Journal of Environmental Radioactivity*, 35(1), 53–69.

Riise, G., H. E. Bjørnstad, H. N. Lien, D. H. Oughton, and B. Salbu. (1990). A study on radionuclide association with soil components using a sequential extraction procedure. *Journal of Radioanalytical and Nuclear Chemistry*, 142(2), 531–538.

Roed, J. (1987). Run-off from and weathering of roof material following the Chernobyl accident. *Radiation Protection Dosimetry*, 21(1/3), 59–63.

Roed, J. (1988). *The Distribution on Trees of Dry Deposited Material from the Chernobyl Accident*. Paper NKA/AKTU-245 (88)2, presented at the *Joint CEC/OECD (NEA) Workshop on Recent Advances in Reactor Accident Consequence Assessment*, Rome, Italy, 25–30 January 1988.

Roed, J. (1990). *Deposition and Removal of Radioactive Substances in an Urban Area*. Final Report of the NKA Project AKTU-245. Nordic Liaison Committee for Atomic Energy at Risø, Denmark, ISBN 87 7303 514 9.

Roed, J., and K. G. Andersson. (1996). Clean-up of urban areas in the CIS countries contaminated by Chernobyl fallout. *Journal of Environmental Radioactivity*, 33(2), 107–116.

Roed J., K. G. Andersson, A. N. Barkovsky, C. L. Fogh, A. S. Mishine, S. K. Olsen, A.V. Ponomarjov, H. Prip, V. P. Ramzaev, and B. F. Vorobiev. (1998). *Mechanical Decontamination Tests in Areas Affected by the Chernobyl Accident*. Risø-R-1029. Risø National Laboratory, Roskilde, Denmark, ISBN 87-550-2361-4.

Roed, J., and P. Jacob. (1990). Deposition on urban surfaces and subsequent weathering. *Proceedings from the Seminar on Methods and Codes for Assessing the Off-Site Consequences of Nuclear Accidents*, Athens, May, 1990.

Roed, J., C. Lange, K. G. Andersson, H. Prip, S. Olsen, V. P. Ramzaev, A. V. Ponomarjov, A. N. Barkovsky, A. S. Mishine, B. F. Vorobiev, A. V. Chesnokov, V. N. Potapov, and S. B. Shcherbak. (1996). *Decontamination in a Russian Settlement*. Risø-R-870, Risø National Laboratory, Roskilde, Denmark, ISBN 87-550-2152-2.

Rosén, K. (2006). Doses and countermeasures for kitchen gardens. In: *URBHAND – Decision Support Handbook for Remediation of Contaminated Inhabited Areas* (Eds K. G. Andersson, M. Ammann, S. Backe and K. Rosén). Version 1-for end-user discussion, Working document from the Nordic project URBHAND supported by NKS-B.

Rulik, P., I. Bucina, and I. Malatova. (1989). Aerosol particle size distribution in dependence on the type of radionuclide after the Chernobyl accident and in the NPP effluents. *Proceedings of the XVth Regional Congress of IRPA*, Visby, Gotland, Sweden, pp. 102–107, ISBN 3-88585-668-9, TÜV Verlag Rheinland GmbH, Köln.

Salbu, B. (2000). Source-related characteristics of radioactive particles: A review. *Radiation Protection Dosimetry*, 92(1–3), 49–54.

Salbu, B., T. Krekling, and D. Oughton. (1998). Characterisation of radioactive particles in the environment. *Analyst*, 23, 843–849.

Salbu, B., D. H. Oughton, A. V. Ratnikov, T. L. Zhigareva, S. V. Kruglov, K. V. Petrov, N. V. Grebenshakikova, S. K. Firsakova, N. P. Astasheva, N. A. Loshchilov, K. Hove, and P. Strand. (1994). The mobility of Cs-137 and Sr-90 in agricultural soils in the Ukraine, Belarus and Russia, 1991. *Health Physics*, 67(5), 518–528.

Sandalls, F. J. (1987). *Removal of Radiocaesium from Urban Surfaces Contaminated as a Result of a Nuclear Accident*. Series AERE-R-12355. Harwell Laboratory, Oxfordshire, ISBN 0-7058-1205-7.

Sandalls, F. J., and S. L. Gaudern. (1988). Radiocaesium on urban surfaces in West Cumbria five months after Chernobyl. *Journal of Environmental Radioactivity*, 7, 87–91.

Sandalls, F. J., M. J. Segal, and N. Viktorova. (1993). Hot particles from Chernobyl: A review. *Journal of Environmental Radioactivity*, 18, 5–22.

Sato, K., T. Hashida, H. Takhashi, and N. Yamasaki. (1997). Strengthening of hydrothermal hot-pressed concrete waste by the addition of slag. *Journal of the Ceramic Society of Japan*, 105(3), 262–264.

Sawhney, B. L. (1967). Caesium sorption in relation to lattice spacing and cation exchange capacity of biotite. *Soil Science Society of America Proceedings*, 31, 181–183.

Sawhney, B. L. (1970). Potassium and cesium ion selectivity in relation to clay mineral structure. *Clays and Clay Minerals*, 18, 47–52.

Sawhney, B. L. (1972). Selective sorption and fixation of cations by clay minerals: A review. *Clays and Clay Minerals*, 20, 93–100.

Sawhney, B. L., and G. K. Voigt. (1969). Chemical and biological weathering in vermiculite from Transvaal. *Soil Science Society of America Proceedings*, 33, 625–629.

Schimmack, W., K. Bunzl, F. Dietl, and D. Klotz. (1994). Infiltration of radionuclides with low mobility (^{137}Cs and ^{60}Co) into a forest soil. Effect of the irrigation intensity. *Journal of Environmental Radioactivity*, 24(1), 53–63.

Schnitzer, M., and S. U. Kahn. (1978). *Soil Organic Matter*. Elsevier, Amsterdam.

Schultz, R. K. (1965). Soil chemistry of radionuclides. *Health Physics*, 11, 1317–1324.

Schultz, R. K., R. Overstreet, and I. Barshad. (1959). On the soil chemistry of caesium-137. *Soil Science*, 89, 16–27.

Schuller, P., A. Ellies, and G. Kirchner. (1997). Vertical migration of fallout Cs-137 in agricultural soils from southern Chile. *The Science of the Total Environment*, 193, 197–205.

Shainberg, I., and W. D. Kemper. (1967). Ion exchange equilibria on montmorillonite. *Soil Science*, 103, 4–9.

Shrivastava, O. P., and R. Shrivastava. (1998). Immobilisation and solidification of cesium on 11 angstrom calcium silicate hydroxy hydrate column. *Indian Journal of Chemistry Section A – Inorganic Bio-Inorganic Physical Theoretical and Analytical Chemistry*, 37(8), 726–729.

Sohier, A., and F. Hardeman. (2006). Radiological dispersion devices: Are we prepared? *Journal of Environmental Radioactivity*, 85(2-3), 171–181.

Solovitch-Vella, N., and J.-M. Garnier. (2006). Comparative kinetic desorption of ^{60}Co, ^{85}Sr and ^{134}Cs from a contaminated natural silica sand column: Influence of varying physicochemical conditions and dissolved organic matter. *Environmental Pollution*, 141, 98–106.

Sprung, S., and E. Siebel. (1991). Assessment of the suitability of limestone for producing Portland limestone cement. *Zement-Kalk-Gips*, 44(1), 1–11.

Squire, H. M., and L. J. Middleton. (1966). Behaviour of Cs-137 in soils and pastures – A long time experiment. *Radiation Botany*, 6, 413–423.

Strazzera, B., M. Dondi, and M. Marsigli. (1997). Composition and ceramic properties of tertiary clays from Southern Sardinia (Italy). *Applied Clay Science*, 12(3), 247–266.

Szerbin, P., E. Koblinger-Bokori, L. Koblinger, I. Végvári, and Á. Ugron. (1999). Caesium-137 migration in Hungarian soils. *The Science of the Total Environment*, 227, 215–227.

Tamura, T., and D. G. Jacobs. (1961). Improving cesium selectivity of bentonites by heat treatment. *Health Physics*, 5, 149–154.

Tessier, A., P. G. C. Campbell, and M. Bisson. (1979). Sqeuential extraction procedure for the speciation of particulate trace metals. *Analytical Chemistry*, 51, 844–851.

Tomasek, M., K. Rybacek, and L. Wilhelmova. (1995). Chemical fraction of radioactive cesium in atmospheric aerosol in Prague after the Chernobyl accident. *Journal of Radioanalytical and Nuclear Chemistry*, 201(5), 409–416.

Tschiersch, J., and B. Georgi. (1987). Chernobyl fallout size distribution in urban areas. *Journal of Aerosol Science*, 18(6), 689–692.

Velasco, R. H., J. P. Toso, M. Belli, and U. Sansone. (1997). Radiocaesium in the northeastern part of Italy after the Chernobyl accident: Vertical soil transport and soil-to-plant transfer. *Journal of Environmental Radioactivity*, 37(1), 73–83.

Wahl, F. M. (1965). High temperature phases of three layer clay minerals and their interactions with common ceramic materials. *Journal of the American Ceramic Society*, 44, 676–681.

Warming, L. (1982). *Weathering and Decontamination of Radioactivity Deposited on Asphalt Surfaces.* Risø-M-2273. Risø National Laboratory, Roskilde, Denmark, ISBN 87-550-0903-4.

Warming, L. (1984). *Weathering and Decontamination of Radioactivity Deposited on Concrete Surfaces.* Risø-M-2473. Risø National Laboratory, Roskilde, Denmark, ISBN 87-550-1068-7.

Wauters, J., L. Sweeck, E. Valke, A. Elsen, and A. Cremers. (1994). Availability of radiocaesium in soils: A new methodology. *Science of the Total Environment*, 157, 239–248.

Wilkins, B. T. (1987). The retention behaviour of radiocaesium on common building materials under natural outdoor conditions. *Radiation Protection Dosimetry*, 21(1/3), 69–73.

Willard, W. K. (1975). *Radiocobalt Cycling in a Small Mammal Food Web.* Report CONF-750503-8. Oak Ridge National Laboratory, Tennessee. Presented at Fourth National Symposium on Radioecology, 12 May 1975, Corvallis, OR.

Winkelmann, I., H. J. Endrulat, S. Fouasnon, P. Gesewsky, R. Haubelt, P. Klopfer, H. Köhler, R. Kohl, D. Kucheida, C. Leising, M. K. Müller, P. Neumann, H. Schmidt, K. Vogl, S. Weimer, H. Wildermuth, S. Winkler, E. Wirth, and S. Wolff. (1987). *Radioactivity Measurements in the Federal Republic of Germany after the Chernobyl Accident.* Report ISH-148/90. Bundesamt für Strahlenschutz, Neuherberg, Germany.

CHAPTER 6

Estimation of Doses in Inhabited Areas

Kasper G. Andersson[1,*], J. Arthur Jones[2] *and* Thomas W. Charnock[2]

Contents

*Corresponding author. Tel.: +45 46 77 41 73; Fax: +45 4677 5330
E-mail address: kasper.andersson@risoe.dk

[1] Risoe National Laboratory for Sustainable Energy, Technical University of Denmark, DK-4000, Roskilde, Denmark
[2] Health Protection Agency, Radiation Protection Division, Chilton, Didcot, Oxon OX11 0RQ, UK

Radioactivity in the Environment, Volume 15
ISSN 1569-4860, DOI 10.1016/S1569-4860(09)00406-9

1. Introduction

People in inhabited areas can be exposed to radiation from a variety of pathways following a release of radioactive material to the atmosphere, whether the release occurs inside or outside the area. Following the release, a plume of material is formed which then disperses away from the initial source. People can inhale material from the plume, leading to internal exposure, as discussed in Chapter 4. Some of the material in the plume could deposit onto people's skin or clothing, leading to external exposure, as also discussed in Chapter 4. Material from the plume can deposit on different surfaces in the inhabited area, as discussed in Chapters 2 and 3, where natural processes will redistribute/deplete it, as discussed in Chapters 3 and 5. These occurrences can lead to external exposure from the material on the surfaces. The material can also deposit on foodstuffs being grown or kept in the area. This leads to internal exposure after the food is consumed. Deposited material can be resuspended, as discussed in Chapters 2 and 3, and subsequently inhaled, leading to internal exposure.

This chapter describes the methods commonly adopted to calculate doses from the different pathways. Several models have been developed for particular aspects of the problem of calculating doses from material in inhabited areas, although none of the current available models addresses all of the pathways considered above. The chapter deals primarily with models used to calculate doses when radioactive material has contaminated a relatively large area, as might occur following an accident at a nuclear site or an event such as a fire or explosion involving radioactive material, including incidents caused by malevolent actions. However, the methods described are also appropriate for modelling other situations leading to a more localised contamination of part of the inhabited area.

2. Why Models are Needed

Dose rates in an area from existing radioactive contamination, from an emergency or a routine release, can be determined by making appropriate measurements. However, models are needed to predict how dose rates might change in the future and to analyse situations, for example in emergency planning and response or setting appropriate discharge limits for continuous releases. Different applications require models with different degrees of complexity. This section considers the type of model that is needed for different applications, with particular reference to the modelling of external exposure from deposited material.

Contamination densities on various surfaces in the area will differ, both because of differences in the initial deposition to the surfaces and differences in the behaviour of material after deposition. The dose rate from material on different surfaces in the area depends on a number of factors such as the different distances between the people and the surfaces and the size of the various surfaces. Overall dose rate therefore depends on the amount of activity on different surfaces, as a function of time following the deposition, and the differing dose rate contributions per unit deposition on the various surfaces.

A number of models have been developed to calculate doses and dose rates in inhabited areas, with different approaches taken. There are relatively simple models that calculate the dose rate in an inhabited area from the dose rate over a uniform soil surface using correction factors to allow for the differences between doses in rural (open) and inhabited areas. Other, more complex, models consider the deposition to the various surfaces, the behaviour of the material on each surface and the dose rates from unit deposition on each surface. Both types of models take into account the shielding properties of the buildings in the area in some way. The more complex models do this explicitly through the dose calculations (as described later in this chapter), while simpler models use location factors, which may have been derived from a more complex model or from experimental data.

Models are needed for application to both routine and accidental releases. The simpler models are generally used to calculate doses from routine releases to the atmosphere. The more detailed models have been developed for application to accidental releases, primarily to provide an input into emergency planning, evaluation of countermeasure strategies and to provide a tool for training.

Models for routine releases need to calculate the dose or dose rate from a discharge which is continuous over an extended period at a reasonably constant rate. Such models generally predict doses over a uniform surface of infinite extent, with simple correction factors to relate them to the doses in an inhabited area. The National Radiological Protection Board (now the Radiation Protection Division of the Health Protection Agency in the UK) and GSF (now Helmholtz Zentrum Muenchen-German Research Centre for Environmental Health in Germany) have developed such models. The NRPB approach (Mayall et al., 1997) explicitly models the movement of radioactive material in the soil column and the shielding by overlying soil. The GSF approach uses attenuation factors derived from observations to describe the time variation of dose rate due to migration of material down the soil column (Jacob et al., 1994).

Other models have been developed which use explicit descriptions of the deposition to different surfaces, the behaviour of material with time after its deposition and the dose rates from the different surfaces. Models of this type include EXPURT, developed by NRPB (UK) (Crick and Brown, 1990; Jones et al., 2006b) and URGENT, developed by Risø, Denmark

(Andersson, 1996a), and ERMIN, developed by several organisations as part of the EC EURANOS project (Jones et al., 2006a). This type of model is generally used to assess dose and the effect of countermeasures following accidents or malicious dispersion incidents, for both emergency planning and responding in the event of an actual release. In particular, compared to the simple models described above, they give more detailed information on the time variation of the dose rate following the deposition. This type of model also gives information on the relative contribution to dose rate of material on different surfaces, which is needed for investigations of the effects of countermeasures to reduce the doses, as discussed further in Chapters 8 and 10. The approach is being incorporated in the European decision support systems ARGOS (2008) and RODOS (2008).

Following a contaminating incident, information on dose rates at particular points or deposition densities on particular surfaces will become available. Models for use after a contaminating incident must also be able to use such information to improve their description of the current situation and their prediction of future doses. Complex models employing data assimilation methods are likely to be required for this type of application (Kaiser et al., 2008).

Models are also required for use in applications such as probabilistic risk assessment, where the average dose to people living in an area is required. These applications do not need a detailed evaluation of particular countermeasures, but can describe them in terms of the dose reduction factor that can be achieved by generic countermeasures. Here relatively simple models and generic assumptions can be used.

The following sections describe the various aspects considered in the models. The chapter ends with comments on the adequacy of the current models and areas where they might be improved.

3. External Dose Rate from Contaminated Surfaces

This section describes the modelling of external beta and gamma dose rates from contaminated surfaces, discusses various features of the area that affect the dose and gives an example of dose rate calculations for a contaminated inhabited area. Models generally distinguish the following surfaces: external walls and roofs of buildings, paved areas (including roads, pavements and areas such as patios and drives), soil, vegetation such as grass, trees and plants, and internal surfaces in buildings, although not all existing models consider all these surfaces (Thiessen et al., 2008). Inhabited areas are very complex, with some features that influence the dose rate varying over short distances. Models can either attempt to calculate dose rate at specific points, using a detailed description of the environment around that point, or adopt a more generic description of those features and so aim to calculate a dose rate that represents typical values in the area.

3.1. Initial deposition to different surfaces

The first step in the calculation is to determine the initial deposition to the different surfaces considered. The basic input to this part of the calculation is likely to be either a dispersion model prediction of deposition to a single surface or measurements of deposition to one or more surfaces, or a combination of both. This must be combined with information on relative deposition to different surfaces to give an initial deposition on all surfaces considered in the model being used.

Information from which values for the relative deposition to different surfaces can be derived was discussed in Chapter 2. Measurements were made in specific locations, and so may refer to particular types of material for the surfaces in the inhabited area. These must be used to derive 'representative' values for the relative deposition on the different categories of surfaces considered in the dose models. Some simplification is necessary as data are not available to describe all possible parametric influences. Typically applied values are given in Table 1, which shows the relative depositions (in terms of deposition density per unit area) to different surfaces, in dry conditions, for elemental iodine and small particles used in the ERMIN model, and based on Cline et al. (1965), Chamberlain (1959), Hull (1966), Katagiri et al. (1997), Freer-Smith et al. (2003), Heinemann and Vogt (1980), Horn et al. (1988), McMahon and Deninson (1979),

Table 1 Relative deposition to different surfaces in dry conditions.

Surface	Deposition to surface relative to that to reference surface		
	Elemental iodine	Aerosol with AMAD	
		$<2\,\mu m$	$2-5\,\mu m$
Paved areas	0.2	0.3	0.7
Roofs	1.5[b]	1.0	4
External walls of buildings	0.14	0.05	0.1
Trees and shrubs	0.4	2.5	5
Grass[a]	1.0	1.0	1.0
Small plants	0.8	1.5	1.5
Soil	0.6	0.3	0.3
Soil and grass[a]	1.0	1.0	1.0

[a]Short cut grass (mowed lawn).

[b]Note that first rain shower will remove practically all elemental iodine from a roof (see Roed, 1987a).

Nakamura and Ohmomo (1980), Nicholson (1989), Roed (1985a, 1987b, 1988, 1990), Sehmel (1973) and Watterson and Nicholson (1996).

The values for trees and shrubs represent the ratio of the activity remaining on the tree to that on the underlying ground. The deposition to these depends on a number of factors, such as the size of the tree and the type of tree (deciduous or evergreen) and, for deciduous trees, whether the tree is in leaf. The deposition to a deciduous tree in winter is only a few percent of that to the same tree in summer. Similar comments are appropriate for deposition to small plants.

Material depositing in rain is treated in a similar way, but the calculation must also consider the extent to which material remains on the surface or runs off with rainwater to drains or sewers. The calculation is carried out in two stages: (a) selecting values for the amount of material depositing in rainwater on the different surfaces and (b) allowing for the fraction of the deposit that runs off with rainwater. Typical values are given in Table 2, which shows the relative deposition in rain and the fraction running off with rainwater, as used in the ERMIN model. The values are based on results from Andersson (1991), Gravenhorst and Höfken (1982), Jacob et al. (1987, 1990), Karlberg (1986, 1992), Müller and Pröhl (1993), Roed (1985b, 1987a, 1990) and Roed and Jacob (1990), amongst others.

Similar comments to those given above for dry deposition to trees and vegetation apply also for wet deposition.

Table 2 Relative deposition to different surfaces in wet conditions.

Surface	Deposition to surface relative to that to reference surface, before run-off			Fraction of deposit running off with rain water		
	Elemental iodine	Aerosol with AMAD		Elemental iodine	Aerosol with AMAD	
		<2 μm	2–5 μm		<2 μm	2–5 μm
Paved areas	1.0	1.0	1.0	0.975	0.55	0.55
Roofs	0.85	0.85	0.85	0.99	0.5	0.55
External walls of buildings	0.01	0.01	0.01	0	0	0
Trees and shrubs	1.0	1.0	1.0	0.99	0.5	0.75
Grass	1.0	1.0	1.0	0.99	0.9	0.9
Small plants	1.0	1.0	1.0	0.99	0.8	0.7
Soil (bare soil)	1.0	1.0	1.0	0	0	0
Soil and grass	1.0	1.0	1.0	0	0	0

Tables 1 and 2 give estimates of the relative deposition densities on the different surfaces. Information is also required on the relative areas of the different surfaces. Different values would be found in different parts of an inhabited area, or in different types of inhabited area. For example suburban areas are likely to have more of the surface covered by soil, grass or trees than city centres. Therefore, an inhabited area could be divided into different sub-areas such that appropriate values for relative amounts of different surfaces can be chosen to represent the various sub-areas. The different sets of values, for each sub-area, would then lead to different predicted dose rates in each area for the same deposition to the reference surface. Within each of the sub-areas, there will also be variations in the relationships between surfaces – particularly the extent to which gardens have trees and the balance between soil and grass areas and paved areas such as patios. Such variations would lead to variations in dose rate over relatively small areas. This variation in dose rate is unlikely to be described in a model for the inhabited area.

3.2. Behaviour of material following deposition

Following its deposition, material is transferred by a number of processes which modify the relative deposition to different surfaces. Material initially deposited on soil travels down the soil column over a period of time following deposition. The dose rate per unit activity in the soil reduces as material passes down the soil column because of the shielding effects of the overlying soil. Material initially deposited on the various impermeable surfaces weathers primarily as a result of the action of wind and rain. These processes remove material from the surface on which it was originally deposited (see Section 4 of Chapter 5 for details). In some cases, this removes the activity from the inhabited area via the sewers. In other cases, it could lead to some of the activity transferring to a different surface, for example if rainwater subsequent to the initial deposition carries material from walls to paved areas near the building, or from some paved areas to soil.

The information on the weathering processes and rates for the different impermeable surfaces is discussed in Chapter 5. Calculations of doses in inhabited areas require information on the amount of material on each surface at all times, and therefore the data described in Chapter 5 must be used to derive equations describing the retention of material. The retention of contamination on a surface is commonly described using a double exponential form

$$A(t) = A_0\left\{a_1 \exp\left(-\ln 2\, t/\tau_1\right) + a_2 \exp\left(-\ln 2\, t/\tau_2\right)\right\} \quad \text{where } a_1 + a_2 = 1 \quad (1)$$

Table 3 Retention times of contaminants on urban surfaces.

Surface	Fraction a_1	Retention time τ_1 (years)	Fraction a_2	Retention time τ_2 (years)
Roads	0.7	0.4	0.3	3
Roofs	0.5	4	0.5	30
Walls	1	7	–	–

where $A(t)$ is the deposition at time t, A_0 the deposition at time zero and fractions a_1 and a_2 of the initial deposit clear with retention half-lives of τ_1 and τ_2, respectively (as demonstrated in Equations (1) and (2)).

Values for the retention parameters for some surfaces, as adopted in the ERMIN model, are given in Table 3. These values are based on the following, and other, studies: Andersson (1991), Andersson et al. (1996a, 2002), Gallego (2006), Jacob et al. (1990), Karlberg (1992), Karlberg and Sundblad (1986), Roed (1987a), Roed et al. (1998), Warming (1982, 1984). As discussed in Chapter 5, a problem is that the available measured data practically exclusively relate to radiocaesium contamination in cationic form.

Rain is one of the main causes of weathering, and much of the activity that is removed from the impermeable surfaces is carried with run-off water to the sewer system. Sewers are generally sufficiently far underground that there would be no exposure of people in the inhabited area from material once it has entered the sewer, and this is generally not considered in inhabited area dose models. Some of the activity may be carried from the surface it originally deposited on to a different surface, and could continue to irradiate people in the area. Models generally either ignore this material after it has been transferred or assume that it becomes uniformly distributed over the surface to which it has transferred. Neither assumption is correct; however, material weathered to a different surface is unlikely to make a significant contribution to the dose in the inhabited area and so approximate treatments of this contribution are acceptable in inhabited area dose modelling.

Material initially deposited on soil surfaces is subject to processes which cause it to move down the soil column over a period of years. This reduces the dose rate from this material as it is shielded by the soil layers above. Calculations of the temporal variation of dose rate in the inhabited area must allow for this shielding effect, and so must consider the rate at which material penetrates into the soil. Therefore, models must calculate the profile of activity as a function of time after the initial deposition. As mentioned in Section 3.3 of Chapter 5, this may be done either using a diffusion – advection equation or by solving a compartment model.

Material also deposits on trees and other plants in the inhabited area. The behaviour of material after its deposition is included in some models for inhabited areas, but a number of assumptions and approximations must be made. These generally reflect a lack of knowledge of the behaviour of this material. Material would be lost from trees because of weathering of material originally deposited on leaves and wooden parts of the tree. Material absorbed into the leaves would also be lost when the latter fall. This would continue to irradiate people if the leaves are left in the area, either on the ground or if placed on a compost heap.

3.3. External dose rate from gamma irradiation

Gamma rays travel for many tens of metres in air, and can pass (with some attenuation) through typical urban structures. Therefore, people indoors and outdoors in an inhabited area would be irradiated by material deposited on the external walls and roofs of nearby buildings and on paved areas, vegetation or soil between those buildings. People would also be irradiated by material that had travelled down the soil column after its initial deposition. People, particularly while inside buildings, are also exposed from material deposited inside the building.

As noted above, deposition density to the various surfaces in the area would be different and the material initially deposited on different surfaces behaves differently. There are many objects in the area that would provide some shielding from the dose from deposited material. For example the dose rate indoors from material outdoors is reduced because of shielding by the building walls, etc. Similarly, dose rate at some locations outdoors from particular surfaces would be reduced by shielding from buildings in the area. Therefore, calculations of dose rate need to consider the amount of material on the different surfaces and the dose rate for a unit deposition on each of the surfaces and allow for the size of each surface, the positions of the surfaces and of shielding material in the area.

There are two methods to calculate dose rates from contaminated surfaces allowing for the presence of shielding material, namely the point kernel and the Monte Carlo methods. These are briefly described below, together with their advantages and disadvantages.

The *point kernel method* is based on a formula for the gamma flux, F (photons $m^{-2}\,s^{-1}$), at a distance r (m) from a point source of strength q (s^{-1}):

$$F = \frac{qB(E, \mu r)e^{-\mu r}}{4\pi r^2} \tag{2}$$

where μ is the linear attenuation coefficient (m^{-1}), B the energy deposition build-up factor and E the initial gamma ray energy (MeV).

In this expression, the term $e^{-\mu r}$ describes the shielding by material in the direct line between the source and the point where dose is being calculated, while the term B describes the contribution to the dose rate of photons which originally left the source travelling away from the target and were scattered back to the target. The gamma flux is then converted to dose using an appropriate conversion factor, which is energy dependent. This is extended to an area source by considering the source to be an array of point sources and integrating over that array. This model allows for the shielding effects of material in the direct line between the part of the source considered and the dose point. The build-up factor, *B*, allows for multiple scattering in an infinite homogeneous medium, but the model does not include a description of the effects of scattering from finite volumes of material. Such models may not give a good description of dose rate indoors because of the complexity of modelling the shielding effects of walls and windows when calculating the dose indoors from material outside. The main advantage of this method is that the equation can be evaluated very rapidly if an appropriate computer code is available. The equation could be handled using some general purpose shielding codes, but some may not be able to handle the complex arrangements of shielding materials (e.g. walls and ceilings) that may need to be considered.

The *Monte Carlo method* is based on randomly generating individual photons, tracking their paths from the source and describing their scattering in collisions with atoms in the media through which they pass in a probabilistic fashion. Doses are calculated by defining a volume around the point at which dose is required, and calculating the energy deposition in the volume (often called the target or detector). Some of the photons that leave the source do not reach the detector while others pass through the detector without interacting with atoms within it; these photons do not contribute to the predicted dose rate. Monte Carlo codes use random numbers to determine the tracks of the photons considered, whether and where they interact with atoms and the outcome of the interactions. They also consider photons or particles produced in the interactions, and include the dose from those particles as they also travel through the volume of interest. The main advantage of this method is that it can describe any arrangement of shielding material and any distribution of the source material on the surfaces in the inhabited area. As the method uses random numbers to track photons or particles, there is a statistical uncertainty in the results of the calculations. To reduce these to an acceptable level, very large numbers of emitted photons must be considered, and therefore the calculations can require substantial computer resources. This is the main disadvantage of the method.

Dose models for inhabited areas use values for the dose rate in different locations per unit deposition density on the different surfaces in the area, calculated by the methods described above. The values are derived for a location that is assumed to represent all or part of the inhabited area, in terms of the size and position of buildings, and their composition, width of streets, locations and sizes of trees and other features. An example of a simplified building scenario actually modelled in some dose calculations is shown in Figure 1.

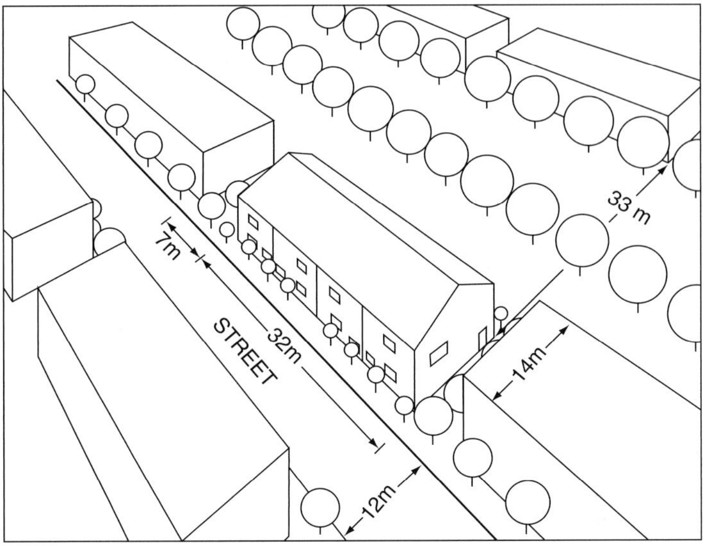

Figure 1 An environment with rows of large two-storey terrace houses as modelled for Monte Carlo radiation transport simulations (from Jacob and Meckbach, 1987).

3.4. External dose rate from beta irradiation

Beta particles travel only a few metres in air, and are stopped by thin layers of other materials. Therefore, people are exposed to beta irradiation only from material deposited very close to them (or on them, as described in Chapter 4) and only if there is no shielding by any surfaces between the person and the radioactive material. In many applications, particularly deposition in the inhabited environment of mixed fission products following a reactor accident, the dose from beta irradiation is much less than that from gamma irradiation. However, for other situations where only a single radionuclide is deposited, such as an accident with a single source or some terrorist events, beta irradiation could be an important exposure pathway (e.g. Andersson et al., 2008b). The modelling of doses from beta irradiation is not considered in many dose models for inhabited areas.

4. Ingestion Dose from Food Contaminated in Inhabited Areas

In many countries, food products are largely produced in agricultural and natural areas, and not in inhabited areas. However, a fraction of the consumed food is, to a varying extent in different areas of the world, produced in small single-household ground lots located in inhabited areas (commonly termed 'kitchen gardens'). For instance, in Sweden, nearly half of the potatoes, and some 10% of the berries, carrots and onions consumed are home-grown (Andersson et al., 2008a). In large village areas of the former Soviet Union, the diet is primarily made up of the yield of kitchen gardens in inhabited areas and food collected in natural areas (Tsalko, 2004).

If an inhabited area is contaminated, consumption of food products grown there might lead to significant dose contributions to the population. Much has been written about doses received from consumption of food items from agricultural and other 'traditional' food-producing areas (Müller and Pröhl, 1993; Nisbet and Woodman, 2000), and this exposure pathway does not as such fall within the scope of this book but is dealt with in some detail in Volume 14 of this book series. As mentioned in Chapter 5, however, it should be noted that food production in kitchen gardens in inhabited areas would be subject to very different conditions than that in agricultural areas. For instance, fertilising, irrigation and tilling practice as well as environmental impact would be different in such small lots in inhabited areas. This inevitably leads to different soil characteristics, and, over time, different vertical distributions and plant uptake availability of deposited contaminants in the soil. The influence of such differences on radionuclide transfer to dietary components has largely been ignored to date. It has further been demonstrated that kitchen garden soils often receive comparatively high quantities of vegetation wastes and other residues of different types, such as ashes from forest fires and slag (Davies, 1978; Norra and Stüben, 2003; Peltola and Astrom, 2003). Such routine human actions are responsible for a significant part of the contamination of kitchen gardens with pollutants like lead and cadmium compared to agricultural areas, but also other urban pollution sources contribute considerably (Douay et al., 2007). As mentioned in Chapter 5, the retention of radioactive contaminants by soil organic matter may be sensitive to changes in pH brought about, for example by the presence in the soil of various pollutants. Fertilisation with ashes containing various pollutant substances, including ^{137}Cs, is in fact restricted by legal action in some countries (SSI, 2005).

Little can be said in general about the magnitude of doses received due to consumption of contaminated kitchen garden products, as this production style adds a host of new locally varying factors to those applying to larger scale food production areas usually considered when assessing dose from ingestion of contaminated food. These factors include specific product types, small-scale management practice, pollution aspects and dietary significance of home-grown food. However, in those contaminated areas where kitchen garden products play an important role in the diet, the resulting doses may be highly significant compared with other dose contributions received due to contamination in an inhabited area (Balonov, 1997).

Another pathway of food contamination in inhabited areas relates to the direct contamination of uncovered food items inside a dwelling, possibly near an open window. The deposition velocity of particles to a plate of hot and moist food can be high, due to turbulence created by convection close to the hot surface and increased adhesion to the moist food (Andersson et al., 2004). At least for some cases, it has been demonstrated that the ingestion dose received through this pathway might be of the same order of magnitude as the inhalation dose received during the passage of the contaminated plume (Andersson et al., 2008b). This problem would obviously also apply to professional food preparation lines in inhabited areas, such as restaurants and bakeries.

5. Other Possible Dose Contributions in the Inhabited Environment

Internal doses from radioactive material inhaled in the early phase during the passage of the contaminated plume may in many types of scenarios be comparatively high, and can be estimated by multiplying time-integrated air concentrations by a representative breathing rate and a dose conversion factor appropriate to the characteristics of the person and the contaminant (see Chapter 4). If the exposure occurs indoors, the air concentrations need to be multiplied by the ratio of indoor to outdoor air concentrations, which depends on the type of contaminant as well as on the dwelling characteristics. The corresponding generic formula is given in Section 6.1.8.

Also after the original contaminated plume has passed, inhalation doses may be received, as deposited radioactive matter can resuspend in the air due to wind or various kinds of human activity, both outdoors and indoors. However, as discussed in Chapter 2, the dose rate component due to this exposure pathway will rapidly decline following the initial deposition of the contamination. Chapter 2 gives an account of how the resuspension can be estimated on the basis of knowledge of deposited contamination levels and appropriate resuspension factors or rates.

External doses from radiation emitted from a passing plume can be estimated from knowledge of the activity concentrations in air of the contaminant (assuming these are reasonably homogeneous within some distance) and the use of dose conversion factors, as illustrated in Section 6.1.9.

As mentioned in Chapter 4, both gamma and beta doses from direct contamination of skin and clothes may be surprisingly high, considering that the contaminants usually only remain on such surfaces for short time. Again, a viable conceptual approach is illustrated in Section 6.1.7. It should be noted that it cannot be ruled out that deposition on skin may under some circumstances lead to limited uptake of contaminants into the body, particularly if the skin is not intact and if the contaminants are in simple ionic form (Schaefer and Lademann, 2001; Tinkle et al., 2003; Koprda et al., 2000). The requirements for calculating how much of a contaminant material can be transferred by contact are discussed in Chapter 4.

For completeness, it should be mentioned that food that is produced in contaminated areas *outside* the inhabited environment may contribute greatly to the dose to inhabitants after an airborne contaminant release. However, the consequences of contamination in agricultural, natural and aquatic environments, where much food comes from, lie outside the scope of this publication.

6. Examples of Calculated Dose Rates

6.1. Illustrative calculations of dose components for a dry deposition case

This section presents a *simplified* example of calculated doses in an inhabited area. The example is aimed at illustrating some of the considerations that may be made in estimating doses received through different pathways. The considered scenario involves a contamination in absence of precipitation in a village in the Bryansk region of Russia, with contaminants released in the spring of 1986 following the accident at the Chernobyl Nuclear Power Plant some 150 km away. For simplicity, the doses are given relative to a dry deposition on the ground of 1 MBq m^{-2} of ^{137}Cs. The example will demonstrate a 'rough' methodology to calculate typical doses received by members of the public over the first 20 years after the Chernobyl accident.

Doses from contamination on outdoor surfaces would here generally be governed by the contamination with ^{137}Cs and ^{134}Cs, since these radionuclides are long-lived gamma-emitters, and the caesium contamination, which was deposited in comparatively high concentrations over the area, is retained over long periods of time after deposition on most outdoor surfaces in a living environment (see Chapter 5). The two radionuclides are in identical physicochemical forms and thus followed the same deposition pattern. The major part of this dose contribution over the period of interest is from ^{137}Cs, primarily due to the very long physical half-life of this radionuclide. Reliable attenuation coefficients for urban dwellings are generally only available for radiation from ^{137}Cs. However, the difference in energy between important photons emitted by ^{134}Cs and ^{137}Cs is on average limited, and the attenuation of the radiation is thus similar. Nevertheless, the dose rate per unit of unshielded contamination for ^{134}Cs is considerably higher than that for ^{137}Cs (Lauridsen, 1982), and this needs to be taken into account. Dose contributions from beta radiation from contamination on these surfaces have not been included in the evaluation, but will generally be of comparatively little importance.

6.1.1. Contamination on streets

Based on *in situ* measurement series made in various parts of Europe after the Chernobyl accident (Tables 1 and 3), the following expression can be deduced for the decrease in external dose rate contribution, $D'(t)$, with time from deposition of radiocaesium (^{137}Cs or ^{134}Cs) on streets (Andersson et al., 2002):

$$D'(t) = \Gamma C_{\rm ref} 0.3 \exp\left(-\frac{\ln 2}{T_{1/2}}t\right)\left(0.7\exp\left(-\frac{\ln 2}{120\ \text{days}}t\right) + 0.3\exp\left(-\frac{\ln 2}{3y}t\right)\right) \quad (3)$$

where $T_{1/2}$ is the physical half-life of the radionuclide, and $C_{\rm ref}$ the initial contamination level of the radionuclide on a grassed reference surface. The initial contamination level on streets after the Chernobyl accident has on average been found to be about half of that value (Andersson et al., 2002). Γ is the conversion factor from contamination level to dose rate. This factor was based on the estimates of Meckbach et al. (1988) for relatively small, detached houses adequately representative of those typical of villages in the Bryansk region. Based on a number of independent surveys in Europe and in California (Jenkins et al., 1992; Andersson, 1996b; Long et al., 2001; Kousa et al., 2002), it seems that the assumption that people on average spend some 15% of their time outdoors is generally reasonable. It was assumed in the calculations that half of this time was spent on paved surfaces (streets, walkways, etc.) and the other half in garden areas.

6.1.2. Contamination on roofs

For roof pavings, the initial dry deposition of radiocaesium after the Chernobyl accident was generally found to correspond approximately to that on the grassed reference surface (Table 1). A measurement series performed after the Chernobyl accident showed that the decrease in contamination level on a roof through natural weathering processes is dependent on the material type (Andersson et al., 2002). The prevailing roofing type in many villages in the Bryansk region is asbestos sheets, and this is assumed to apply in the given case. The change with time (Table 3) in the external dose rate contribution from a radiocaesium concentration on an asbestos roof has been found to follow the expression (Andersson et al., 2002):

$$D'(t) = \Gamma 0.5 C_{\rm ref} \exp\left(-\frac{\ln 2}{T_{1/2}}t\right)\left(\exp\left(-\frac{\ln 2}{4\ \text{years}}t\right) + \exp\left(-\frac{\ln 2}{30y}t\right)\right) \quad (4)$$

in which $C_{\rm ref}$ is the initial contamination level on a grassed reference surface, and $T_{1/2}$ is the physical half-life of the radionuclide (^{137}Cs or ^{134}Cs). For clay or concrete roofs, the weathering process is significantly faster. It has been found that covers of moss and algae on a roof may increase the retention of radiocaesium additionally (Fogh et al., 1999). Also for roofs the Γ factor was based on the estimates of Meckbach et al. (1988). The roofs in the Bryansk area often have a light construction, and the critical population group is residents of single-storey or top floors of multi-storey buildings, as these by far get the highest dose contribution from the roof. It was assumed here that inhabitants spend on average about half of their time in a top floor flat, whereas the rest of the time is spent either outdoors or on other floors of buildings in the area.

6.1.3. Contamination on walls

In situ measurement series carried out after the Chernobyl accident have shown that natural weathering processes only very slowly lead to a reduction in the contamination level on walls, probably primarily due to the vertical orientation (Andersson et al., 2002). The possible influence of differences in material characteristics is indeterminable in the available data sets but, in general, the natural removal of radiocaesium contamination on walls seems to occur with a half-life of about 7 years (Table 3), as indicated in the formula below for the temporal decline in the dose rate from ^{137}Cs or ^{134}Cs contamination on walls:

$$D'(t) = \Gamma 0.05 C_{\text{ref}} \exp\left(-\frac{\ln 2}{7\ \text{years}} t\right) \exp\left(-\frac{\ln 2}{T_{1/2}} t\right) \quad (5)$$

where $T_{1/2}$ is the physical half-life of the radionuclide, and the conversion factor from contamination level to dose rate, Γ, is again based on the findings of Meckbach et al. (1988) for relatively small brick or wooden houses characteristic of the Bryansk villages. The difference in shielding provided by these two types of dwellings is substantial. As indicated by the formula, the radiocaesium dry deposition per unit area on a wall is based on the Chernobyl experience (Table 1) expected to be of the order of 5% of that on the grassed reference surface.

6.1.4. Contamination on open (grassed) soil areas

Bunzl et al. (1997) modelled the influence on dose of the mobility of ^{137}Cs in soil. They found one time-function for the areas closest to the Chernobyl reactor, where the contamination is to a great extent associated with large, insoluble fragmentation particles and another for contamination at greater distances, where caesium was primarily transported in the air in the form of small condensation particles. It is clearly the latter situation that would be of relevance to this case study. The local dose rate shortly after deposition depends to some extent on the surface roughness and possibly penetration with rainwater early after the deposition. For an area where deposition took place in rain, Bunzl et al. (1997) estimated this value to be of the order of 60% of that from contamination on top of a smooth surface. For a dry deposition scenario, the initial contaminant penetration into the soil would be expected to be less, and based on the findings of Roed (1990), it is estimated that a value of the order of 80% would be reasonable in this case, and this was applied in the expression of Bunzl et al. (C_{ref} is the initial contamination level on the surface and Γ is the dose conversion factor for surface contamination):

$$D'(t) = \Gamma C_{\text{ref}}\left(0.46 \exp\left(-\frac{t \ln 2}{3.3y}\right) + 0.34 \exp\left(-\frac{t \ln 2}{21y}\right)\right) \tag{6}$$

However, the local soil type and conditions would of course influence the long-term behaviour of the contaminants in the soil and thus the long-term dose. The time constants found by Bunzl et al. in a different area might thus not adequately describe the migration processes in typical soil in the Bryansk region. Therefore, a test was carried out to examine the local adequacy to the case study of the equation above. In 1995 and 1997, a large number of soil core samples were taken at Novo Bobovichi and Guta Muravinka in the Novozybkov area (Roed et al., 1996, 1998). These samples were sectioned at ca. 2 cm intervals to determine the vertical distribution profile of Chernobyl ^{137}Cs at the time. Based on typical undisturbed soil (sandy loam) profiles from these investigations and MCNP Monte Carlo photon transport calculations (Briesmeister, 1993) for infinite, homogeneous isotropic plane ^{137}Cs sources at various depths in soil with representative density, the reductions in dose rate over, respectively, 9.3 and 11.3 years following the Chernobyl accident could be estimated. The dose rate was estimated to have been reduced to 33% in the early autumn of 1995 and 28% in the early autumn of 1997. In comparison, the above formula gives the values 31% for the early autumn of 1995 and 25% for the early autumn of 1997, which is in reasonable agreement, suggesting that the parameters in the above formula are applicable for this case. Similarly, for ^{134}Cs the following modified formula was used:

$$D'(t) = \Gamma C_{\text{ref}}\left(0.46 \exp\left(-\frac{t \ln 2}{3.3y} - \frac{t \ln 2}{2.06y}\right) + 0.34 \exp\left(-\frac{t \ln 2}{2.06y}\right)\right) \tag{7}$$

It should be stressed that the above time-functions were originally established for very large open areas. In the earliest phase after a dry deposition of ^{137}Cs and ^{134}Cs, most of the dose rate would, due to, for example ground roughness, be expected to arise from contamination within the nearest 10 m (Andersson, 1996b; Nielsen, 1977). Later on, as the contamination migrates down, the influence of the contamination at large distances will become less. As the dwellings in villages of the Bryansk area typically have only one storey and are surrounded by large open garden areas, the error in using a model for an infinite area will often be small.

The dose conversion factor, Γ, was based on the modelling of Meckbach et al. (1988) for relatively small brick or wooden houses characteristic of the villages. As mentioned above, it was assumed that people spend 15% of their time outdoors, evenly distributed between pavements and garden areas. A large fraction of the dose contribution from these open areas is generally received during the little time spent outdoors.

6.1.5. Contamination on trees and shrubs

Due to the efficient filtration through the very large surface provided by leaves or needles, dry contamination of trees and shrubs can be highly significant. After the Chernobyl accident, it has been found that the contamination level on a tree per unit area of the ground covered by the tree was about 2.5 times as high as that on the grassed reference surface (Table 1). The behaviour of the contamination on the trees will vary according to, for example the tree species, climate, season and environmental conditions, but radiocaesium contamination on common spruce, which is typical of the area of interest, has been reported to be removed from tree canopies with a half-life of the order of 100 days (Bunzl et al., 1997), and this is what has been assumed in the model. In spite of this short half-life, the greatest contributions to dose from contamination in tree canopies are clearly from the radiocaesium isotopes, and only these have been taken into account in the calculations. In the expression below, Γ is as above the dose conversion factor (based on the work of Meckbach et al., 1988), $T_{1/2}$ is the physical half-life of the radionuclide (^{137}Cs or ^{134}Cs) and C_{ref} is the contamination level on the grassed reference surface. The shielding provided by dwellings against radiation from contamination on trees and shrubs will often be limited, as much unscattered radiation from contamination on the needles/leaves will enter the dwelling through its thin windows.

$$D'(t) = \Gamma 2.5 C_{\text{ref}} \exp\left(-\frac{\ln 2}{100\ \text{days}} t\right) \exp\left(-\frac{\ln 2}{T_{1/2}} t\right) \tag{8}$$

If the contamination in the needles or leaves is allowed to accumulate on the ground or on roofs, this will naturally add further to the dose.

6.1.6. Contamination on indoor surfaces

Due to factors such as the long physical half-lives and association with small particulates after a major nuclear power plant accident (generally leading to relatively long-term retention on indoor surfaces), the main contributors to dose from contamination on indoor surfaces would be expected to be ^{137}Cs and ^{134}Cs, and only these have been considered here. Tracer experiments have shown that deposition velocities for aerosol in the relevant size range are about one order of magnitude higher to the floor than to the walls and ceiling of a room (Lange, 1995). Investigations have indicated that the retention half-life of Chernobyl ^{137}Cs on a floor (assuming almost daily vacuuming) was of the order of 1 month (Allott et al., 1994). The retention half-life of the contaminants on walls and ceiling would, based on the findings of Andersson (1991), be estimated to be of the order of 2–3 years, since routine cleaning procedures are generally not applied to such surfaces (Chapter 3). Based on the findings of Roed (1990), the following time-function can be deduced for the dose rate from contamination on an indoor surface:

$$D'(t) = \Gamma \frac{V_{d,s}}{V_{d,ref}} f \frac{\lambda_v}{\lambda_v + \lambda_d} C_{ref} \exp\left(-\frac{\ln 2}{T_{1/2,s}} t\right) \quad (9)$$

where λ_v is the rate coefficient of ventilation (the fraction of air that is exchanged per unit of time), λ_d is the rate coefficient of deposition (the fraction of aerosols in the dwelling that deposits per unit of time) and f is the filtering factor (the fraction of aerosols entering the dwelling which are not retained in cracks and fissures of the building structure). $V_{d,s}$ is the deposition velocity for the aerosol to the indoor surface in question, $V_{d,ref}$ is the deposition velocity to the grassed reference surface, C_{ref} is the initial contamination level on this reference surface, Γ is the dose conversion factor and $T_{1/2,s}$ is the effective half-life of the contamination (^{137}Cs or ^{134}Cs) on the indoor surface in question, taking into account radioactive decay as well as physical removal. Typical values of V_d, V_{ref}, f, λ_v and λ_d are given for various conditions by Andersson et al. (1996b), who also described a method for calculations of the dose conversion factor, Γ. Simple calculations using this method formed the basis of the rough estimates of Γ used in this study. As above, it was assumed that people on average spend about 85% of the time indoors. Dose contributions from beta radiation from contamination on these surfaces have not been included in the evaluation, but are of minor importance.

6.1.7. Contamination on humans

The retention half-life on humans of contaminants from the Chernobyl accident is believed to have been short (Andersson et al., 2002). This means that short-lived radionuclides may have an important bearing on the dose contribution. On human skin, an upper limit of the retention half-life is set by the shedding of the stratum corneum, which occurs over a few weeks. For supermicroneous particles, the retention half-life is considerably shorter (Andersson et al., 2002). On clothes, the retention half-life is typically a day or two (Fogh et al., 1999). Contamination on hair has for simplicity not been considered in the model, since it is likely to be of minor importance for dose (Andersson et al., 2002). It was assumed in the calculations that people were staying indoors during the passage of the contaminated plume. The various radionuclide concentrations assessed by Kryshev (1996) determine the source term for both this case study and the calculations of Andersson et al. (2002), which are thus here directly applicable. The calculations were made according to the formula:

$$D = \Gamma \frac{1}{\lambda_{\text{physical}} + \lambda_{\text{clearance}}} V_{\text{d,surface}} \frac{f\lambda_{\text{v}}}{\lambda_{\text{v}} + \lambda_{\text{d}}} \int_{t=T_{\text{o}}}^{T_1} C_{\text{o}}(t)dt \qquad (10)$$

where Γ is the conversion factor from surface contamination level to either gamma or beta dose to the affected body parts, $\lambda_{\text{physical}}$ the radioactive decay constant (governing the physical half-life) and $\lambda_{\text{clearance}}$ the rate constant for clearance of contamination from the surface. $C_{\text{o}}(t)$ is the outdoor contaminant air concentration time-function, f, λ_{v} and λ_{d} are defined above and $V_{\text{d,surface}}$ the deposition velocity to the human surface in question (see Chapter 4). The time-integrated outdoor contaminant air concentrations were scaled to a level that would result in a deposition of $1\,\text{MBq}\,\text{m}^{-2}$ of ^{137}Cs on a lawn. The radionuclides that were found to have had the greatest influence on both gamma and beta doses from the contamination on humans are ^{131}I, ^{137}Cs, ^{134}Cs, ^{106}Ru and ^{103}Ru. Dose contributions from contaminants inadvertently transferred onto humans by contact and from deposition to humans of resuspended contaminants in the indoor environment have not been included here, but are likely to be of little importance in the particular case.

6.1.8. Contamination inhaled during the plume passage

The doses received through inhalation of contaminants as the polluted plume passed over the area were calculated on the assumption that people stayed indoors during virtually the whole period where the air concentrations were highest. The committed doses can then be calculated from the following expression:

$$D = \text{BR}\Gamma \frac{f\lambda_{\text{v}}}{\lambda_{\text{v}} + \lambda_{\text{d}}} \int_{t=T_{\text{o}}}^{T_1} C_{\text{o}}(t)dt \qquad (11)$$

where BR is the breathing rate, which may here be taken to be $3.3\,10^{-4}\,\text{m}^3\,\text{s}^{-1}$ here (ICRP, 1994), Γ the inhalation dose coefficient (ICRP, 1995) and the other factors in the expression are as defined above. The greatest contributor to the inhalation dose was found to be ^{137}Cs. Dose contributions from inhalation of resuspended contaminants in the indoor environment have not been considered here, but would be of minor importance.

6.1.9. External irradiation from the contaminated plume

This estimation was carried out by multiplying the time-integrated contaminant concentrations of the radionuclides in air assessed by Kryshev

(1996) by calculated dose conversion factors for contamination in a semi-infinite space of air. The applied conversion factors for gamma dose rate to the human body and conversion factors for beta dose rate at a depth in human tissue corresponding to the epidermal basis layer were taken from Hedemann Jensen (1992).

6.1.10. Contamination in locally produced food

Doses from ingestion of contaminated food in a particular area are difficult to estimate adequately, as a proper investigation would require much knowledge of the fractions of the main dose-contributing dietary constituents that were actually produced in the area. Food produced locally in kitchen gardens could constitute a major (but highly variable) part of the diet in these areas. As discussed above, contamination levels may vary considerably, and also agricultural practices and food preparation procedures will greatly influence the ingestion doses. As mentioned in Section 4, it has been claimed (Balonov, 1997) that in the Bryansk region internal and external irradiation contribute about equally to the long-term dose.

6.1.11. Discussion of dose calculations

Table 4 shows 'best estimates', based on the methodologies and parameters described above, of the various contributions to dose that may have been received over 20 years after the Chernobyl accident by people living in a dry contaminated area in the Bryansk region. As can be seen, a large dose contribution is received from contamination in soil areas, due to the generally very efficient retention of radiocaesium contamination in the upper layers of soil. In urban centres, however, where these areas have much more limited dimensions, the contributions of other surfaces to dose would be of much greater relative importance. Contamination on roofs can give significant doses, but only to people living on the top floor of a building. In an area of the type considered it is not expected that a very large part of the time spent outdoors is spent on paved areas, and this is reflected in the dose estimate. Also, natural weathering processes rather rapidly reduce the contamination level on these surfaces. Walls generally carry little contamination, but if the walls are thin, this contribution to dose may be significant. The contribution from contamination on trees varies greatly according to the shielding provided by the dwelling (e.g. the window fraction of the wall areas). Contamination on freely exposed human skin can result in significant gamma doses to the body and beta doses to the skin over a short period of time. Also, as can be seen, the inhalation dose received during the plume passage may be significant compared with other contributions, stressing the importance of sheltering people during the earliest phase of an accident. However, the external dose received from

Table 4 Summary of estimates of dose contributions for the case study.

Dose contribution from:	Best estimate of magnitude
Contaminated streets	0.7–1.5 mSv in a town area. A major part of the dose contribution is generally received during the time spent outdoors, and the influence of dwelling type is thus small.
Contaminated roofs	7–10 mSv expected to inhabitants of a top floor flat with a very light roof construction, which is representative of some dwellings in the area. Significantly less with thicker attic construction. Generally significantly <1 mSv to persons not residing on top floor.
Contaminated walls	0.3–1 mSv for brick walls, and 1–2 mSv for very thin wooden walls.
Contaminated soil areas	16–22 mSv for brick walls. May be some four times higher in areas with low shielding wooden walls.
Contaminated trees and shrubs	2–3 mSv with a not very well shielded house, and <0.5 mSv with a well-shielded building.
Contaminated indoor surfaces	0.5–2 mSv depending on a variety of factors.
Contamination on humans	2–3 mSv gamma dose to the body plus a beta dose of 0.5–1 Sv to freely exposed and 0.04–0.08 Sv to skin covered by 0.4 mm cotton (many radionuclides contribute).
Inhalation (plume passage)	6–8 mSv (many radionuclides contribute).
Contaminated plume (external)	0.3–0.7 mSv gamma dose to the body plus a beta dose of ca. 0.2–0.3 mSv to freely exposed skin and much less where covered with clothing (many radionuclides contribute).
Ingestion of contaminated food	Highly variable, but possibly comparable to external dose contribution.

Doses are given per MBq m^{-2} of ^{137}Cs on the grassed reference surface.

contaminants in the air during the plume passage seems to be of lesser significance. As can be seen from the table, remarkably many different dose contributions here are of the same order of magnitude. This would not have been the case if the contamination had occurred in strong rain, which would have left comparatively little contamination on other outdoor surfaces compared to permeable horizontal surfaces (grass/soil). Also, near ground air concentrations might well have been so depleted by the rain that inhalation doses and doses from deposition of contaminant particles on indoor surfaces and humans would have been comparatively small. Further, it should be noted that, in connection with other scenarios, for instance some types of explosions, a considerable fraction of the contaminant particles could be much larger than assumed in the above case study, which would give an enhanced overall dry deposition velocity and a comparatively higher contamination level on horizontal surfaces. Such large particles would also weather off surfaces much more rapidly than assumed in the example above. Similarly, it has been found that the modelling of weathering functions requires special care in connection with studies of the Pripyat area, very near to the Chernobyl power plant, where a large part of the contamination was associated with large insoluble core fragment particles (Thiessen et al., 2008).

6.2. Example of external dose calculations made with a complex model

This section presents some examples of calculated external gamma dose rates in urban areas and it illustrates the variations between dose rates in different types of areas and the relative importance of different surfaces in contributing to dose rates.

The model used to generate the examples is the ERMIN model (Jones et al., 2006b). The model calculation can be divided into four stages. ERMIN was applied to a scenario created by the Urban Remediation Working Group of the IAEA Environmental Models for Radiation Safety (EMRAS) programme. The scenario was based on the town of Pripyat that was evacuated following the Chernobyl accident. The deposition specified in the scenario was applied to several of the environments in the ERMIN database (Charnock et al., 2009; Thiessen et al., 2008). Some of the environments could correspond to real environments in the Pripyat settlement, whereas others are less appropriate. Figures 2 and 3 show the predicted dose rates and cumulative dose both outdoors and where appropriate indoors. The variation between environments seen in the figures is due to different configuration of urban surfaces.

The 'ideal' environment represents an area of grass with no trees, buildings, or paved surfaces. The 'open grassy area' environment represents an area of mostly grass but with some trees and paved surfaces. The 'terraced houses' environment represents a street of terraced houses or similar low buildings with a relatively high density (similar to that shown in Figure 1). The 'multi-storey block of flats opposite parkland' environment and the 'multi-storey block of flats amongst other house blocks' environment use a similar building construction but the buildings are set in very different surroundings: either next to a park or surrounded by other similar buildings. For this example, the 'multi-storey block of flats

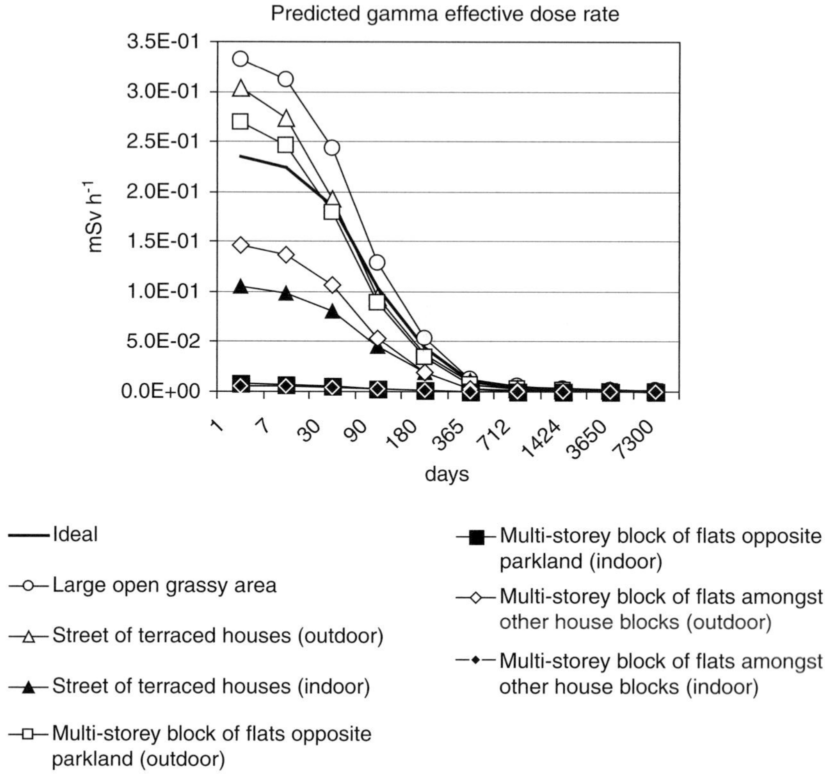

Figure 2 Predicted gamma dose rate for different ERMIN environments based on the Pripyat scenario.

amongst other house blocks' has been specified as having no trees and a high proportion of paved areas in order to represent a very heavily built-up environment; one not actually present in the Pripyat scenario.

In order to understand the predicted differences, it is necessary to decompose the overall dose and dose rates into the individual components from different urban surfaces. The dose rate from a surface depends on the deposition to the surface, the area of surface and the geometry of that surface in relation to the population. In ERMIN, the geometry and area of surface are combined in a parameter called the unit dose rate. The ERMIN database contains unit dose rates calculated using Monte Carlo modelling code for each surface in each environment. The unit dose rate is the dose rate per $Bq\,m^{-2}$ on the surface. Two unit dose rates are calculated for each radionuclide and environment; a representative dose rate inside the buildings (if there is an interior) and a representative outdoor dose rate. Figure 4 shows the unit dose rates for each surface in each environment,

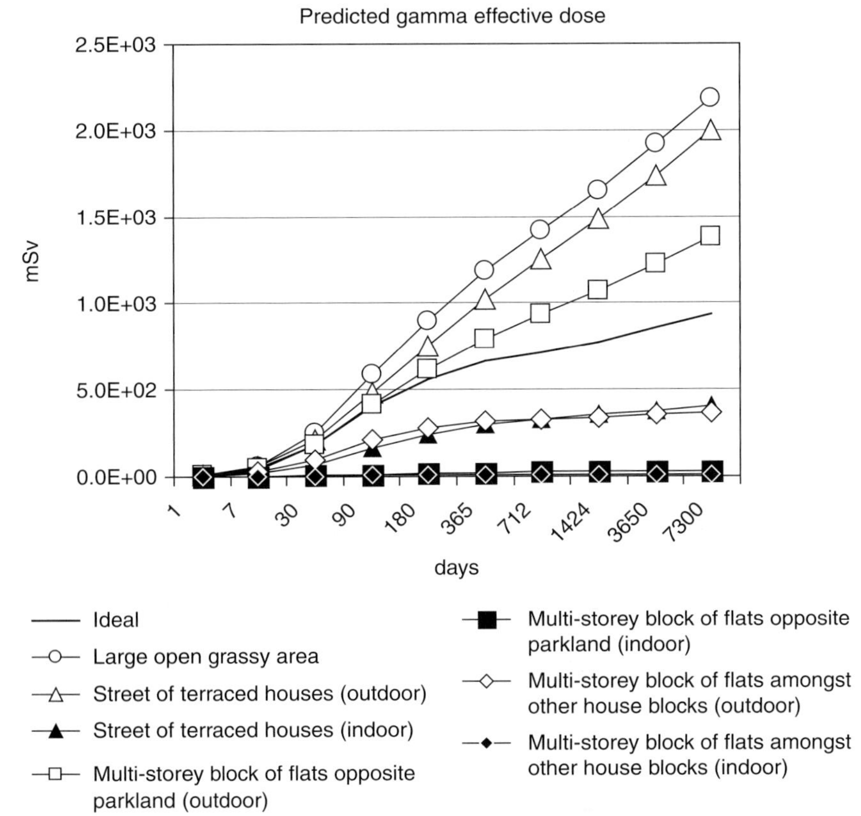

Figure 3 Predicted cumulative dose for different ERMIN environments based on the Pripyat scenario.

weighted by the amount of surface in the default configuration of that environment. The unit dose rates from the ERMIN database are combined with predicted surface activities to give the predicted representative dose rates and doses outdoors and indoors in the ERMIN environments shown in Figures 2 and 3.

Figure 5 shows the predicted outdoor dose rate from each surface in each environment. As can be seen, very different surfaces contribute significantly to the dose rate in the different types of environment at different times. Similar figures can be created to examine the estimated outcome in terms of dose of countermeasure strategies or single countermeasures. Visualisations such as these, verified with suitable measurements, allow decision-makers to target appropriate countermeasures to particular surfaces and environments.

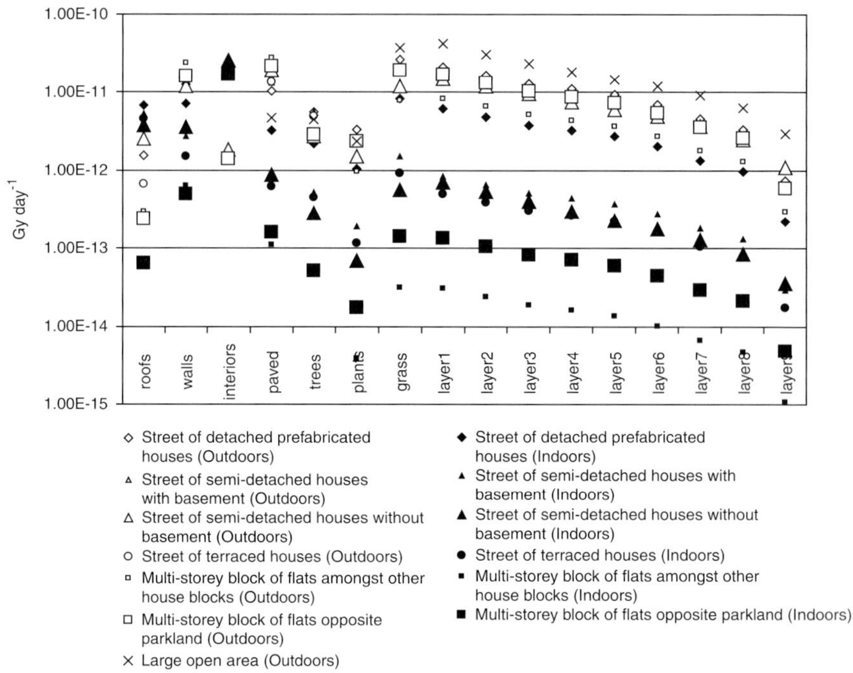

Figure 4 Unit dose rates for all ERMIN environments weighted by the area of each surface within the default configuration of the environment.

7. Doses from Non-Anthropogenic Sources

Essentially all materials contain small quantities of radioactive material, and this leads to exposure of people through a variety of exposure pathways. Soil and building materials contain small quantities of radioisotopes of uranium (^{234}U, ^{235}U, ^{238}U) and thorium (^{232}Th) and their decay products, and of other isotopes such as ^{40}K. People are exposed to external beta and gamma irradiation from these nuclides, both when indoors and when outdoors. The external dose rates from such materials depend on the activity concentration in various materials and the shielding properties of buildings. The decay products of these nuclides include isotopes of radon, which is a noble gas and escapes from soil or building materials and can become trapped inside buildings, leading to internal doses following the inhalation of the radon or its daughter products. People are also irradiated by these nuclides following inhalation of resuspended material or by consumption of foods, as the nuclides enter the food chain through crop root uptake. People are exposed to natural radiation in other ways, such as from cosmic radiation, though this is not considered in detail here.

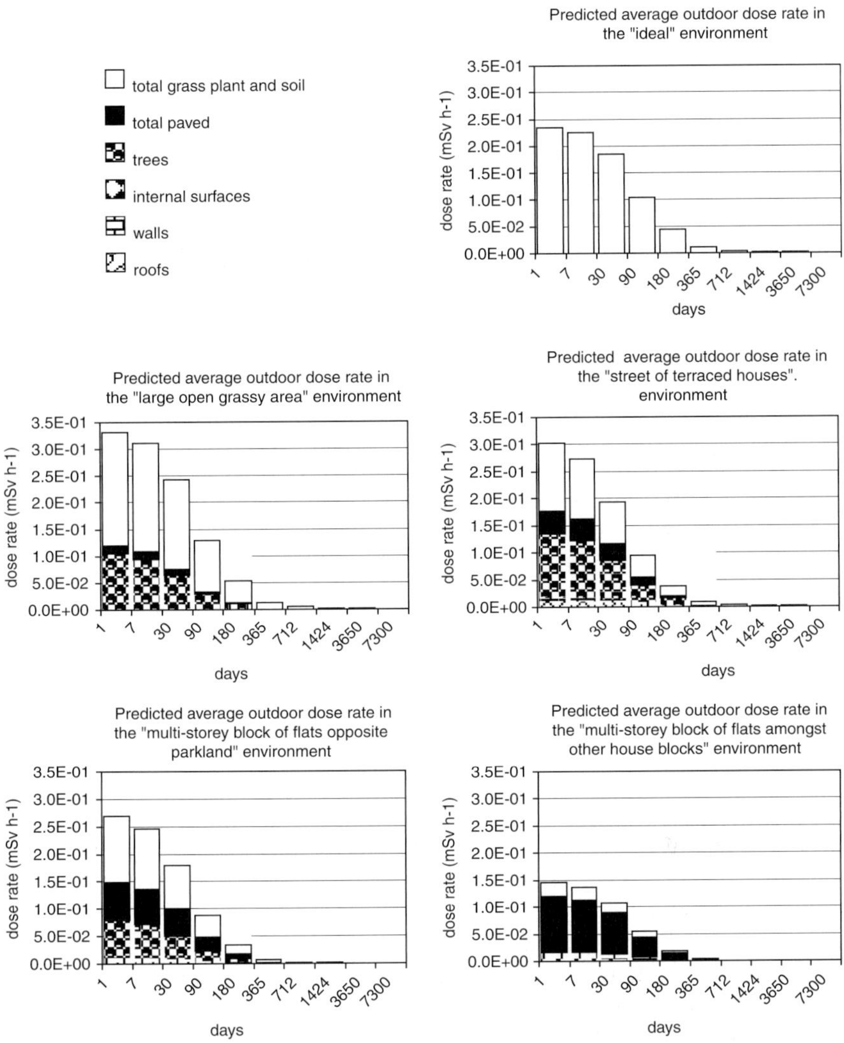

Figure 5 The predicted contribution to outdoor dose rate from each surface in each environment.

The concentrations of these radionuclides in soil vary considerably, both within different parts of the same country and between different countries. Many different materials are used in the construction of buildings, again with differences within and between countries. The concentration of radon isotopes inside buildings depends on many things, such as the construction of the building (particularly whether there is an air space below the floor), the ventilation rate within the building and the material from which the building is made. These differences can lead to considerable variations in the dose rates from natural material between different areas. Dose rates from all sources of exposure have been reviewed in a large number of publications, for example Watson et al. (2005) for the UK and UNSCEAR (2000) for all countries. The following summarises some of the data presented in those reports.

Watson et al. show that the average dose rate in the UK from natural sources is 2.23 mSv per year, with about half of that coming from radon exposure indoors. They give some information on the range of doses from radon, which is discussed later in this section. UNSCEAR presents a histogram of the fraction of the world population with dose rates from natural material in different ranges, and quotes an annual average dose of 2.4 mSv per year. UNSCEAR presents detailed information on the range of dose rates from different exposure pathways in many countries of the world; some of this information is summarised below.

UNSCEAR gives information on the ranges of concentrations of ^{40}K, ^{238}U and ^{226}Ra in soil in different countries; the range for ^{238}U in Germany is $11-330\,Bq\,kg^{-1}$, with similar ranges in other countries. UNSCEAR gives a range of dose rates from terrestrial gamma radiation, giving values for the indoor and outdoor dose rates; in most countries the dose rate is greater indoors than outdoors. As examples, the range quoted for Japan is $21-77\,nGy\,h^{-1}$, both indoors and outdoors, whereas for Switzerland the indoor dose rate ranges from 20 to $200\,nGy\,h^{-1}$ while the outdoor dose rate ranges from 15 to $120\,nGy\,h^{-1}$.

The external dose rates from natural terrestrial gamma radiation tend to be larger indoors than outdoors, the opposite of the situation for external gamma irradiation from material discharged from nuclear sites or from other sources. This reflects the differences in the distribution of the radioactive materials in the two cases, and the effects of shielding by buildings. Material released from nuclear activities is mainly deposited outdoors, and buildings provide shielding from the material. Natural radioactive material is incorporated into soil and building materials. While buildings provide some shielding against the doses from material outdoors, they also cause exposure from radioactive material in the building structures themselves.

Watson et al. (2005) show that the average concentration of ^{222}Rn in indoor air in the UK is $20\,Bq\,m^{-3}$, corresponding to a dose rate of 1 mSv per year. A survey of concentrations in UK houses showed that around 0.5% of UK houses have concentrations in excess of $200\,Bq\,m^{-3}$; 2 houses were found with concentrations in excess of $10{,}000\,Bq\,m^{-3}$ corresponding to a dose rate in excess of 500 mSv per year.

This brief summary has been included to give some context to the examples of doses from deposited material presented above. The original reports, Watson et al. (2005) and UNSCEAR (2000), or other similar documents, should be consulted if further details are required.

8. Current and Future Inhabited Area Dose Model Trends and Needs

This section considers trends and needs of models for calculating the dose from deposited material in inhabited areas. The needs and trends of models for atmospheric dispersion in inhabited areas are discussed in Chapter 2.

There is no generally accepted criterion for how accurate (in terms of the comparison between model predictions and measurements) a model should be for particular applications and indeed it is difficult to validate models against measurements robustly. There is uncertainty in the model predictions, with some parts of a model being more uncertain than others, and so reducing the uncertainty in one part of a model may not necessarily improve the overall uncertainty in the model prediction. Deciding whether a model is adequate for a particular application is, therefore, somewhat subjective.

The assumptions used in models and the data on which the parameter values are based were summarised in previous sections and described in more detail in earlier chapters of this book. This section comments on where current models for different applications are weak, where improvements are most needed, and which predictions are likely to be particularly sensitive to the values assigned to particular parameters.

8.1. Initial deposition

The information on the relative dry deposition to different surfaces is mostly based on measurements made after the Chernobyl accident, and therefore relates to a situation where the contamination is reasonably constant over the area from which most of the dose is received. This may not be the case for some situations where the source is inside the inhabited area.

As mentioned in Chapter 2, dry deposition is governed by a number of different factors, such as the roughness of the surface, wind speed and surface moisture. Particle size is an important factor for aerosols, while chemical reactions between the gas and the surface are important for gases and vapours. Where possible, the values of relative deposition to different surfaces should be taken from measurements made on different surfaces in a single area, rather than separate sets of measurements, to reduce the effect of variation between deposition rates in different conditions. As noted earlier, there is a body of information on dry deposition to different surfaces which is reasonably consistent. However, much of the information comes from measurements made after the Chernobyl accident, and this may not be applicable to other situations. Certainly, particles with other physicochemical characteristics, for instance released in connection with an explosion, could have very different deposition velocities to the different surfaces in the inhabited environment.

Dry deposition on trees may be particularly important for two reasons. Firstly, dry deposition on trees, particularly when the trees are in leaf, is likely to be considerably greater than that to the underlying surface – Jacob and Meckbach (1987) suggest that it is an order of magnitude greater. Secondly, material on trees could be opposite windows in buildings, and so there is less shielding by building walls for material on trees than for material on the ground.

It is generally assumed that deposition of aerosols is governed by the particle size rather than the particular element considered. There is some information on the deposition velocities of different elements (VAMP, 1994) suggesting an order of magnitude variation in the deposition velocities of some elements to paved surfaces. These differences may simply reflect the limited database from which the values are derived.

The indoor contamination is governed by the factors discussed in Chapter 3. Indoor deposition is important because people spend a large part of their time inside houses, and there is little shielding between people and material deposited in the same room. There are very few measurements of relative deposition to indoor and outdoor surfaces, and the deposition indoors has to be calculated from assumptions about the indoor/outdoor ratio of activity concentration in air. This ratio is highly case-specific.

The contamination of surfaces following wet deposition is determined by the amount of rain falling on the surfaces and the amount of radioactive material carried away by run-off water. The data for relative deposition to different surfaces are reasonably consistent, although much of the data comes from measurements in Gävle, Sweden, or southern Germany after the Chernobyl accident. However, the fraction of material initially running off a surface also depends on the characteristics of the contaminant.

The current models reflect the understanding of deposition on surfaces in inhabited areas. However, this is based on limited (though reasonably consistent) data, which could be improved. In particular the information on dry deposition rates indoors and to single trees in inhabited areas (rather than to forests) are limited. There is very little information on deposition in foggy conditions, deposition in snow and deposition to lying snow, and these conditions are generally ignored in current models; they could be of concern in some parts of Europe.

8.2. Weathering

Information on weathering can be obtained from laboratory or field experiments or measurements made at different times after the Chernobyl accident. Information from the different sources is reasonably consistent.

Information on weathering at long times after deposition can only be obtained from experiments or surveys carried out over long periods.

Until rather recently, published data on the weathering from roofs were only available for a period of about 2 years after the deposition. Andersson et al. (2002b) have published information on the retention over a period of around 15 years after the deposition (see Section 4 of Chapter 5). The results showed that, at least for some types of roof material, the retention is greater than that obtained by simply extrapolating the earlier results to longer times. This shows that estimates of doses over long periods could be uncertain as the greater than expected retention could contribute more to the dose rate at times of several years or more after the initial deposit.

Some experiments suggest that material is only removed from surfaces such as walls, roofs or paved areas by rain, with the fraction removed being proportional to the rainfall (Wilkins, 1987). Warming (1982, 1984) measured dose rate over a few months after an experimental deposition on roads, and also reported the daily rainfall rates over the same period. This showed instances with rain and dose rate reductions, but also periods with dose rate reductions in the absence of rain and periods of rain with little reduction in dose rate, suggesting that rainfall is not the only mechanism responsible for natural weathering. Roed (1990) showed that weathering processes on roads are strongly dependent on the surface dust loading. Jacob et al. (1987) investigated retention at five locations in Munich and suggested that street cleaning may have an impact on the removal from paved areas. Investigations in different locations in Gävle, Sweden (Andersson, 1991) showed that in some areas with a high level of traffic, the weathering on streets was significantly faster than in other areas. Measurements on roof pavings contaminated by the Chernobyl accident indicate that the contaminant weathering half-lives can vary significantly according to roof material type (Andersson et al., 2002). These studies suggest that weathering could be dependent on specific (local) features, and therefore that the use of the available values for retention functions in locations other than where the measurements were made leads to uncertainty in the model predictions.

There is little information on retention of material on indoor surfaces, and this is generally represented in models by one retention half-life. Models suggest that deposition on internal surfaces may be an important contributor to doses over the first few months after deposition. Therefore, uncertainty in the modelling of retention indoors could be a major contributor to the uncertainty on the total dose for nuclides with half-lives of a few months. Dose rates in normal living following countermeasures to reduce the activity on outdoor surfaces, and hence the effectiveness of such countermeasures in reducing normal living doses, will also be sensitive to this modelling.

The generally applied weathering parameters are to a very great extent derived from studies of Chernobyl radiocaesium and cannot be expected to be adequate to model for instance the sometimes much larger particles that would be expected in connection with other contaminating incidents such as fires and explosions (Andersson et al., 2008b). Also for other scenarios, where radiocaesium is not the dominant dose contributor, the use of such parameters would be problematic.

8.3. Calculating dose rate from contamination on different surfaces

The point kernel and Monte Carlo methods of calculating dose rates from material on surfaces in inhabited areas were described in Section 3.3. The methods are well established, and the calculations can be undertaken easily if the arrangement of buildings and the composition of their walls are specified.

However, to cover the various sources that might be used in a malevolent incident, data libraries of dose rate conversion factor are needed for a number of radionuclides that are not traditionally considered, as they would not release or would not constitute a problem following a release from a nuclear power plant accident. As the contaminant might be a single pure beta emitter, beta dose rate conversion factors are needed for adequate geometries. For instance, the beta dose rate from a contaminated surface near to an uncovered part of the body (e.g. a pillow or a bed) could be much higher than that received by walking on a contaminated floor.

The current models use libraries of dose calculations for typical or standardised arrangements of inhabited area surfaces. However, the dose in an actual location will reflect the precise arrangement of surfaces (particularly trees); the arrangement of buildings and composition of materials will affect the shielding produced. The doses in the idealised locations are then applied to a range of situations which are considered to be sufficiently similar that the calculations are appropriate. However, there have been few studies of the sensitivity of calculated doses to changes in the assumed separation of the buildings. Some results from one such study (Jones, 2004) are shown in Table 5.

This compares dose rates indoors and outdoors for various spacings of semi-detached houses with those calculated for a grid where each semi-detached house is 20 m from the adjacent houses. The results show that the doses for streets of houses differ from those for the square grid of houses by factors of around 2. However, the doses for the streets of houses differ by around 10% for the different spacings considered here. This suggests that the dose rate in an idealised setting could reasonably be used to represent the dose rate in a range of real situations. The results in Table 5 are from one study. More similar studies should be carried out to determine how well the doses in real environments are represented by those in idealised locations; this would also give information on which characteristics of the

Table 5 Ratio of dose rates for different building arrangements and spacings to those for a square grid of brick buildings at an energy of 0.5 MeV.

House arrangement	Dose rate from surface	Ratio of dose rate indoors	Ratio of dose rate outdoors
Square grid of brick buildings, 20 m apart	Roof	1.0	1.0
	Walls	1.0	1.0
	Paved areas	1.0	1.0
Street of brick houses, 20 m street width, 2 m between adjacent houses	Roof	2.70	0.58
	Walls	0.41	0.12
	Paved areas	2.39	0.83
Street of brick houses, 15 m street width, 2 m between adjacent houses	Roof	2.72	0.48
	Walls	0.42	0.19
	Paved areas	2.02	0.76
Street of brick houses, 20 m street width, 4 m between adjacent houses	Roof	2.67	0.54
	Walls	0.41	0.12
	Paved areas	2.48	0.84

environment need more detailed specification and which could be modelled in a more generic way. Such studies could guide the collection of information (such as the building geometries, construction materials, number and size of trees) which would characterise inhabited areas in the region to which a model would be applied.

8.4. Behaviour of people

Models calculate doses at particular locations, and predict doses to people spending all their time at that location. Doses to people living or spending time in an area need to take account of the behaviour of people, in particular the time spent in different locations. Therefore, models need to calculate an appropriate average of doses at different locations where people may spend their time. This could simply be an average over different locations in a single idealised environment or it could be an average over doses in different environments if the movement of people around a larger, non-uniform area must be considered.

The detail required for information on the population behaviour depends on the use for which the model results are intended. For example if the aim is to estimate average doses over a long period, then long-term averages of the time spent indoors and outdoors may be sufficient. This is likely to be adequate for calculations for routine releases, probabilistic risk assessment and emergency planning. However, if the aim is to calculate doses received by people in the short term after an actual accident, then more detailed information would be required. The variation between seasons of the year or days of the week may need to be considered for some applications. Information on the time spent in different areas within a larger region may also be needed for some applications. Information on the long-term averages of time spent indoors and outdoors is available. However, more detailed information may not be available.

Another aspect of people's behaviour is that it is important to know whether a significant part of the diet consumed in the area is produced locally, for example in kitchen gardens. In some areas of the world, this is a deep-rooted tradition, and it may not be possible to substitute dietary components with an uncontaminated supply produced elsewhere within a reasonable time. Specific factors governing the radiological significance of consumption of foods produced in contaminated inhabited areas are currently not well understood.

8.5. The choice of model

This chapter has described the range of models available for the assessment of doses, the type of information that they need and the applications for which models of different complexity might be appropriate. One of the problems of using the more complex models is that, in general, a large amount of detailed information is needed. There are gaps in our knowledge on some of the processes that govern the doses received and some data, particularly on environmental characteristics and peoples' behaviours, are not readily available. The choice of model for a particular application depends on a number of factors, such as:

- Whether the data required are available.
- How accurate the dose calculations need to be.
- What period the doses are to be calculated over.
- Whether the model is to be applied to accidental or routine discharges.

The development of more complex models does not make simpler models superfluous, as different models are appropriate or adequate for different purposes and detailed models should only be used if the required input data are available for the actual situation of interest. However, the decision to use a simple model should be justified by considering whether it contains sufficient detail to represent the situation being modelled in an appropriate way.

Simple models can be adequate for some applications, particularly if averaging is involved, such as giving a rough large-scale overview after an accidental release, in probabilistic risk assessment and for routine releases. It may be better to use a simple model to calculate the average dose within large, inhomogeneous areas, where the required information, particularly on population behaviour, for a detailed model may not be available.

However, a simple model cannot help assess the effects of countermeasures in reducing doses, as this type of application must consider the contribution of different surfaces to the dose rate and its time variation. Many of the simpler models rely on the use of appropriate multiplying factors to describe the difference between doses over an infinite plane surface and doses in inhabited areas – these factors can be obtained by suitable measurements or from calculations with a complex model.

8.6. Future work needed

There are gaps in the data required to support models for some applications, particularly for types of releases other than those from large power plant accidents. There are also considerable gaps in relation to population behaviour and how to represent real areas in terms of idealised situations. There would be benefits in undertaking an analysis of the uncertainty in the predictions of models for doses in inhabited areas, and identifying further research needed to improve the models.

REFERENCES

Allott, R. W., M. Kelly, and C. N. Hewitt. (1994). A model of environmental behaviour of contaminated dust and its application to determining dust fluxes and residence times. *Atmospheric Environment*, 28(4), 679–687.

Andersson, K. G. (1991). Contamination and decontamination of urban areas. PhD Thesis, Risø National Laboratory, Roskilde, Denmark.

Andersson, K. G. (1996a). Modelling external radiation doses in contaminated urban areas: Implications for development of decontamination strategies. *Proceedings of the IRPA9 International Congress on Radiation Protection*, Vienna, pp. 3-265–3-268, ISBN 3-9500255-4-5.

Andersson, K. G. (1996b). *Evaluation of Early Phase Nuclear Accident Clean-up Procedures for Nordic Residential Areas.* NKS Report NKS/EKO-5(96)18, ISBN 87-550-2250-2.

Andersson, K. G., M. Ammann, S. Backe, and K. Rosén. (2008a). *Decision Support Handbook for Recovery of Contaminated Inhabited Areas.* NKS Report NKS-175. Nordic Nuclear Safety Research (NKS), Roskilde, Denmark, ISBN 978-87-7893-241-9.

Andersson, K. G., T. Mikkelsen, P. Astrup, S. Thykier-Nielsen, L. H. Jacobsen, L. Schou-Jensen, S. C. Hoe, and S. P. Nielsen. (2008b). Estimation of health hazards resulting from a radiological terrorist attack in a city. *Radiation Protection Dosimetry*, 131(3), 297–307.

Andersson, K. G., J. Roed, M. A. Byrne, H. Hession, P. Clark, E. Elahi, A. Byskov, X. L. Hou, H. Prip, S. K. Olsen, and T. Roed. (2004). *Airborne Contamination in the Indoor Environment and Its Implications for Dose.* Risoe-R-1462(EN). Risoe National Laboratory, Roskilde, Denmark, ISBN 87-550-3317-2.

Andersson, K. G., J. Roed, and C. L. Fogh. (2002). Weathering of radiocaesium contamination on urban streets, walls and roofs. *Journal of Environmental Radioactivity*, 62(1), 49–60.

Andersson, K. G., J. Roed, P. Jacob, and R. Meckbach. (1996a). Weathering of ^{137}Cs on various surfaces in inhabited areas and calculated location factors. In: *Deposition of Radionuclides, their Subsequent Relocation in the Environment and Resulting Implications* (Ed. J. Tschiersch). EUR 16604 EN, ISBN 92-827-4903-7, pp. 47–57.

Andersson, K. G., J. Roed, H. G. Paretzke, and J. Tschiersch. (1996b). Modelling of the radiological impact of a deposit of artificial radionuclides in inhabited areas. In: *Deposition of Radionuclides, their Subsequent Relocation in the Environment and Resulting Implications* (Ed. J. Tschiersch). EUR 16604 EN, ISBN 92-827-4903-7, pp. 83–94.

ARGOS. (2008). ARGOS CBRN Whitepaper. Decision support for emergency management, www.pdc.dk/argos

Balonov, M. I. (1997). Discussion on semi-natural ecosystems. In: *Health Impacts of Large Releases of Radionuclides* (Eds J. V. Lake, G. R. Bock, and G. Cardew). Wiley, Chichester, UK. ISBN 0-471-96510-3, 41pp.

Briesmeister, J. (Ed.) (1993). *MCNP-A general Monte Carlo N-Particle Transport Code.* Users Manual. Report LA-12625. Los Alamos National Laboratory, Los Alamos, NM.

Bunzl, K., P. Jacob, W. Schimmack, R. M. Alexakhin, N. P. Arkhipov, Y. Ivanov, and S. V. Kruglov. (1997). ^{137}Cs mobility in soils and its long-term effect on the external radiation exposure. *Radiation and Environmental Biophysics*, 36, 31–37.

Chamberlain, A. C. (1959). Deposition of iodine-131 in Northern England in October 1957. *Quarterly Journal of the Royal Meteorological Society*, 350–361.

Charnock, T. W., J. A. Jones, L. N. Singer, K. G. Andersson, J. Roed, S. Thykier-Nielsen, T. Mikkelsen, P. Astrup, J. C. Kaiser, H. Müller, G. Pröhl, W. Raskob, S. C. Hoe, L. H. Jacobsen, L. Schou Jensen, and F. Gering. (2009). Calculating the consequences of recovery, a European model for inhabited areas. *Radioprotection*, ISSN 0033-8451 (submitted for publication).

Cline, J. F., D. O. Wilson, and F. P. Hungate. (1965). Effect of physical and biological conditions on deposition and retention of ^{131}I on plants. *Health Physics*, 11, 713–717.

Crick, M. J., and J. Brown. (1990) *EXPURT – A model for evaluating exposure from radioactive material deposited in the urban environment.* NRPB-R235. Chilton, UK.

Davies, B. E. (1978). Plant-available lead and other metals in British garden soils. *Science of the Total Environment*, 9, 243–262.

Douay, F., H. Roussel, H. Fourrier, C. Heyman, and G. Chateau. (2007). Investigation of heavy metal concentrations on urban soils, dust and vegetables nearby a former smelter site in Montagne du Nord, Northern France. *Journal of Soils and Sediments*, 7(3), 143–146.

Fogh, C. L., K. G. Andersson, A. N. Barkovsky, A. S. Mishine, A. V. Ponomarjov, V. P. Ramzaev, and J. Roed. (1999). Decontamination in a Russian Settlement. *Health Physics*, 76(4), 421–430.

Freer-Smith, P. H., A. A. El-Khatib, and G. Taylor. (2003). Capture of particulate pollution by trees: A comparison of species typical of semi-arid areas (Ficus Nitida and Eucalyptus Globulus) with European and North American species. *Water, Air and Soil Pollution*, 155(1–4), 173–187.

Gallego, E. (2006). MUD: A model to investigate the migration of ^{137}Cs in the urban environment and drainage and sewage treatment systems. *Journal of Environmental Radioactivity*, 85(2–3), 247–264.

Gravenhorst, G., and K. D. Höfken. (1982). Concentration of aerosol constituents above and beneath a beech and a spruce forest canopy. In: *Deposition of Atmospheric Pollutants* (Eds H. W. Georgii and J. Pankrath). D. Reidel, Dordrecht, The Netherlands.

Hedemann Jensen, P. (1992). *Atmospheric Dispersion and Environmental Consequences – Exposure from Radioactive Plume Pathways.* Risø National Laboratory, Risø-M-2849(EN), ISBN 87-550-1613-8.

Heinemann, K., and K. J. Vogt. (1980). Measurements of the deposition of iodine onto vegetation and the biological half-life of iodine on vegetation. *Health Physics*, 39, 463–474.

Horn, H. G., M. Maqua, and H. Bonka. (1988). *Nasse und trockene Ablagerung Radioaktiver Stoffe auf die Vegetation und den Erdboden*. Schriftenreihe Reaktorsicherheit und Strahlenschutz, BMU 199–195, ISSN 0724-3316, in German.

Hull, A. P. (1966). Environmental monitoring of ^{131}I in small concentrations and some comparisons with meteorological calculations. *Health Physics*, 12, 1317–1325.

ICRP. (1994). *International Commission on Radiological Protection, Publication 66: Human Respiratory Tract Model for Radiological Protection*. Pergamon Press, Oxford, New York, Tokyo, ISBN 0-08-041154-1.

ICRP. (1995). *International Commission on Radiological Protection, Publication 72: Age-Dependent Doses to Members of the Public from Intake of Radionuclides: Part 5 – Compilation of Ingestion and inhalation Dose Coefficients*. Pergamon Press, Oxford, New York, Tokyo, ISBN 0 08 042737 5.

Jacob, P., and R. Meckbach. (1987). Shielding factors and external dose evaluation. *Radiation Protection Dosimetry*, 21(1/3), 79–85.

Jacob, P., R. Meckbach, and H. M. Müller. (1987). Reduction of external exposure from deposited Chernobyl activity by run-off, weathering, street cleaning and migration in the soil. *Radiation Protection Dosimetry*, 21(1/3), 51–57.

Jacob, P., R. Meckbach, H. M. Müller, and K. Meimberg. (1990). *Abnahme von künstlicher Radioaktivität nach Deposition in städtischer Umgebung*. Abschlussbericht zum Forschungsvorhaben, Förderkennzeichen St. Sch. 1051, GSF, Neuherberg, Germany.

Jacob, P., R. Meckbach, H. G. Paretzke, I. Likhtarev, I. Los, L. Kovgan, and I. Komarikov. (1994). Attenuation effects on the kerma rates in air after cesium depositions on grasslands. *Radiation and Environmental Biophysics*, 33, 251–267.

Jenkins, P. L., T. J. Phillips, E. J. Mulberg, and S. P. Hui. (1992). Activity patterns of Californians: Use of and proximity to indoor pollutant sources. *Atmospheric Environment*, 26A(12), 2141–2148.

Jones, A., T. Charnock, L. Singer, J. Roed, K. G. Andersson, S. Thykier-Nielsen, T. Mikkelsen, P. Astrup, J. C. Kaiser, H. Müller, G. Pröhl, W. Raskob, S. Hoe, L. H. Jacobsen, L. Schou-Jensen, and F. Gering. (2006a). *Description of the Modelling of Transfer and Dose Calculations within ERMIN*. Draft of deliverable D5C2R2 of the CEC-EURANOS project, EURANOS(CAT2)-TN(05)-04.

Jones, J. A. (2004). *Investigations of the Effects of Building Spacings and Configurations of Trees for Dose Rates in the CONDO v3.1 Project*. NRPB-EA/11/2004. Chilton, UK.

Jones, J. A., L. N. Singer, and J. Brown. (2006b). The EXPURT model for calculating external γ doses from deposited material in inhabited areas. *Journal of Environmental Radioactivity*, 85(2–3), 314–329.

Kaiser, J. C., F. Gering, K. Andersson, T. Charnock, S. Hoe, and L. H. Jacobsen. (2008). Correction of model predictions for the radioactive contamination of the environment with data assimilation. *Proceedings of the International Conference on Radioecology & Environmental Radioactivity* (Eds P. Strand, J. Brown and T. Jølle), 15–20 June 2008, Bergen, Norway, Oral and Oral Poster Presentations, Part 1, Norwegian Radiation Protection Authority, Østerås, Norway, ISBN 978-82-90362-25-1, pp. 130–133.

Karlberg, O. (1986). *Avrinning och retention av Chernobyl nedfallet i stadsmiljö – Fält-mätningar med gamma spectrometer (in Swedish)*. Technical Note (6 pages), Studsvik AB, Nyköping, Sweden.

Karlberg, O. (1992). *The Environmental Behaviour of Chernobyl Deposition in a High Fallout Region of Sweden – Measurements and Analysis of the Urban Programme 1986–1989*. Report NS-92/1, ISBN 91-7010-184-1. Studsvik Nuclear, Stockholm, Sweden.

Karlberg, O., and B. Sundblad. (1986). *Weathering and Migration of Chernobyl Fallout*. Studsvik Arbetsrapport – Technical Note, NP-86/78.

Katagiri, K., T. Shimizue, Y. Akatsu, and H. Ishiguro. (1997). Study on the behaviour of ^{129}I in the terrestrial environment. *Journal of Radioanalytical and Nuclear Chemistry*, 226, 23–27.

Koprda, V., M. Harangozó, and Z. Kassai. (2000). Transfer of radionuclides across skin barriers of animal skin models in vitro. *Journal of Radioanalytical and Nuclear Chemistry*, 246(3), 505–509.

Kousa, A., J. Kukkonen, A. Karppinen, P. Aarnio, and T. Koskentalo. (2002). A model for evaluating the population exposure to ambient air pollution in an urban area. *Atmospheric Environment*, 36, 2109–2119.

Kryshev, I. I. (1996). Dose reconstruction for the areas of Russia affected by ^{131}I contamination. *Radiation Protection Dosimetry*, 64(1/2), 93–96.

Lange, C. (1995). Indoor deposition and the protective effect of houses against airborne pollution. PhD Thesis, Risø National Laboratory, Roskilde, Denmark.

Lauridsen, B. (1982). *Table of Exposure Rate Constants and Dose Equivalent Rate Constants*. Risø-M-2322. Risø National Laboratory, Roskilde, Denmark. ISBN 87-550-0898-4.

Long, C. M., H. H. Suh, P. J. Catalano, and P. Koutrakis. (2001). Using time- and size-resolved particulate data to quantify indoor penetration and deposition behaviour. *Environmental Science and Technology*, 35(10), 2089–2099.

Mayall, A., T. Cabianca, C. Attwood, C. A. Fayers, J. G. Smith, J. Penfold, D. Steadman, G. Martin, T. P. Morris, and J. R. Simmonds. (1997). PC CREAM-97 (PC CREAM-98 code update) NRPB, NRPBSR-296 (EUR 17791), National Radiological Protection Board, Chilton, UK, ISBN 9-28280-320-1.

McMahon, T. A., and P. J. Deninson. (1979). Empirical atmospheric deposition parameters – A survey. *Atmospheric Environment*, 13, 571–585.

Meckbach, R., P. Jacob, and H. G. Paretzke. (1988). Gamma exposures due to radionuclides deposited in urban environments. Part 1: Kerma rates from contaminated urban surfaces. *Radiation Protection Dosimetry*, 25(3), 167–179.

Müller, H., and G. Pröhl. (1993). ECOSYS-87: A dynamic model for assessing radiological consequences of nuclear accidents. *Health Physics*, 64(3), 232–252.

Nakamura, Y., and Y. Ohmomo. (1980). Factors used for the estimation of gaseous radioactive iodine intake through vegetation. 2. Uptake of elemental iodine by spinach leaves. *Health Physics*, 38(3), 315–320.

Nicholson, K. W. (1989). The dry deposition of small particles – A review of experimental measurements. *Journal of Radiological Protection*, 9(2), 113–119.

Nielsen, S. P. (1977). *In Situ Measurements of Environmental Gamma Radiation Using a Mobile Ge(Li) Spectrometer System*. Denmark, Risø Report 367, Risø National Laboratory, Roskilde, Denmark, ISBN 87-550-0501-2.

Nisbet, A. F., and R. F. M. Woodman. (2000). Soil-to-plant transfer factors for radiocesium and radiostrontium in agricultural systems. *Health Physics*, 78(3), 279–288.

Norra, S., and D. Stüben. (2003). Urban soils. *Journal of Soils and Sediments*, 3(4), 230–233.

Peltola, P., and M. Astrom. (2003). Urban geochemistry: A multimedia and multi-element survey of a small town in northern Europe. *Environmental Geochemistry and Health*, 25, 397–419.

RODOS. (2008). Realtime online decision support system for nuclear emergency management, www.rodos.fzk.de

Roed, J. (1985a). *Dry Deposition of Urban Surfaces*. Risø Report R-515. Risø National Laboratory, Roskilde, Denmark, ISBN 87-550-1069-5.

Roed, J. (1985b). *Run-Off from Roofs*. Risø Report Risø-M-2471. Risø National Laboratory, Roskilde, Denmark, ISBN 87-550-1070-9.

Roed, J. (1987a). Run-off from and weathering of roof material following the Chernobyl accident. *Radiation Protection Dosimetry*, 21(1/3), 59–63.

Roed, J. (1987b). Dry deposition in rural and urban areas in Denmark. *Radiation Protection Dosimetry*, 21(1/3), 33–36.

Roed, J. (1988). The distribution on trees of dry deposited material from the Chernobyl accident, paper NKA/AKTU-245 (88)2. Presented at the Joint CEC/OECD (NEA) Workshop on Recent Advances in Reactor Accident Consequence Assessment, Rome, Italy, 25–30 January 1988.
Roed, J. (1990). *Deposition and Removal of Radioactive Substances in an Urban Area.* Final Report of the NKA Project AKTU-245, Nordic Liaison Committee for Atomic Energy, Roskilde, Denmark, ISBN 87 7303 514 9.
Roed, J., K. G. Andersson, A. N. Barkovsky, C. L. Fogh, A. S. Mishine, S. K. Olsen, A. V. Ponomarjov, H. Prip, V. P. Ramzaev, and B. F. Vorobiev. (1998). *Mechanical Decontamination Tests in Areas Affected by the Chernobyl Accident.* Risø-R-1029. Risø National Laboratory, Roskilde, Denmark, ISBN 87-550-2361-4.
Roed, J., and P. Jacob. (1990). Deposition on urban surfaces and subsequent weathering. *Proceedings of a Seminar on Methods and Codes for Assessing the Off-Site Consequences of Nuclear Accidents.* Athens, Greece, EUR13013, Vol. 1, pp. 335–356.
Roed, J., C. Lange, K. G. Andersson, H. Prip, S. Olsen, V. P. Ramzaev, A. V. Ponomarjov, A. N. Barkovsky, A. S. Mishine, B. F. Vorobiev, A. V. Chesnokov, V. N. Potapov, and S. B. Shcherbak. (1996). *Decontamination in a Russian Settlement.* Risø-R-870. Risø National Laboratory, Roskilde, Denmark, ISBN 87-550-2152-2.
Schaefer, H., and J. Lademann. (2001). The role of follicular penetration. *Skin Pharmacology and Applied Skin Physiology*, 14, 23–27.
Sehmel, G. A. (1973). Particle eddy diffusitivities and deposition velocities for isothermal flow and smooth surfaces. *Aerosol Science*, 4, 125–138.
SSI. (2005). The Swedish Radiation Protection Authority's regulations and general advice on the handling of ashes contaminated by caesium-137, Document SSI FS 2005:1, SSI, Stockholm, Sweden, www.ssi.se/forfattning/pdf_eng/2005_1e.pdf
Thiessen, K. M., B. Batandjieva, K. G. Andersson, A. Arkhipov, T. W. Charnock, F. Gallay, S. Gaschak, V. Golikov, W. T. Hwang, J. C. Kaiser, S. Kamboj, M. Steiner, J. Tomás, D. Trifunovic, C. Yu, R. L. Zelmer, and B. Zlobenko. (2008). Improvement of modeling capabilities for assessing urban contamination: The EMRAS Urban Remediation Working Group. *Applied Radiation and Isotopes*, 66, 1741–1744.
Tinkle, S. S., J. M. Antonini, B. A. Rich, J. R. Roberts, R. Salmen, K. De Pree, and E. J. Adkins. (2003). Skin as a route of exposure and sensitization in chronic beryllium disease. *Environmental Health Perspectives*, 111(9), 1202–1208.
Tsalko, V. (2004). Statement by the Chairman of the Chernobyl Affairs Committee at the Council of Ministers of the Republic of Belarus at the International Conference 'Chernobyl Forum', www.belarusembassy.org/humanitarian/tsalko.htm
UNSCEAR. (2000). *Report of the United Nations Scientific Committee on the Effects of Atomic Radiation to the General Assembly.* UN, New York, Vol. 1.
VAMP. (1994). *Modelling the Deposition of Airborne Radionuclides into the Urban Environment.* First Report of the VAMP Urban Working Group. IAEA-Tecdoc-760, Vienna.
Warming, L. (1982). *Weathering and Decontamination of Radioactivity Deposited on Asphalt Surfaces.* Risø-M-2273. Risø National Laboratory, Roskilde, Denmark.
Warming, L. (1984). *Weathering and Decontamination of Radioactivity Deposited on Concrete Surfaces.* Risø-M-2473. Risø National Laboratory, Roskilde, Denmark.
Watson, S. J., A. L. Jones, W. B. Oatway, and J. S. Hughes. (2005). Ionising radiation exposure of the UK population: 2005 Review. Chilton HPA-RPD-001, http://www.hpa.org.uk/radiation/publications/hpa_rpd_reports/2005/hpa_rpd_001.htm
Watterson, J. D., and K. W. Nicholson. (1996). Dry deposition and interception of 4-22 μm diameter particles to a lettuce crop. *Journal of Aerosol Science*, 27(5), 759–767.
Wilkins, B. T. (1987). The retention behaviour of radiocaesium on common building materials under natural outdoor conditions. *Radiation Protection Dosimetry*, 21(1–3), 69–73.

CHAPTER 7

Measurement and Screening of Contaminated Inhabited Areas

Peter Zombori*

Contents

*Corresponding author. Tel./Fax: +36 1 392 2500
E-mail address: zombori@aeki.kfki.hu

KFKI Atomic Energy Research Institute, H-1525 Budapest, Hungary

Radioactivity in the Environment, Volume 15
ISSN 1569-4860, DOI 10.1016/S1569-4860(09)00407-0

1. Introduction

There are two major approaches to measuring radioactive environmental contamination (whether in inhabited areas or in the natural environment). One approach is *sampling and laboratory analysis* of environmental samples. This is more precise and accurate, more specific, regarding the radionuclide composition of the contamination and provides higher sensitivity (i.e. the detection of lower levels of contamination). Nevertheless, the laboratory analysis is time consuming and labour intensive, it requires more equipment (sometimes complicated sample treatment, shielded counting facilities), the amount of samples processed is limited and the interpretation of the measured data is often questionable due to the problems of representativity of sampling. The other approach is focusing on the *in situ instrumental techniques for the measurement* of radiation levels and radioactive contaminations. This is much faster, more data characterising the spatial and temporal variation of the contamination can be obtained within a certain assessment period and the instruments and methods are simpler than those used in laboratory analyses. Although the direct instrumental field measurements have their limitations and disadvantages (lower sensitivity, limited analytical capabilities, influence of the fluctuating natural background radiation, etc.) and the sampling and laboratory sample analysis is expected to complement the monitoring programme, the direct *in situ* measurements should play the decisive role in the measurement and screening of contaminated inhabited areas. This chapter is aimed at providing an overview of the requirements and features in developing a targeted strategy for measurement and screening of radioactively contaminated inhabited areas following an airborne release.

2. Main Issues to be Considered When Designing Contamination Monitoring Capabilities

Contamination of inhabited areas will certainly require difficult decisions to be made. These decisions may affect the everyday life and activity of many people. It is, therefore, very important to take all factors into consideration, not least the actual situation concerning the contamination. It is the process of contamination monitoring which is there to feed the decision-making with reliable, factual data.

Ideally, the decisions should be based on as many good quality, consistent and coherent data as possible. The coverage in time should begin from before the release starts (threat analysis, possible scenarios and predictive model calculations) to the point in the late phase of an accident when the contamination is not a concern any more. It should include arrangements and systems used in routine monitoring, as well as the special procedures of emergency operations. It should combine systems of fixed monitoring stations, mobile monitoring units, sampling and sample analysis techniques. The dynamics and the actual phase of the accident scenario has an influence on the selection of the monitoring techniques; less sophisticated but more rapid instrumental methods will be preferred in the early phase whereas the time-consuming, tedious laboratory analysis will be better suited to the later phase. Treatment of data (in the field and in the control centre), model calculations and presentation of the results are also important parts of the whole complex process of contamination monitoring.

When building up the capabilities for radiation monitoring the following issues must be considered:

- The objectives (Why do we need a radiation monitoring capability?)
- The scope and extent (What should be covered and what should not be covered by our monitoring programme?)
- The quantities (What specific quantities do we want to measure?)
- The methods (What techniques do we choose to monitor?)
- The instrumentation (What equipment do we use to perform the measurements?)
- The plan (What are the elements of the monitoring programme and how are we going to implement them?)
- The logistics (What are the requirements to be met and conditions to be provided to make the monitoring work?)
- The use (What do we do with the data obtained by the contamination monitoring?)

The issues above should be considered thoroughly when building up a national strategy for an efficient nuclear or radiological emergency preparedness and response system (not forgetting that radiation monitoring is just a part of such a system). The list above may be a useful guide, however, even in cases when the scope of the monitoring is more limited, for example when contamination monitoring is intended to serve the radiological safety and dose assessment of the population.

3. Objectives and Scope of Contamination Measurements and Screening

Using instrumental techniques for measuring the contamination in inhabited areas has three main objectives:

- Determining the levels of contamination as functions of space and time in inhabited areas, in order to make decisions about the necessity and the best means of intervention.

- Providing input for the assessment of the dose via different pathways.
- Determining the efficiency of the intervention where it was performed.

Beyond the objectives mentioned above, contamination measurements can serve important scientific and technical research and development purposes, as well.

The selection of the proper equipment and the best measurement strategy is an optimisation procedure. It has to take into account the availability of the instrumentation and manpower, time and other resources. Cutting-edge technologies in measuring surface activity concentration may have been developed in special laboratories but may not yet be available on the market, or if they are on the market they may turn out to be just not affordable. A balance has to be found between the amount of data and the level of detail, the time needed to get a result and the lower limit of detection, etc. In many cases, the measurement simply has to be made with the available instrumentation, there being no possibility (time, money, supplier, etc.) of selecting the technically optimal solution.

The results of the contamination measurements are the most important inputs for the assessment of doses through different pathways. These assessment methods are discussed in detail in other chapters.

4. Instrumentation

The selection of the devices intended for the measurement of radionuclide concentrations in different media and on various surfaces must take into account the levels to be dealt with. High levels of contamination can be measured by simpler devices and within shorter periods of time. In such cases, the main problem is the upper limit of the detection method used. Some devices show saturation effects (e.g. direct gamma spectrometric measurements in high contamination areas) or simply stop functioning (e.g. GM-tube based detectors). When lower levels of contamination are to be determined one has to consider the lower limits of detection achievable with the instrumentation and methods applied. The conditions (the type of the radiation, the type and size of the detector, the radiation background, the measurement time, etc.) may prevent the detection of minor amounts of radioactive material contaminating the investigated surface or media. These limitations must be taken into consideration when choosing the proper instrumentation for the contamination measurements.

Measurement of the various types of radiation – due to the differing transport and interaction properties – may require different devices and methods. For every type of radiation (mainly alpha, beta and gamma), the energy of the particles also has a strong influence on the measurement performance (detection limits, accuracy, fluctuation, etc.). Not all components of a mixed radionuclide composition can be determined with the same instrument and method. On the other hand, most of the nuclides emit more than one particle in the process of nuclear transformation, offering more choices for their detection. One has to consider many factors to find the best solution for the quantitative assessment of the contamination (Knoll, 2002).

4.1. Measurement of pure alpha emitters

Alpha particles are usually emitted when radionuclides of high-Z elements undergo nuclear transformation. When released by the parent nuclide these particles have a rather high kinetic energy (several MeV) which is gradually degraded during its transport and interaction in the surrounding matter. This fact makes the detection of alpha radiation rather difficult, especially in field conditions. Thin layers of solid matter can totally absorb the alpha radiation, preventing it from reaching any detector material. Even a couple of centimetres of air can cause the same effect. The thickness of air needed for the total absorption is dependent on the initial energy of the alpha particle.

The accurate quantitative measurement of pure alpha emitter contamination is possible only in laboratory conditions. This requires careful sampling of the investigated surfaces, tedious sample preparation with often lengthy and laborious chemical procedures, and finally measuring with specially shielded counting devices. These methods – due to the specificity of the applied chemical procedures – can provide isotope-specific information about the alpha-emitting radionuclide composition of the sampled surface.

Despite the difficulties mentioned above, there are some instruments available for the *in situ* measurement of pure alpha emitters on different surfaces. A possible detector is a ZnS scintillator, combined with a plastic substrate, which can detect higher energy alphas (>1 MeV) with an efficiency of more that 40% over 2π. The measurement performance is dependent on many factors (detector size and geometry, protection grid transparency, radiation background, counting time, etc.) that all have to be taken into account when determining measurement efficiency and the lower limit of detection values. Depending on the detector specification and the measuring circumstances, a lower limit of detection below $0.01\ \mathrm{Bq\ cm^{-2}}$ can easily be achieved.

When the alpha decay is accompanied by the simultaneous release of gamma radiation then the accurate *in situ* measurement of the contamination becomes much easier, due to the relatively easy detection of gamma particles, as discussed below.

4.2. Measurement of pure beta emitters

Beta decay is the typical transformation mode of medium-Z radionuclides but this type of decay can be observed throughout the nuclide chart, from the light-Z materials up to the heavy transuranium elements. The emitted β^- (electron) and β^+ (positron) particles have different energies: for a given radionuclide the beta energies follow a certain energy distribution characterised by the maximum and average energies (E_{max} and E_{avg}) of the emitted electrons/positrons. These energies determine the detectability of the (pure) beta-emitting radionuclide contamination: since beta particles are charged, their energies are gradually diminished during the subsequent interactions with the surrounding matter, thus limiting the range they can pass through. Consequently, the contamination of low-energy emitters (e.g. ^3H with $E_{max} = 18.6$ keV and ^{14}C with $E_{max} = 156$ keV) are hard to detect by direct field measurements, whereas others with higher beta energies (e.g. ^{90}Sr with $E_{max} = 546$ keV) are much easier to measure.

The accurate quantitative measurement of pure beta emitter contamination requires sampling and laboratory analysis. This is especially true for the low-energy beta emitters. After careful sampling, sample preparation (sometimes involving chemical procedures) and measuring with specially shielded counting devices the isotope-specific contamination values can be determined.

There are many devices readily available for the direct measurement of pure beta emitters. Several types of detectors are in use, a complete overview of their specification would exceed the scope of this chapter. Only for illustration purposes, a beta contamination monitor using thin plastic scintillator can detect betas above 30 keV with an efficiency of more that 40% over 2π. As for the alpha survey meters, their measurement performance is dependent on many factors and it is difficult to characterise them with a single efficiency or lower limit of detection value. Depending on the detector specification and the measuring circumstances a lower limit of detection well below 0.1 Bq cm^{-2} can be achieved.

As mentioned in the discussion of the alpha measurements, beta decay is often accompanied by the simultaneous release of gamma radiation. In this case, quantitative *in situ* measurement of the contamination becomes relatively easy by the detection of these gamma particles.

4.3. Measurement of gamma emitters

Gamma radiation is usually a 'by-product' of the nuclear transformation; it is normally the result of the de-energising process taking place (in one or several steps) in the daughter nuclide. This process leads to the release of (one or more) gamma photon(s). Emission of gamma radiation may or may not follow the emission of alpha or beta particles, depending on the nuclear characteristics of the parent and daughter radionuclides. Since the detection of gamma radiation of medium energy (50–3,000 keV) is relatively easy and there are many ways of doing it, determination of the contamination by measurement of the gamma radiation has become the prevailing method whenever the radionuclide in question emits medium energy gamma photon(s).

Nevertheless, gamma measurement is a technique with several difficulties. The strong energy dependence of gamma photon interaction with matter makes the quantitative assessment of contamination rather problematic. Low-energy gammas may be considerably attenuated before reaching the detector; higher-energy gammas may travel through thin detector material without being detected; the detection process can be disturbed by the strong interference of the background radiation, etc. All these factors pose limitations against the measurement of minor surface contamination. In such cases, sampling, sample preparation in a laboratory and counting in low-background gamma spectroscopy systems may extend the range of measuring capabilities.

There are many types of detectors usable for the measurement of gamma-emitting radionuclide concentration on surfaces. Gas-filled detectors (ionisation chambers, proportional counters, GM-tubes) as well as solid-state detectors (scintillators and semiconductors) are in general use for this purpose. Detectors of the first category (gas-filled counters) are simpler, cheaper and easy-to-use but the information obtainable is more limited (not least because of their poorer energy resolution). Their response time is longer and the discrimination between gamma rays coming from the measured surface versus those coming from a potentially contaminated environment (background) is rather difficult. Nevertheless, with proper calibration and customisation these detectors can be used in most cases when a large number of measurements has to be performed (e.g. for screening purposes).

Instruments based on solid-state detectors are typically more complex, more expensive and require more skill and understanding of the detection process. In return, they give the user much more information about the type of the contamination, most importantly about its radionuclide composition. Both scintillators and semiconductors are usable for spectrometric measurements, with, however, differing efficiency and energy resolution. The bulky NaI(Tl) scintillation detectors have a superb efficiency but a limited energy resolution. Contamination survey instruments based on this measuring technique offer fast response and are easy-to-use but their analytical capabilities (isotope identification) are limited to the measurement of a contamination composed of one or only a few gamma-emitting radionuclides (provided their medium energy gammas are not overlapping in the spectrum). On the other hand, portable spectrometers based on cooled high-purity semiconductor detectors (*HPGe*) have excellent energy resolution but typically lower measurement efficiency. With the on-site application of these instruments (so-called *in situ* gamma spectrometry), the radionuclide composition of contamination on larger areas can be determined with high accuracy and low limit of detection (below $100\,\mathrm{Bq\,m^{-2}}$).

5. Contamination Monitoring Techniques, Basic Elements of a Comprehensive Monitoring Programme

The present publication is focusing on different scenarios where airborne contamination of surfaces in inhabited areas is a major issue. The characteristics of the atmospheric transport and of the deposition processes can lead to a contamination pattern which is hard to predict. The Chernobyl reactor accident is a typical example of this feature: no simulation model calculation – despite gradual model refinement, a growing number of model parameters and the ever-increasing computing power – can fully reconstruct the actual contamination pattern (not to mention the comparison with real-time predictive model calculations).

The basic input to any dose assessment for all the pathways considered in this book is the radionuclide composition and the isotope-specific average contamination of the relevant environmental media (volume activity concentration in the air and surface activity concentration building up during the deposition). These are the data that should be obtained before the details of contamination on different surfaces are studied. This type of information will influence the further monitoring and (if needed) intervention strategies: when and what to measure in detail, what methods can be used, what intervention is required, where to decontaminate and with what technique, etc.

5.1. Air contamination monitoring

The contamination will first appear in the ground-level atmosphere and its monitoring while still airborne will give the investigator timely and valuable information. Pumping air through a vessel of well-defined volume or a filter can provide a representative air sample that can be measured either directly at the site of the sampling or in a laboratory. It is very important to keep the procedure under strict control from the point of view of metrology requirements to obtain good quality air activity concentration values. If the measurement (either *in situ* or in the laboratory) is carried out by a spectrometer (semiconductor or scintillator), the isotopic composition of the air contamination can easily be determined.

Monitoring the fluctuation and the integrated value of the airborne radionuclide concentration is important for the assessment of the inhalation dose, especially when the release of radioiodine isotopes is involved. The air concentration may give some hint about the expectable surface contamination but the latter is dependent on many other factors, not least the physico-chemical form of the radionuclides, the aerosol particle size distribution, the form of the deposition (wet or dry), etc. This feature of the deposition is well demonstrated by the contamination patterns in the Nordic countries after the Chernobyl accident: highly contaminated air masses travelled through Finland without leaving too much surface contamination because of the relatively dry weather conditions while the deposition in certain parts of Sweden was really high due to the heavy rain during passage of the contaminated cloud.

5.2. Large area contamination monitoring

The airborne contamination may be important from the inhalation dose point of view and also for the preliminary assessment of the resulting deposition but its monitoring is strongly limited in time. After the active phase of an atmospheric release the air will rapidly clear by deposition or dilution, the air concentration of radionuclides will diminish to a level which is hard to detect and insignificant from dosimetric aspects. In the longer term, the doses to the population are determined (directly or indirectly) by the surface concentration of the deposited radionuclides.

The best available method for the early survey of large area contamination is *in situ* gamma spectrometry (ICRU, 1994). If well calibrated, this method gives a reliable average value for all or most of the gamma-emitting radionuclides deposited. It is important to note that – when using the original, *un-collimated* set-up (Figure 1) – *in situ* gamma spectrometry typically provides an averaged value of contamination within a flat surface of about 30–50 m radius (this area is dependent on the energy of the gamma radiation and on the actual distribution of the radioactive material on/in the ground). Therefore, it is important to select places for the measurement which are large and flat enough (parks, gardens, wide road and squares, large public area like sports grounds, etc.). In case of emergency when urgent protective actions may be warranted, preference should be given to public places where larger numbers of people tend to spend more time. If such large areas are not available or more detailed mapping of the contamination is needed, then the application of collimators can restrict the area measured under and around the detector.

After major contamination events, rapid mapping of large areas is a high-priority task. This can be done by applying airborne *in situ* gamma spectrometry (*Note*: In this case, 'airborne' refers to the equipment and not to the contamination!). Both helicopters and fixed-winged airplanes can be

Figure 1 Equipment for *in situ* gamma spectrometry.

used for this purpose; the method is an extension of ground-based *in situ* spectrometry (IAEA, 1999; ICRU, 1994). This method is excellent in the rapid assessment of contamination contours of medium- and higher-energy gamma emitters on the ground. The sensitivity and analytical performance of the procedure is strongly influenced by the flight parameters (speed, height, flight routing, etc.) and has some serious limitations (spatial and time resolution, inefficiency for low-energy gammas, etc.).

5.3. Sampling and measurement of the soil concentration of radionuclides

In situ gamma spectrometry can (and in many cases should) be complemented by *sampling* of the surfaces and by *laboratory analysis* of these samples. This additional assessment can reveal other possible components of the contamination not detectable by the direct measurement (e.g., pure alpha and beta emitters).

Soil sampling is also important to determine the depth distribution of the radionuclides. Care must be taken at core sampling and also during the preparation of the samples to preserve the layer structure. Tube core samplers in use nowadays are designed so that they take a soil column of 20–40 cm depth and the integrity of the core can be maintained until the different layers can be separated into further sub-samples. If equidistant division (1–2 cm each) is not practicable then the division of the core should take into account the fact that the top layer has much higher importance from the external radiation dose point of view. Therefore, a 1–2 cm layer thickness is to be achieved for the top 10 cm, while the deeper layers can be divided into 5 cm, then 10 cm thick layers.

In any case, preserving the integrity of a core sample and cutting it into thin slices is rather difficult. Special techniques have been developed to achieve the required stratified sampling which use deep-freezing of the core and cutting the hardened soil column to high precision using a diamond saw. This method is described in detail by Roed and Togawa (1996).

The sub-samples are prepared for gamma spectrometry according to standard laboratory practice and are measured (preferably with HPGe spectrometers) for the counting times necessary for collection of statistically significant peak areas.

5.4. Surface contamination monitoring

When the isotopic composition of the contamination is determined and the 'big picture' of the deposition pattern is known, a more detailed investigation of the distribution of the radioactive substances on different surfaces can start. This work requires less analytical work (the isotopic composition is not supposed to vary too much) but much more direct field measurement.

In the event that the contamination is mainly composed of gamma emitters, the most direct means of analysis of isotopic composition is the application of *collimated in situ gamma spectrometry*. This is the technique which can be applied in an environment with differing contamination values on various surfaces. The inhabited areas are typical examples of this situation, because of the structured landscape and the great variety of materials used and surface positions (horizontal, vertical and tilted) applied during the building construction. The collimator helps making sure that only photons emitted from the surface 'seen' by the detector are measured. With proper calibration, nuclide-specific surface concentration values can be obtained immediately following the measurement (for single-line gamma emitters or simple spectrum composition, NaI(Tl) detectors are adequate; for complex spectra, detectors of better resolution – typically HPGe – should be used). Figure 2 shows a typical arrangement for the measurement of contaminated walls. The design of the collimator must take into account the fact that the contamination on the studied surface is much lower than on the underlying horizontal surfaces (e.g. pavement or soil) so the structure must provide shielding against this undesired background interference (Andersson et al., 2002).

Spectrometers are not always available and their use is rather complicated. In general, the contamination measurement can be achieved with a properly selected *portable surface contamination monitor*. If the preliminary sample

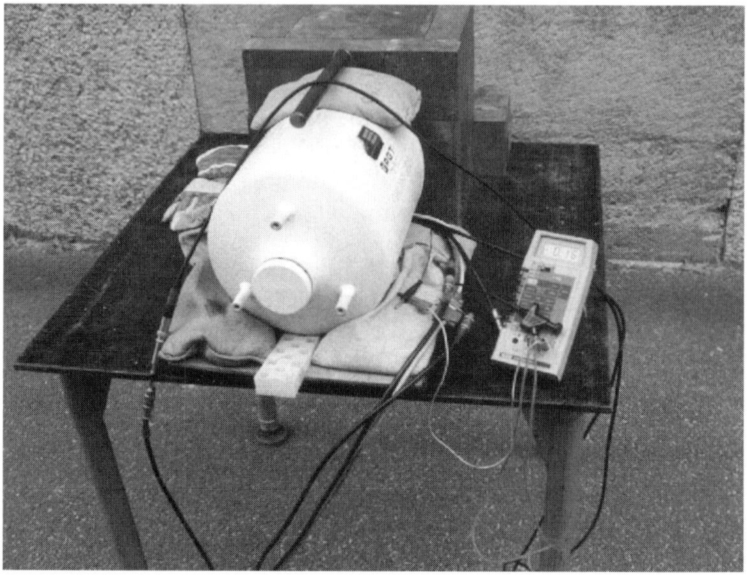

Figure 2 Wall surface contamination monitoring with a collimated *in situ* spectrometer.

analyses indicate that the contamination is largely composed of pure alpha and/or pure beta-emitting radionuclides then preference is to be given to the specific (alpha and beta) contamination monitors; in the case of gamma emitters, simpler devices can be used (Figure 3).

Although most contamination monitors leave the factory with certain preset calibration factors (that may or may not be built into the measurement algorithms), a recalibration may be necessary based on the actual radionuclide composition of the deposited matter. It is rather easy in the case of a single-radionuclide contamination but becomes more complicated with an increasing number of radionuclides involved. The re-calibration can be done theoretically by calculating the weighted average of the calibration factors for the radionuclides present in the contamination mix or (preferably and more practically) can be determined experimentally.

After such preparation, the measurement of the contamination on different surfaces in an inhabited area can be carried out. Assuming that the relocation of the radioactive materials from different surfaces (by natural processes or by decontamination) does not change the isotopic composition considerably, direct measurement will be a reliable means of contamination measurement and screening (the composition of the contamination may, however, change with time due to the differing half-lives of the radionuclides. For this reason, the need for re-adjustment of the calibration factors should be taken into account.)

Figure 3 Surface contamination monitoring with a portable contamination meter.

5.5. Characterisation of the contamination by dose rate measurements

Lacking specific surface contamination monitors, any simple *dose rate meter* can be used – with limited validity, however – for the assessment of the contamination. These devices can be used only for gamma radiation measurement and the quantitative analysis is very difficult. But it may be applicable to the characterisation of the contamination pattern in simple geometries (large, flat areas with constant gamma-emitter composition, separated rooms, etc.); efficient mapping has already been carried out with such devices (Figure 4).

Figure 4 Contamination measurements with a portable dose rate meter.

The general requirement to know the measurement properties of the equipment used is even more imperative if using dose rate meters for contamination measurements. The instrument's background readings, the efficiency, the directional dependence of its response, etc. are important factors to be taken into consideration when using them for the indicated purpose.

On the other hand, dose rate measurements have a great significance from the point of view of assessing the population dose from external sources. If well calibrated, these instruments directly show what one would like to know: the ambient dose rate at a given location.

Dose rate meters have a special role in the early indication of an event leading to contamination of larger areas. Many countries operate networks of fixed dose rate monitoring stations (*early warning systems*) that are automatically collecting dose rate data and transmitting them to a central location, based on a pre-defined operational cycle. These systems have the advantage of being the first to provide information about the arrival of the contamination (whether still airborne or already deposited) and to indicate the geographic location of the high-priority areas. The more stations involved in the network the better use of its information can be guaranteed (in this regard preference should be given to inhabited areas). The interpretation of the measurement data needs some special attention, however; it is difficult (if not impossible) to differentiate between the airborne and the deposited component of the contamination, the potential contamination of the detector and its immediate environment can distort the measurement, etc.

With the passing of time, the significance of the data obtained from these stations diminishes and more valuable data will be obtained from other types of monitoring (mobile monitoring units, *in situ* gamma spectrometry, sampling and laboratory analysis of samples, etc.).

Dose rate meters can be mounted on a vehicle. If data collection and storage is automated and combined with the simultaneous collection of signals of a GPS device, the system can perform a special mapping function, so-called *route monitoring*, which is one of the most powerful methods of screening large contaminated areas.

5.6. Personal monitoring

Radioactive contamination of inhabited areas may lead to the exposure (both external and internal) of humans living and/or working in the given environment. The primary objective of all the monitoring, intervention and decontamination efforts discussed throughout this publication is to diminish this exposure to a level which is well below the threshold of the deterministic effects, making sure that the remaining exposure is as low as reasonably achievable.

Monitoring the radioactive contamination and exposure of the human body is a complex task. It involves the measurement of the contamination of external surfaces (skin and clothing), the internal contamination (by direct whole body measurement of the gamma emitters and by excreta analysis for the pure alpha and beta emitters), as well as continuous monitoring of the external exposure. Doses received via the different pathways should be individually assessed and then summed up to estimate the total effective dose received by the given person.

In an emergency situation, members of the public seldom wear dosimeters that would allow the experts to determine the external dose to the individual. Determination of the committed dose (and some reconstruction of the previously obtained dose) is possible by measurement of the external surface and internal contamination, but such complex assessments are rarely done, only if there are special indications of high contamination/exposure in the case of some individuals.

For the professionals who are working on the implementation of intervention, monitoring and decontamination during an emergency, the situation is different, these occupationally exposed workers being obliged to wear personal dosimeters that enable them and their employers to assess the exposure they may suffer. Both integrating-type passive detectors (e.g. films, TLD) and self-reading electronic dosimeters can be chosen and ideally the two types should be used together.

5.7. QA measurements

Monitoring is based on the readings of certain nuclear instruments which are calibrated for dose rate and/or surface contamination measurement. The principle of quality assurance (QA) assumes that the devices have a valid certificate of quality and that they have been calibrated with sources that can be traced back to national or international activity standards. For several reasons (low battery, faulty components, ambient air temperature, spectral changes of the radiation field, etc.), however, the performance of these devices can deteriorate with time. It is, therefore, very important to carry out regular QA measurements of the monitoring instrumentation to check the consistency with the original calibration and with the other instruments.

QA measurements can be done in two ways:

- using a calibration source in a fixed measurement position (some devices are shipped together with such sources),
- performing regular intercomparison measurements with all the devices used for the monitoring, on well-defined sites where the radiation field is assumed to be constant with time.

These measurements can reveal changes in the calibration status of the detectors and any systematic deviation between the different instruments.

6. Scenarios

Contamination of inhabited areas can occur as a consequence of various events. It is worthwhile to look briefly into the possible scenarios that can lead to such a situation for, beyond the common characteristics, there may be specific features which can have an influence on the contamination monitoring strategy.

There are several events that can lead to the release of substantial amounts of radioactive materials into the environment. For many scenarios there is empirical experience, while others are only hypothetical (nevertheless credible) situations. The following is a detailed (though not comprehensive) list of the possible starting events (see Chapter 1 for examples):

- Accidents associated with the nuclear fuel cycle
  - Accidents in fuel fabrication facilities
  - Accidents in reactors (power/research/training)
  - Accidents in fuel reprocessing plants
  - Accidents involving spent nuclear fuel (transport, storage)
- Nuclear weapons detonation
- Nuclear weapons accident

- Accidental dispersion of high-activity radioactive sources (melting, fire, explosion, etc.)
- Intentional dispersion of high-activity radioactive sources ('dirty bomb' or RDD)
- Re-entry of space satellite containing nuclear or radioactive power source

Knowing the starting event can help shape the relevant contamination monitoring strategy.

Accidents in fuel fabrication facilities will lead to the dispersal of the nuclear fuel material (typically enriched U) in an area defined by the release and possible transport process. In such a case, the contaminated area will be limited due to the rapid gravitational deposition of the uranium particulates near the release site. The monitoring should focus on the detection of the dispersed uranium isotopes. The basic method is surface contamination measurement by any portable surface contamination monitor which is equipped with an alpha detector. Depending on the level of the contamination, a gamma monitor (surface contamination probe or gamma dose rate meter) can also be used to detect the 185.7 keV energy gamma line of ^{235}U.

Accidents in reactors (whether power plants, research reactors or training facilities) have the potential of contaminating large areas with a multitude of radionuclides. The scale of such an event is hardly predictable; it is strongly dependent on many factors (design, power, type of accident, containment structure, environmental characteristics, etc.). In principle, the higher the power, the more radioactive material can be dispersed. The accidents in the Three Mile Island NPP and in the Chernobyl Power Plant fall into this category. Responding to such an accident requires the full range of monitoring capabilities, from the early warning systems, through mobile monitoring units to *in situ* measurements and laboratory analysis of environmental and food samples. The earliest possible determination of the isotopic composition is a high-priority task, it is essential to choose the proper monitoring method and to make predictions about the expectable trends in the radiological situation. There are many relatively short-lived radionuclides in the contamination mix but the actual composition depends on many factors. In the early phase of the emergency, rapid instrumental methods (air- and car-borne *in situ* gamma spectrometry) will be best suited for quick mapping and determining the focal areas of intervention (screening). If the isotopic composition is known, then simpler methods (dose rate measurements, portable contamination monitoring) can be used, assuming that the composition is constant in space and the temporal variation is governed by the law of radioactive decay. Even the most rapid contamination monitoring will, however, probably be preceded by model calculations for the transport (trajectory, dispersion and deposition) of the released material. These predictions should be used to orient the monitoring and to direct the focus on the high-priority areas (potential hot spots and deposition plumes) but should not be considered as the ultimate truth. As soon as monitoring data start to arrive the transport model predictions must be reviewed and – when necessary – revised (*data assimilation*).

Accidents in fuel reprocessing plants will disperse a different mixture of radionuclides compared to reactor accidents. Most probably the short-lived nuclides are not present anymore. The actual composition (and the corresponding best monitoring technique) is dependent on a number of factors (which component of the reprocessed fuel is involved, the mode of dispersion, weather conditions, etc.) so it is difficult to give a generic guidance for contamination monitoring. The response described for the reactor accidents may be appropriate in this case, too, although the area of contamination is probably more limited and the time factor may be less pressing due to the lack of short-lived *radionuclides*. Although there have been accidents in fuel reprocessing facilities, the number of such plants in the world is rather small and specialised teams capable of performing the special contamination monitoring, should there be a need for it, are employed at these installations.

Accidents involving spent fuel either during transport or storage are of rather low probability because this kind of shipment is done relatively rarely and under extremely careful conditions and, when disposed of, there is very little activity around the stored spent nuclear material. In spite of this circumstance, accidents can occur and monitoring of environmental contamination may be required. The monitoring strategy must be similar to that described for the fuel reprocessing plants. The isotopic composition is characterised by the long-lived radionuclides (^{90}Sr, ^{137}Cs, transuranics, etc.), so the use of both gamma detectors and alpha/beta monitors should be considered.

Nuclear weapons detonation is probably the worst possible scenario, not least because of the destruction caused by the heat and pressure wave generated by the explosion. In the vicinity of the epicentre, this primary destruction is the higher priority concern, radiation monitoring coming after the first response to the emergency is implemented. With increasing distance, the primary destruction becomes less significant and the radiological contamination and its short- and long-term health consequences will be the main problem. There are limited experimental data and a great need for model calculations regarding the dispersion and deposition of nuclear fallout after a nuclear explosion. The isotopic composition is rather well known, containing a mixture of alpha-, beta- and gamma-emitting radionuclides with various initial activities and half-lives. The range is even wider than in the case of an accidental release from a nuclear reactor. The monitoring strategy must be similar to that developed for a reactor accident: rapid instrumental methods (air- and car-borne *in situ* gamma spectrometry) in the first phase, to determine the isotopic composition, and simpler but faster methods (dose rate measurements, portable contamination monitoring) for mapping purposes. In a nuclear explosion scenario, probably all contamination resources will have to be mobilised.

Nuclear weapons accidents (i.e. dispersion of a nuclear warhead by traditional explosives) have happened in the past. Some devices were unintentionally dispersed in conventional explosions during air transportation (Palomares, Thule). Fortunately, these cases occurred in rural areas but there is no guarantee that this cannot happen again in populated locations. There is, however, a big difference between these cases and the nuclear explosion: the contaminated area is much more limited and the isotopic composition is much simpler. After a nuclear weapons accident, the monitoring will focus on the localisation and recovery of the scattered debris, composed of the dispersed fissile nuclear material (enriched uranium or plutonium), which – through its alpha radiation – poses a threat via internal exposure if inhaled. The technique mentioned in the discussion of fuel fabrication facility accidents should be applied: use of portable surface contamination monitors equipped with an alpha detector and (for higher levels of contamination) surface contamination gamma probes or gamma dose rate meters can also be used (to detect the 185.7 keV energy gamma line of ^{235}U).

Accidental dispersion of high-activity radioactive sources is the scenario with the widest possible spectrum of initiating events. The common cause of these accidents is the loss of operational or regulatory control over the sources which is – under normal circumstances and in accordance with the licence conditions – supposed to be continuously maintained. Negligence, inappropriate training, incompetence, component failure, etc., can lead to a situation when the source gets 'lost' at least from the supervision of the operator. Then the source can get into a scrap yard and then to a melting furnace, which has been a typical scenario for the accidental dispersion of many radioactive sources. Radioactive sources can also get involved in high-intensity fire or explosions which would then result in the release of radioactive material into the atmosphere. The range of possible transport mechanisms can extend to extremely strange situations, like the contamination of the large section of a city by the population itself (Goiânia). This latter case demonstrates how difficult it is to give any prediction about the distribution of the dispersed radioactive material and it is, therefore, left mostly for the monitoring to find the contamination 'hot spots'. From the technical point of view the task may be simpler: as soon as the type of the source is known (typically one or only a couple of radionuclides) the search can focus on that specific nuclide.

Intentional dispersion of a high-activity radioactive sources, often termed a 'dirty bomb' or radiation dispersion device (RDD) is – up to the date of publication of this book – also a hypothetical scenario, although some attempts to construct and detonate such a device have been revealed. It is assumed that the usually exaggerated aversion of the population against ionising radiation would make these devices a tempting means of generating more fear and terror, especially if they are combined with classical terrorist techniques (bombing). Although the contamination caused by detonating a 'dirty bomb' would most likely not result in deterministic health effects, the psychological effects like panic, fear, the disruption of normal living conditions and the recovery costs (especially in an inhabited environment) could be enormous. The size of the contaminated area is dependent on the explosives used and also on the physical and chemical characteristics of the radioactive material involved. Fresh spent fuel would be among the most dangerous radioactive materials to be used for this purpose but these substances are usually under strict control and they are hard to obtain and handle. Industry, scientific research and medicine use high-activity sources in a large number of applications and these sources have a higher chance of getting out of operational and regulatory control. Radionuclides like ^{60}Co, ^{137}Cs, ^{192}Ir, ^{90}Sr are potential candidates that are likely to be chosen for malicious purposes. Alpha-emitting radionuclides (like Am, Pu or U isotopes) can also be put into an RDD to enhance the 'fear factor' and to complicate detection and response. Metallic and heavy element components, larger particulates released during the explosion will scatter and deposit in an elliptical ground pattern determined by the blast dynamics. Fine particulates and volatile elements may travel a greater distance (see Section 5 of Chapter 2 and Section 10 of Chapter 5), their transport being governed by the ground-level weather conditions (wind speed and direction). These two dispersal processes may lead to different deposition patterns of the different components, complicating contamination monitoring. The gamma-emitter contamination can be measured by the gamma detectors discussed above (*in situ* gamma spectrometry, dose rate monitoring) for the alpha contamination one has to use the special portable surface contamination monitors (with alpha detection option) or take samples and analyse them in a specialised laboratory.

Re-entry of a space satellite containing nuclear or radioactive power source, although rare, is a scenario that requires due consideration and preparation. Nuclear reactors and thermoelectric generators (RTGs) have often been used to provide a power source in the circumstances of deep space exploration when even solar energy is scarce. The RTGs contain varying amounts of radionuclides (e.g. ^{90}Sr, ^{210}Po or ^{238}Pu). If a reactor is used onboard, about 30 kg of highly enriched uranium can generate around 100 kW_{th}. Accidents during launch are low-probability (but not totally excludable) events and the design of the radioactive batteries should provide enough protection to maintain the integrity of the batteries during impact on the ground. If, however, unused satellites (especially those produced in the earlier years of space exploration) return from near-surface orbits, the disintegration of the power source is highly probable. There are some techniques to influence the return trajectory (if control over the satellite is not completely lost) or at least to predict it in most cases. During re-entry, the debris (composed of larger fragments, small particles and dust) will be dispersed over a huge area amounting to thousands of square kilometres (the debris of Cosmos 954 covered an area of 100,000 km^2 in Canada in 1988). One of the main challenges of the scenario is to determine the contours of this vast contaminated area and to collect (if possible) the larger fragments (*hot particles*) scattered around. If the area is inhabited, preference should be given to places where the population density is higher to optimise the monitoring and intervention efforts from the collective dose point of view and to avert external doses. Advance knowledge of the type of the power source can help select the appropriate monitoring technique. Since an RTG contains a single radionuclide, the detection of the contamination may be more difficult, if special nuclide-specific techniques are required. ^{90}Sr is a pure beta emitter but the beta energy is rather high and the brehmstrahlung is detectable by gamma detectors. It is more difficult to determine contamination by ^{238}Pu which has only alpha activity, poses high risk of inhalation dose to those inhaling it but resists monitoring efforts if only gamma probes are used. In this case, surface contamination monitors with special alpha detectors can be used or the lengthy and tedious sampling and analysis method. Operating reactors will disperse a variety of fission products plus uranium isotopes. The monitoring method should be similar to that used in case of reactor accidents on the ground (*in situ* gamma spectrometry, dose rate monitoring). Because of the great extent of the area, mobile measuring techniques (both air-borne and ground-based) can play a major role in such a case.

7. Measurement of Dose Rates

The idea of using dose rate meters for the assessment of an inhabited area is rather obvious. These instruments (whether in fixed stations, mobile units or hand-held devices) are a relatively simple, easy-to-use means of getting fast information about the temporal and spatial variation of the radiation environment. One has to bear in mind, however, the limitations and difficulties when we want to use the direct readings of the instruments to produce contamination values that comply with strict metrological requirements.

The basic problem is that the dose rate meter picks up the signal from a complex environment and it is rather difficult to relate the readings of the instrument to a specific concentration value. In the early phase of an airborne radioactive contamination scenario, the air surrounding the detector is filled with the radioactive substance and the detected radiation comes from this contaminated air, the contribution of the nearby surfaces to the signal being negligible. The relationship between the average volume concentration and the dose rate is dependent on many factors like the nuclear characteristics (radiation type, energy and branching ratio) of the radionuclide(s), the detection efficiency (energy and directional dependence) and the distribution of the radionuclide(s) in the air. For the simple case of an evenly contaminated atmosphere, calculations can be made by assuming a semi-infinite half-space geometry and the result of such an assessment can be seen in Table 1 (Lahtinen, 2006). (*Note*: These geometries are more accurate for large open areas (agricultural areas, pastures, meadows, etc.) and are probably less accurately applicable to inhabited areas where the terrain is more structured by different buildings, streets, etc.)

The data in Table 1 illustrate well how difficult it is to say anything about the possible air concentration values, based simply on the measured dose rate values, without knowing the type of radioactive contamination. Considering the typical background gamma dose rates (around 0.1 μSv h^{-1}), the additional dose rate of 0.1 μSv h^{-1} is just on the brink of detectability. As a rule of thumb, it takes about a kBq m^{-3} concentration of medium energy gamma-emitters to produce this excess dose rate in the readings of a standard dose rate meter. On the other hand, alpha-emitters of particular concern in several emergency scenarios (U, Pu) will not be detected until more than 1,000 times this concentration is present in the

Table 1 Concentrations of some radionuclides in air causing a rise of 0.1 μS h^{-1} in external dose rate levels.

Nuclide	Half-life	Concentration (Bq m^{-3})	Inhalation dose resulting from an exposure of one hour (μSv)
I-131	8.04 days	1,700	11
I-132	2.30 h	270	0.03
I-133	20.8 h	1,000	1.5
Cs-134	2.06 years	400	8.0
Cs-137[a]	30.1 years	1,200	47
U-235[b]	7.04×10^8 years	4,100	1200
U-238[b]	4.47×10^9 years	6,000,000	5.1×10^7
Pu-23 8[b]	87.7 years	6,900,000	7.6×10^8
Pu-23 9[b]	24,131 years	7,000,000	8.4×10^8

[a]In equilibrium with ^{137m}Ba.

[b]Alpha emitters. Intensity of gamma and X-rays is very low and thus detection with ordinary dose rate monitoring devices is difficult (with the exception of ^{235}U).

atmosphere. The data in the column showing the inhalation doses (for adults) resulting from an exposure of 1 h demonstrate the even more pronounced importance of identifying the radionuclide that produces the 0.1 $\mu Sv\,h^{-1}$ excess dose rate in our meter.

It is worth emphasising that 1 $kBq\,m^{-3}$ is a considerable concentration. For comparison, the daily average activity concentration of ^{137}Cs in ground-level air of Budapest, Hungary, during the days following the Chernobyl accident reached 1 $Bq\,m^{-3}$ only on two occasions (1 May and 6 May 2006). These air contaminations, hardly detectable by the dose rate meters while airborne, led to significant surface contamination (5 $kBq\,m^{-2}$) after being deposited on different urban and rural areas, causing a situation that required attention for a long time.

Following the cessation of the release of radioactive substances, the air contamination will diminish rapidly. By dispersion and deposition the air becomes clean; on the other hand, the surface contamination will build up. In this later phase, the dose rate measured by the meters originate totally from the deposited surface contamination. In the long run, ground-level air becomes contaminated again only by the resuspension of the deposited radionuclides (from the monitoring point of view this component is negligible, however).

In the late phase of an airborne release scenario, the dose rate is solely determined by the contribution of the deposited gamma-emitting radionuclides. For the case of evenly contaminated plain surfaces, calculated factors relating the dose rates to the surface activity concentrations are available in the literature (ICRU, 1994; IAEA, 2000). From these factors, the surface concentrations needed to produce the same 0.1 $\mu Sv\,h^{-1}$ increment in the dose rate can be calculated. A few examples for some typical radioactive contaminants are shown in Table 2 (the values should be treated with some caution: deviation from the ideal plain surface used in the model calculations is very likely, especially in inhabited areas where

Table 2 Concentrations of some radionuclides on the ground causing a rise of 0.1 $\mu S\,h^{-1}$ in external dose rate levels (IAEA, 2000).

Nuclide	Half-life	Concentration ($kBq\,m^{-2}$)
Co-60	5.27 years	12
I-131	8.04 days	77
I-132	2.30 h	13
I-133	20.8 h	48
Cs-134	2.06 years	19
Cs-137	30.1 years	48
Ce-144	285 days	500
U-235	7.04×10^8 years	192
Am-241	432.7 years	1030

buildings, structures, various paved and unpaved surfaces will define an environment different from those used in the calculations).

Bearing in mind the limitations discussed in the previous paragraphs, the dose meter is a potentially useful instrument for contamination monitoring. This easy-to-use, relatively cheap, widely available type of monitoring equipment can be used very efficiently for the mapping of a larger area. If carried in hand, attention should be paid to maintaining a standard measuring position (typically 1 m above the ground).

8. Screening of Contamination Level

For the qualitative and quantitative determination of the surface contamination of larger areas (plots, meadows, pastures, gardens, etc.), *in situ* gamma spectrometry can be used. If the total inventory of relatively fresh fallout is to be determined, then HPGe spectrometry is the best solution. If well calibrated, it can provide the activity concentration of all the radionuclides deposited on the ground after the accident. The accuracy of the method is limited by the statistical and calibration errors, complexity of the terrain, contaminated and/or shielding objects and, last but not least, the penetration of the radionuclides into the soil.

The measurement range of *in situ* spectrometry depends greatly on the detector efficiency. For typical HPGe spectrometers of 20%–30% relative detector efficiency, ^{137}Cs contamination (the predominant long-term contaminant of the Chernobyl accident) ranging from 100 Bq m^{-2} to 10 MBq m^{-2} can be measured. In settlements where monitoring and possible decontamination work is projected, however, this level of contamination is very unlikely to be exceeded.

In situ spectrometric measurements are normally made at a predefined, standard height (normally 1 m) with a downward looking detector (see Figure 1). It is aimed at determining the amount of radionuclide concentration on the surface. If the fallout occurred a long time ago and/or by wet deposition, then it is important to know the concentration depth profile. This profile has an influence on the transport of the photons emanating from the contaminated soil, and should be taken into consideration when converting the measured signal to the deposition value. ICRU Publication 53 (1994) gives the conversion factors for the different radionuclides as a function of the exponential depth profile parameter (*relaxation length*). For the determination of the relaxation length of the actual depth distribution, it is necessary to take samples of each layer of soil (in 2–4 cm steps, if possible, to 10 cm, 10–15 cm, 15–20 cm) and measure the specific activity of the soil layers by gamma spectrometry.

In this case, the samples may be taken only for zones different by dose rate and soil structure (one to three samples for a zone). Should there be significant differences of the depth distribution of activity in the limits of a given zone, it is necessary to take additional soil samples in such a zone and if necessary additionally split the zone into several sub-zones.

Once the nuclide-specific radionuclide concentration is determined, the expectable ambient dose rate at 1 m height can be calculated, based on the conversion factors published in the literature (ICRU, 1994; IAEA, 2000). The calculation is achieved using the following expression:

$$H_{\text{ext}} = \sum_{i=1}^{n} \bar{C}_{g,i} \times \text{CF}_i \tag{1}$$

where, H_{ext} is the ambient dose rate from deposition ($\text{mSv}\,\text{h}^{-1}$), $\bar{C}_{g,i}$ the average deposition (ground) concentration of radionuclide i ($\text{kBq}\,\text{m}^{-2}$), CF_i the conversion factor from Table 3; ambient dose rate per unit concentration ($\text{mSv}\,\text{h}^{-1}$ per $\text{kBq}\,\text{m}^{-2}$) for radionuclide i and n the number of radionuclides.

The calculation should take into account the shielding factors and the partial occupancy: This can be estimated by the following equation:

$$H^{\text{po}}_{\text{ext}} = H_{\text{ext}}[\text{SF} \times \text{OF} + (1 - \text{OF})] \tag{2}$$

where $H^{\text{po}}_{\text{ext}}$ is the ambient dose rate from deposition for the period of concern assuming shielding and partial occupancy ($\text{mSv}\,\text{h}^{-1}$), SF the shielding factor from measurements during occupancy or from Table 4, OF the occupancy fraction; fraction of time the shielding factor SF is applicable, for example the fraction of time spent indoors; it is assumed that, for the rest of the time, there is no shielding; default – 0.6.

More details about the dose assessment can be found in IAEA (1997) and IAEA (2000).

The calculations above form the basis of the assessment of the annual effective dose expectable from external exposure. In the case of a long-lived contaminant, the product of the calculated ambient dose rate from deposition and the number of the hours in a year (8,766 h) gives an upper estimate for the expectable annual exposure. This is already a number which can be compared with the prevailing dose limits during contamination screening.

It is rather difficult to give any definite figures for the screening criteria. The values will be dependent on the radionuclide(s), the use of the place where the monitoring was made, type of the environment and even social factors as, for example how much (real or perceived) risk associated with the contamination a certain society is willing to accept. These preset values should, however, be developed well before the contamination occurs.

Table 3 Conversion factors for exposure to ground contamination for a number of radionuclides.

Radionuclide	Conversion factor CF[a]; ambient dose rate from deposition ($mSv\,h^{-1}/kBq\,m^{-2}$)
Co-60	8.3E-06
Zr-95	2.6E-06
Nb-95	2.6E-06
Ru-103	1.6E-06
Ru-106+Rh-106	7.5E-07
Ag-110m	9.4E-06
Sb-124	6.0E-06
Te-129m	1.3E-07
Te-132	8.0E-07
I-131	1.3E-06
I-132	7.8E-06
I-133	2.1E-06
I-134	8.9E-06
I-135+Xe-135m	5.4E-06
Cs-134	5.4E-06
Cs-136	7.4E-06
Cs-137+Ba-137m	2.1E-06
Ba-140	6.4E-07
La-140	7.6E-06
Ce-141	2.6E-07
Ce-144+Pr-144	2.0E-07
U-235	5.2E-07
Am-241	9.7E-08

[a]Based on 'Dose conversion for exposure to contaminated ground surface' factors from external exposure to radionuclides in air, water and soil (EPA, 1992).

The following recommended dose limits can be considered when developing the criteria (IAEA, 1996) (see also Section 10 of Chapter 2):

- 1 mSv effective dose is the internationally recommended annual limit for the general population in the context of normal practices involving the use of ionising radiation.
- 2–3 mSv effective dose is the typical average annual dose to the general population from natural sources. The range can extend well beyond 10 mSv, however.
- 10 mSv is an existing annual effective dose limit on remediation, considering an appropriately defined critical group. Below this dose (from all sources), remediation is not considered justified. If the annual equivalent dose to any organ exceeds 100 mSv, however, the appropriate clean-up action is justified under any circumstance.

Table 4 Shielding factors (SF) for surface deposition (Jacob and Meckbach, 1987).

Structure or location	Representative SF[a]	Representative range
1 m above an infinite smooth surface	1.0	–
1 m above ordinary ground	0.7	0.47–0.85
One and two story wood-frame house (no basement)	0.4	0.2–0.5
One and two story block and brick house (no basement)	0.2	0.04–0.4
House basement, one or two walls fully exposed		
One-story, less than 1 m of basement wall exposed	0.1	0.03–0.15
Two story, less than 1 m of basement wall exposed	0.05	0.03–0.07
Three or four story structures (500–1,000 m^2 per floor)[b]		
First and second floor	0.05	0.01–0.08
Basement	0.01	0.001–0.07
Multi-story structures (>1,000 m^2 per floor)[b]		
Upper floors	0.01	0.001–0.02
Basement	0.005	0.001–0.15

[a]The ratio of the interior to the exterior doses; away from doors and windows.
[b]The SF values are appropriate if indoor deposition is negligible, that is wet deposition outdoors; for dry deposition these numbers may be higher depending on air exchange rate.

- 30 mSv effective dose per month is the limit above which relocation of the affected population is to be considered. 10 mSv per month is the limit for termination of relocation and return to the contaminated area (this limitation implicitly assumes that the contamination is partly composed of short-lived radionuclides and that the radiation level will gradually diminish with time). This latter limit also implies that living in an environment where the annual effective dose temporarily exceeds 20 mSv is considered acceptable, provided that appropriate remediation actions are initiated (in accordance with the previous recommendation).
- 1,000 mSv effective dose projected for a whole lifetime would justify permanent relocation.

One should also take into account the fact that external exposure originating from the surface contamination is not the only component of the annual dose. Other pathways of exposure (e.g. committed dose due to incorporation of the radioactive contaminants) should also be assessed and considered when applying the screening criteria. Although the dose limits above are based on broad international consensus and are published in many papers, the actual definition of the prevailing screening criteria is normally left to national competence and should be determined for the individual countries separately by the competent national authorities.

In an emergency situation, however, a timely response is crucial and urgent decisions are needed. In such a case the detailed and accurate assessment of doses is simply not practicable. Therefore, it is wise to define certain measurable quantities that are easier to obtain and set the screening criteria accordingly. Such derived intervention levels (or *operational interventional levels, OILs*) can be determined from the dose limits above, using simple models and applying certain assumptions.

An example of using surface contamination values as criteria for zoning contaminated areas is the practice followed in the former USSR after the Chernobyl accident. Different map categories were established depending on the concentrations of ^{137}Cs, ^{90}Sr and $^{238}Pu-^{240}Pu$ on the ground. For example for ^{137}Cs isolines of 1, 5, 15 and 40 Ci km^{-2} (37, 185, 555 and 1,480 kBq m^{-2}, respectively) were defined. The zoning of the area was done according to these contamination categories, as indicated in Table 5. The names of the zones indicate the countermeasures applied to the population living in the given zone.

Table 5 Zoning of the contaminated areas around the Chernobyl nuclear power plant according to the level of radioactive contamination (Konoplya and Rolevich, 1996).

Name of the zone	Surface contamination (kBq m^{-2})		
	^{137}Cs	^{90}Sr	$^{238}Pu-^{240}Pu$
Zone of residence with periodic radiation surveillance	37–185 (1–5 Ci km^{-2})	5.55–18.5	0.37–0.74
Zone with right for resettling out	185–555 (5–15 Cikm^{-2})	18.5–74	0.74–1.85
Zone of subsequent resettling out	555–1,480 (15–40 Ci km^{-2})	74–111	1.85–3.7
Zone of primary resettling out	>1,480 (>40 Ci km^{-2})	>111	>3.7
Zone of evacuation	The territory around the Chernobyl NPP from which the population was evacuated in 1986 (the so-called '30 km zone')		

The contamination values obtained by the wide-scale monitoring programme formed the basis of decisions with far-reaching consequences (evacuation and resettlement of hundreds of thousands of people from several hundred settlements, agricultural countermeasures, decontamination efforts, medical follow-up of the affected population, regular radiation surveillance, etc.). This fact underlines the outstanding significance of reliable and accurate contamination monitoring.

REFERENCES

Andersson, K. G., J. Roed, and C. L. Fogh. (2002). Weathering of radiocaesium contamination on urban streets, walls and roofs. *Journal of Environmental Radioactivity*, 62, 49–60.

EPA. (1992). *US Environmental Protection Agency, Manual of Protective Action Guides and Protective Actions for Nuclear Incidents*. Report US EPA, EPA 400-R-92-001. Washington, DC.

IAEA. (1996). *International Basic Safety Standards for Protection against Ionizing Radiation and for the Safety of Radiation Sources*. Safety Series No. 115. IAEA, Vienna.

IAEA. (1997). *Generic Assessment Procedures for Determining Protective Actions During a Reactor Accident*. IAEA-TECDOC-955. International Atomic Energy Agency, Vienna.

IAEA. (1999). *Generic Procedures for Monitoring on a Nuclear or Radiological Emergency*. IAEA-TECDOC-1092. International Atomic Energy Agency, Vienna.

IAEA. (2000). *Generic Procedures for Assessment and Response During a Radiological Emergency*. IAEA-TECDOC-1162. International Atomic Energy Agency, Vienna.

ICRU. (1994). *Gamma-Ray Spectrometry in the Environment*. ICRU Report 53. International Commission on Radiation Units and Measurements, Bethesda, MD.

Jacob, P., and R. Meckbach. (1987). Shielding factors and external dose evaluation. *Radiation Protection Dosimetry*, 21(1/3), 79–85.

Knoll, G. F. (2002). *Radiation Detection and Measurement*. Wiley, New York, 3rd edn., ISBN 0-471-07338-5.

Konoplya, E. F., and I. V. Rolevich. (Eds) (1996). *The Chernobyl Catastrophe Consequences in the Republic of Belarus*. National Report. Ministry for Emergencies and Population Protection from the Chernobyl NPP Catastrophe Consequences & Academy of Sciences of Belarus, Republic of Belarus.

Lahtinen, J. (2006). *Emergency Monitoring Strategy and Radiation Measurements. Working Document of the NKS Project Emergency Management and Radiation Monitoring in Nuclear and Radiological Accidents (EMARAD)*. NKS-142. Radiation and Nuclear Safety Authority (STUK), Finland.

Roed, J., and O. Togawa. (1996). A method for the determination of vertical distribution profiles of radioactive contamination in soils. *Proceedings of IRPA 9, 1996 International Congress on Radiation Protection*, Vienna, Austria, Vol. 2, pp. 2-666–2-668.

CHAPTER 8

Countermeasures for Reduction of Dose in Contaminated Inhabited Areas

Kasper G. Andersson*

Contents

1. Introduction

As mentioned in Chapter 5, before the Chernobyl accident in 1986 it was considered highly unlikely that any plausible incident leading to airborne dispersion of radioactive contaminants would affect inhabited areas to any significant extent. Therefore, efforts made until then to investigate countermeasures that might be implemented in the event of a large contaminating incident had almost entirely focused on rural land areas, and in

*Corresponding author. Tel.: +45 46 77 41 73; Fax: +45 4677 5330
E-mail address: kasper.andersson@risoe.dk

Risø National Laboratory for Sustainable Energy, Technical University of Denmark, DK-4000 Roskilde, Denmark

Radioactivity in the Environment, Volume 15
ISSN 1569-4860, DOI 10.1016/S1569-4860(09)00408-2

particular on agriculture. The countermeasures that were introduced in the late 1980s in contaminated settlements in Russia and Ukraine were, therefore, largely selected on a 'hit and miss' basis (Anisimova et al., 1994), and the outcome of the implementation was generally poor, as important pitfalls were overlooked (see example in Section 4 of Chapter 10). The Chernobyl accident, however, also provided a unique opportunity to study a wide variety of countermeasures for inhabited areas and test them *in situ*, and the decision-making knowledge base was thus improved considerably. This chapter is aimed at giving an introductory overview of the different countermeasures that might be implemented to reduce doses in an inhabited area, while at the same time giving an account of the types of aspects it would be necessary to consider in evaluating the usefulness and appropriateness of a countermeasure, either for generic emergency preparedness or for a very specific emergency situation. One important aspect to consider is that a number of countermeasures generate contaminated waste that must be managed safely. Therefore, this chapter also includes descriptions of options for management of waste produced by the implementation of countermeasures in inhabited areas.

2. Types of Countermeasures

2.1. Countermeasures for reduction of doses from different exposure pathways

The relative importance of introducing countermeasures for reducing radiation doses arising from different exposure pathways depends strongly on the type of release that led to the contamination, and on the characteristics of the affected inhabited environment. The different contributions to the radiation doses which might all require reduction, depending on the particular case, include the following:

- Internal dose from inhalation during plume passage
- Internal dose from inhalation of resuspended contaminants
- External dose from the passing contaminated plume
- External dose from contamination on outdoor surfaces
- External dose from contamination on indoor surfaces
- External and internal dose from contamination deposited on humans
- External dose from contaminants transferred onto humans by contact
- Internal dose from consumption of food contaminated in inhabited areas (issues relating to food contaminated elsewhere fall outside the scope of the book)

The Chernobyl accident showed that some of the above dose contributions would be expected to be greater than others in an area contaminated after a major nuclear power plant accident. For instance, inhalation doses received during the plume passage (particularly in dry weather) and external doses from deposited contaminants would in practically all perceivable cases be expected to be among the most important, whereas external exposure during the plume passage and from inhalation of resuspended contaminants (particularly in the indoor environment) would generally not be expected to contribute greatly to the total dose (see Chapter 6). As demonstrated in the example in Section 3.5 of Chapter 6, many different dose contributions could nevertheless be significant, and might need reduction through active implementation of countermeasures.

On the other hand, if the contaminating incident were a terror attack involving a 'dirty bomb' explosion, it would be impossible to say that one dose contribution would generally be likely to be greater than another. A terror attack could for instance involve a single alpha emitter, like ^{238}Pu or ^{252}Cf. In that case all external dose contributions can be ignored, as the radiation will not penetrate the human stratum corneum, and thus not cause a dose to living cells via this route. In this case, exposure pathways involving the intake of radionuclides via inhalation or ingestion will be of interest. Different types of 'dirty bomb' attacks would also involve very varying fractions of inhalable contaminant particles. Liquid droplet and particularly vapour condensation particles, that may under some circumstances be produced in a 'dirty bomb' explosion if sufficiently high pressure is generated, would be small and certainly inhalable, whereas the physical fractionation of, for instance, some metals would lead to negligible inhalable particle concentrations (Harper et al., 2007). Inhalation doses could thus have highly varying importance according to the specific scenario. Clearly also, the relationship between indoor and outdoor particle concentrations in the air would vary greatly according to particle size, as two of the governing factors, indoor deposition rate and the filter factor, are highly sensitive to variations in particle size, particularly for particles larger than ca. 5 μm (see Chapter 3). Also doses from deposition on humans would be highly dependent on particle size, as submicroneous particles may lodge in skin cavities like hair follicles for weeks, whereas the typical natural clearance half-life (without washing/scrubbing) on human skin of particles in the 10 μm range has been found to be of the order of 3 h (Hession et al., 2006; Andersson et al., 2004; see Chapter 4).

It is thus clear that an operational preparedness needs to build in many different countermeasures that can reduce doses received through different pathways. Some countermeasures, like different means of removing or burying contamination in areas (Andersson et al., 2003; Nisbet et al., 2004), can reduce both external and internal exposure pathways, whereas others are targeted specifically on a single dose component.

2.2. Countermeasures for different time phases

Any nuclear or radiological release incident will be followed by an initial emergency phase in which urgent countermeasures may be required to protect individuals from short term, relatively high risks. Such countermeasures can include sheltering and evacuation. In the subsequent recovery phase, an increased number of measurements together with consequence assessment modelling can provide a better and more detailed picture of the situation. Here, the overall ultimate aim of introducing countermeasures in a contaminated inhabited area is the return of more normal living conditions, where the inhabitants can live and the local community as a whole can function in the area without more or less constantly focusing on the emergency and its implications. The two phases are not separated by any exact boundaries, but the recovery phase might be defined as starting when the initial atmospheric release has largely ended, and continuing until the radiological situation has again reached an acceptable level (Brown et al., 2007; ICRP, 2007).

The time that would be available to introduce countermeasures in an early emergency phase would strongly depend on the type of incident leading to the contaminant release. In connection with even a large accident at a nuclear power plant, contaminants may be contained for some time before an atmospheric release occurs, and that period may provide sufficient time to issue early warnings and for timely implementation of early measures like sheltering, depending on the distance of the given area from the power plant. In areas at considerable distance from the power plant, an early warning might give time to implement early countermeasures that require more time, including evacuation of populations and possibly respiratory protection, for example through application of vacuum cleaners in a prescribed way to reduce indoor contaminant concentrations. Distribution of simple filter masks has also been suggested for respiratory protection, but if the contaminants are primarily in the form of submicroneous condensation particles, as observed for the Chernobyl radiocaesium, except in a small area very near to the power plant (Reineking et al., 1987; Rulik et al., 1989; Dorrian, 1997), such masks would typically only reduce the inhalation dose contribution by 10–25%, whereas it would be much more efficient for supermicroneous particles.

By contrast to power plant accidents, radiological terror attacks (e.g. dirty bombs) would in general be expected to involve a much less powerful explosion, which would imply that the effective release height of the contaminants would be substantially lower than for instance that observed in connection with the Chernobyl accident (Andersson et al., 2008b). Also, depending on the physicochemical form of the contaminants as well as the construction of the device, a 'dirty bomb' explosion would at least in some cases only lead to production of large particles, which would have a high deposition velocity through gravitational settling. For instance, ceramic contaminants would in general not undergo phase transition in an explosion, and small liquid or vapour condensation particles would thus not be formed-only supermicroneous particles resulting from physical fragmentation (Harper et al., 2007). Because of these two factors, atmospheric contaminant dispersion from a 'dirty bomb' would in general lead to contamination of a much smaller area than would a major nuclear power plant explosion. Because of the short dispersion period, it would be highly questionable whether even simple countermeasures like indoor sheltering with closed doors, windows and forced ventilation systems, could be implemented effectively and consistently after a 'dirty bomb' attack on a timescale that they would be effective. It should be noted that a significant fraction of the contaminants would be sufficiently small to be inhalable, even from an explosion dispersing a ceramic source (Andersson et al., 2008b).

When the contamination has occurred in an inhabited area, simple countermeasures like applying an agent on a surface to suppress resuspension of inhalable contaminants in the air may be considered. Such countermeasures may be effective in the short term, but since the resuspension rate is often described as decreasing in proportion to the reciprocal of time in units of days (Garland and Pomeroy, 1994; see also Chapter 2), they generally rapidly lose effectiveness in reducing doses. Also, a number of simple countermeasures may have great beneficial effect far into the future if they are implemented over a relatively short period of time. For instance, if the contamination took place in dry weather, deposition to a lawn would to a great extent have occurred to the grass cover rather than to the underlying soil. However, when the next heavy rain shower sets in, a large part of this contamination will be washed off the vegetation and transferred to the soil. Therefore, if the lawn is mown before this transfer occurs, and the cut grass is removed from the area, a much more severe long-term soil contamination problem can largely be avoided. Similarly, if contamination occurs in snow or to a snow-covered surface, removal of the contaminated snow before the first thaw can prevent contamination of the underlying surface with much more serious implications. Although early countermeasure implementation can thus be highly advantageous, it must be considered whether extra doses to workers are justifiable in comparison with the doses averted through the implementation of remediation measures. If the dose rate is dominated in the early phase by contributions from short-lived radionuclides, it may be advantageous to postpone remediation for some time. Rapid countermeasure implementation in general also needs to be balanced against the corresponding risk of making wrong, and possibly irreversible, decisions, which may affect the long-term radiological situation and the return to normal living. This risk will decrease with time, as more case-specific information becomes available.

Often the implications of postponing countermeasure implementation would be that the contamination becomes more difficult to remove, and more demanding and expensive techniques may need to be applied. An example of this is the natural migration over time of contaminants to greater depth in a surface, for example soil, demanding that a thicker surface layer be removed to decontaminate the surface. Also, due to the increased fixation in many materials in the inhabited complex, particularly for caesium (see Chapter 5), the effect of simple house or street surface washing techniques will decrease with time, and other techniques, possibly removing an entire outer layer of the surface, may become necessary. Such countermeasures would often be more time consuming, require more skilled personnel and produce waste that must be managed in a safe way, while possibly also necessitating subsequent resurfacing, either to make a surface applicable for use (e.g. making a road safe for vehicle traffic), to reduce surface permeability or for aesthetic reasons. Other types of countermeasures, which could for instance involve the replacement or mixing of thick soil layers, can have practically the same effect in reducing dose rate whether they are implemented early or several years after the contamination, depending on the mobility of the contaminant in the soil.

It should also be noted that some countermeasures may take considerable time to implement, for instance if the equipment or sufficiently skilled workers are not readily available in the area, if the method is very slow to undertake, if weather conditions (e.g. frost) make it impossible to start immediately, or if the measures generate large quantities of waste, for which at least temporary repositories of adequate capacity and safety must first be made available.

2.3. Countermeasures for decontamination or shielding

In many situations, there will be a choice between different solutions for essentially the same contamination problem, and in some cases, the choice will be between applying a countermeasure that entirely removes the contamination from the area (decontamination) and one that reduces dose by placing the contamination under a layer of a shielding material (e.g. uncontaminated soil), whereby both external and internal dose components may also be reduced (the latter by limiting contaminant resuspension that can cause inhalation doses and possibly by reducing ingestion dose if edible kitchen garden crops are subsequently grown).

One advantage of decontamination methods is that much of the contamination is removed from the area, although it must be stressed that no remediation technique is in practice capable of removing 100% of the contamination, even if an entire contaminated surface is removed (Brown et al., 2007). This is due to inadvertent blending and smearing of contamination, or contamination of other surfaces either directly or via resuspension of particulate during the decontamination process. If the contamination is instead placed under a shielding material, the public perception may be that the problem still exists. However, a disadvantage is that decontamination methods also require that the removed contamination is stored or buried somewhere. Management of the waste in the local area might be preferred, since it would often be the least expensive, and it could also be argued that it would be unfair to export this problem to a different area.

An important concern in connection with shielding methods like digging of soil to effectively bury contamination is that it needs to be considered very carefully in advance if the solution is satisfactory and sufficiently effective, so that it will not be necessary at a later date to remove the contamination from the area. Once the contamination is buried at some depth, a very large amount of waste would need to be removed to decontaminate the area – much more than if only a surface layer were removed. Also, if contamination is buried, it would be necessary to instruct inhabitants in the future not to dig beyond the burial depth, which might be seen as a restriction on their ability to return to 'normal life'.

2.4. Countermeasures for different surfaces

Over the years, a wide range of countermeasures have been suggested for treatment of contaminated surfaces, and tested on sufficiently large scale to allow identification of the various types of effects on the inhabited area as a whole (e.g. Roed and Andersson, 1996; Hubert et al., 1996; Fogh et al., 1999; Roed et al., 2006). The more promising of these countermeasures have been described in detail. A total of 59 different countermeasures have in a recent comprehensive study (Brown et al., 2007; Andersson et al., 2008a) been found to have potential for application to reduce doses in the event of airborne contamination of an inhabited area. Except for a few simple countermeasures (such as sheltering, evacuation, issuing stable iodine tablets and closing air ducts) that might be carried out in the pre-release and early emergency phases, if time permits, the countermeasures are targeted for the recovery phase – i.e. for reducing doses from contamination deposited in the inhabited environment. The different surfaces in an inhabited area can broadly be distinguished in three categories:

- Gardens, parks and other open unpaved areas
- Roads and other paved areas
- Buildings.

The relative importance of treating a given surface depends on a variety of factors such as the type of contaminating incident, characteristics of the contaminant(s), surface material and orientation, the weather during and after contaminant deposition, and the time that has elapsed since the contamination took place. Also, any countermeasure that is cost-effective and uncomplicated to carry out may be used as an attractive part of a dose reduction strategy (see Chapter 10), even though it may need to be combined with other methods that can bring about a more substantial dose reduction. Therefore, countermeasure descriptions are in general required for all types of potentially contaminated surfaces.

Simplicity and availability of resources are major concerns in selecting potentially appropriate countermeasures to avoid restricting access to (parts of) an area over longer times. Complicated procedures can have a higher inherent risk of failure as they may be too complex and demanding to be carried out correctly by workers with locally available skills, and the result could be an irreversible worsening of the situation in the area (Anisimova et al., 1994). Some types of sophisticated equipment may have been deemed promising on the basis of small case studies, but if the equipment or the skills to use it cannot be made available over an affected area during the relevant time period, the method cannot be used. Therefore, good operational methods for dose reduction in inhabited areas often contain elements of common practice based on their use for other purposes.

In the following few pages, a selection of short descriptions are given of a number of countermeasures designed to be used in the recovery phase following a contaminating incident to reduce doses in an inhabited area (based on the work in the European EURANOS and STRATEGY projects; www.euranos.fzk.de; www.strategy-ec.org.uk). As can be seen, many of the proposed techniques are indeed based on simple principles.

2.4.1. Gardens, parks and other open unpaved areas

2.4.1.1. Lawn mowing. If deposition occurs without precipitation, much of the contamination on a lawn will, in the earliest phase (days to weeks), be present on the grass rather than on the soil. Cutting the lawn (and removing the grass) can therefore, to some extent, prevent soil contamination. Up to some 90% of the contamination on a lawn can be removed in this way. It is a rapid and relatively inexpensive method (typically a few hundred to a few thousand square metres can be treated per hour), but generates waste, which can have rather high specific activity. For example, for a deposition of 1,000 Bq m^{-2}, the activity concentration in grass waste would be of the order of 10,000,000 Bq m^{-3}.

2.4.1.2. Snow removal. If contaminant deposition occurs to a snow surface, mechanical removal of a snow layer before the first thaw can prevent contamination of underlying (soil or asphalt) surfaces. If carried out carefully, more than 90% of the contamination can be removed from the surface. Time consumption is a factor of 2–3 less than that of topsoil removal if the work is done with similar equipment. It can also be carried out manually with spades, but then typically takes at least 10 times longer. Snow removal may also be considered for paved areas and roofs.

2.4.1.3. Topsoil removal. Removal of a thin top layer containing most of the contamination from a soil surface, using a 'bobcat' mini-bulldozer, tractor with scraper or similar equipment (see Figure 1). This can remove more than 90% of the contamination from the surface, if it is optimised according to the vertical contaminant distribution in the soil. The procedure takes some 50–100 h ha^{-1}. It can also be carried out manually with spades, but then typically takes at least 10 times longer.

2.4.1.4. Triple digging of soil/turf. By manually triple digging, the order of three vertical soil layers is changed: the thin contaminated top layer is buried in the bottom, with the turf facing down, the bottom layer is placed on top of this and the intermediate layer (not inverted) is placed at the top. Thereby shielding from contaminated soil is achieved with a minimised impact on soil fertility. Optimised application over a large area can reduce external dose rate above the surface by 80–90%. However, the method is very time consuming (20–30 min m^{-2}).

2.4.1.5. Ordinary manual digging of soil/turf. Using ordinary manual digging, the contaminated topsoil is at least partially placed deeper in the soil, and some shielding is obtained against the contaminated soil. The method is less effective than triple digging which ensures that contamination is placed deep in the vertical soil profile. Effort should be

Figure 1 Topsoil removal in a contaminated village area in Russia in 1995, using a tractor mounted with a scraper.

made to place the turf in the bottom of the profile to maximise the effect. The procedure can be accomplished at a rate of about 15–20 min m^{-2}, but the effect in terms of reduction of dose from the surface is typically only some 50–75%.

2.4.1.6. Ploughing (park areas). By ploughing much of the contamination on a soil surface will be buried deep in the vertical profile (depending on ploughing depth), so that exposure to the contaminated soil is substantially reduced. A special 'skim-and burial' plough ideally skims off a thin top layer, which is placed in the bottom of the vertical soil profile. A deeper soil layer (to about 45 cm) is placed on top of this (this layer is not inverted). Thereby shielding is achieved with a minimised impact on soil fertility. A reduction in dose rate above the surface of a factor of 10 is achievable here at a rate of 3 h ha^{-1}.

2.4.1.7. Application of clean sand/soil around dwellings and in frequently occupied areas. Sand or soil from a radiologically clean area can be applied around dwellings, to shield against radiation. This would typically be considered for reduction of dose rates from residual radiation after removal of a topsoil layer. Application of 10 cm soil over a large area may reduce dose rate from the surface by 80%. The time consumption is comparable to that of topsoil removal if the work is done with similar equipment.

2.4.1.8. Tie-down (fixation). Resuspension of contaminants from an open area can under some circumstances pose a problem that calls for action, particularly for alpha-emitting radionuclides. Various substances can be sprayed onto the surface to suppress formation of inhalable dust. For instance water, acrylic paint or lignin (a non-toxic by-product from paper production that can be purchased at low cost and in very large quantities) may be used, the preferable substance depending on, for example the duration over which the tie-down is needed. The applied layer may also reduce external beta dose rate significantly. A few hundred to a few thousand square metres can normally be treated in an hour, depending on the type of substance applied.

2.4.1.9. Resurfacing with asphalt. A more permanent method of fixation which also provides a shielding layer against external radiation can be achieved by covering the area with asphalt. However, this would often not be an acceptable solution for garden areas, and it would severely complicate any subsequent removal of the contamination. It would also often be a relatively expensive option to carry out. A 5–6 cm thick layer could be expected to reduce the external dose rate above the surface from a ^{137}Cs contamination by a factor of 2–3. Resuspension will be eliminated.

A team of four persons would typically take 6–7 h to treat an area of 100 m^2.

2.4.1.10. Pruning or removal of trees and shrubs. If deposition occurs without precipitation, shrubs and trees (particularly if in leaf) may receive and retain very high levels of contamination compared with other vegetation. Their pruning or removal from areas such as gardens may, therefore, significantly reduce the dose to inhabitants. The fraction of the initial deposit on the vegetation that can be removed is high. Fallen-off leaves and needles must be removed. The time consumption depends greatly on the amount and type of vegetation.

2.4.1.11. Removal of leaves and needles. Depending on the type of vegetation and the season, contaminated leaves and needles will, through natural processes, fall off over a period of days to years. If these contaminated leaves and needles are left on the ground, they will come closer to people living in the area and the external dose rate will rise. Thorough collection of leaves/needles, for example followed by composting in a safe place, can practically eliminate dose received through this pathway. The time consumption depends greatly on the amount and type of vegetation.

2.4.2. Roads and other paved areas

2.4.2.1. Road planing. Grinding off a thin contaminated top layer of asphalt surfaces with a road planer (rotating grinding equipment used by contractors). Subsequent re-paving is likely to be necessary. Reduction in contamination level by 80–90% is achievable, but the method is relatively expensive, and waste must be managed appropriately. About 1,000 m^2 could be treated in an hour by a team of four workers.

2.4.2.2. Vacuum sweeping roads and walkways. Vacuum sweeping with a type of machine used in many areas of Europe for routine street cleaning (see Figure 2). Machines often have three rotating brushes and a vacuuming attachment. The road dust is accumulated in a vessel behind the operator. If applied early the method can remove 50–70% of the contamination. It is quick to implement (ca. 3,000 $m^2 h^{-1}$) and relatively inexpensive, but generates waste, which can have rather high specific activity. For example, for a deposition of 1,000 Bq m^{-2}, the activity concentration in street dust waste would be of the order of 5,000,000 Bq m^{-3}.

2.4.2.3. Firehosing roads and walkways. Hosing with water to remove loosely bound contamination from surfaces of roads or other horizontal pavings. This is particularly efficient if applied early after contaminant deposition, where 50–75% of the contamination may be removed. It is

Figure 2 A typical ride-on vacuum sweeper with three rotating brushes. Some sweepers apply a water spray to control the resuspension of dust. The dust is collected in a vessel on the vehicle, typically behind the operator's seat. The vessel may need to be shielded from the cab to reduce the external dose to the operator, as the specific activity of contaminated street dust may be high. Vacuum-sweeper vehicles are used by municipal authorities in many countries for routine street cleaning, and the equipment and skills to implement this countermeasure are thus often readily available.

generally not possible to collect the loosened radioactive material, which will end up in drains. It is a relatively rapid method (ca. $50-100\,m^2\,h^{-1}$).

2.4.2.4. Turning paving slabs. As the contamination on a paving slab (e.g. made of concrete) will be attached to the surface, turning it will place the contamination under the slab, whereby a shielding effect is obtained (the amount of shielding depends on the type of radiation and thickness and material of the slab). At the same time, the method effectively prevents against resuspension of contaminants that were deposited on the slab. The procedure is simple to carry out, but is time consuming (ca. $5-10\,m^2\,h^{-1}$).

2.4.2.5. Tie-down (fixation). Resuspension of contaminants from a paved area can under some circumstances pose a problem that calls for action. Various substances can be sprayed onto the surface to suppress formation of inhalable dust. Of particular use for such areas are water, sand, bitumen and lignin, the choice of material depending on, for example the duration over which the tie-down is needed. The applied layer may also reduce external beta dose rate significantly. A few hundred to a few thousand square metres can normally be treated in an hour, depending on the type of substance applied.

2.4.3. Buildings

2.4.3.1. High-pressure water hosing of walls. Water hosing with high-pressure nozzles can remove part (typically 35–80%) of the contamination deposited on walls of buildings, especially if applied early after deposition. The waste is practically impossible to collect and is led to drains. The hosing of 1 m^2 of wall surface is estimated to take about 1 min.

2.4.3.2. Sandblasting of walls. By sandblasting, using high-pressure air with sand injected, a thin layer of the surface of a wall is removed, taking with it much (typically 75–85%) of the contamination. Wet sandblasting is recommended, as both the efficiency and the control of the generated dust is better. The method generates rather large amounts of waste (ca. 3 $kg\,m^{-2}$), which is virtually impossible to collect, and is relatively time consuming (ca. 4 $min\,m^{-2}$).

2.4.3.3. Ammonium treatment of walls (caesium contamination only). The ammonium ion has similarities with the caesium ion, and can (on non-specific sorption sites) exchange with caesium. A solution of ammonium nitrate is sprayed onto the wall at low pressure, ensuring a continuous flow over the wall. The method is generally most efficient soon after contamination has occurred, where up to half of the caesium may be removed. It generates 6 l of liquid waste per square metre, which is virtually impossible to collect. It is relatively time consuming (ca. 5 $min\,m^{-2}$).

2.4.3.4. Mechanical abrasion of wooden walls. On (painted) wooden walls the contamination will initially be located on the outer surface, and abrasion with a hand-held grinder will remove a thin surface (paint) layer, including the contamination. The technique is commonly applied for removing old paint and cleaning surfaces prior to surface treatment (e.g. painting). The method generates very little waste (which must be managed safely), but even if it is carried out early after the contamination took place, the reduction in contamination on the wall is seldom by more than a factor of 2. It is rather time consuming (ca. 0.5 $h\,m^{-2}$).

2.4.3.5. High-pressure water hosing of roofs. Water hosing with high-pressure nozzle can remove part (typically 35–80%) of the contamination deposited on roofs of buildings, if applied early after deposition. Time consumption is some 1–2 $min\,m^{-2}$. The generated waste (estimated 20 $l\,m^{-2}$ liquid and 0.2 $kg\,m^{-2}$ solid) can be collected through drainpipes.

Figure 3 Roof cleaning device based on a rotating brush in an enclosure. The picture on the right shows the device in use on a contaminated asbestos roof in Novye Bobovichi, Russia.

2.4.3.6. Roof cleaning by cleaning device. Various designs exist for roof cleaning and are commercially available, involving rotating brushes or oscillating high-pressure jet systems (See Figure 3). They are typically mounted on an extendible rod and operated from the ground or from the rooftop. The reduction in the roof contamination is similar to that of ordinary high-pressure hosing. These roof cleaning methods generally have the advantage that the cleaning takes place under an enclosure, thus minimising spreading of loosened contamination and, for example asbestos. The time consumption is of the order of $4-10\,\mathrm{min\,m^{-2}}$ for a team of two workers.

2.4.3.7. Roof replacement. Replacement of roofs with new or cleaned tiles or slates can, at a comparatively high cost (time consumption is obviously highly variable), largely eliminate the contamination on a roof. Gutters and drainpipes must also be removed, as these can accumulate high concentrations of contaminants. The method generates much waste, which must be managed safely.

2.4.3.8. Intensive indoor surface cleaning. Dose contributions from indoor contamination may be significant, especially over the first year. Also over longer periods, contamination may be brought into dwellings over longer timescales, for instance attached to the soles of shoes. The method is to provide self-help advice to carefully clean floor surfaces thoroughly, particularly carpets, within a short period after contamination occurs. For instance, the contamination level on a hard floor material can typically be reduced by 35–65% by washing. Different processes take between 0.5 and $2\,\mathrm{min\,m^{-2}}$.

2.4.3.9. Demolition. Demolishing buildings is a very destructive method that might be considered if other techniques fail to solve the problems associated with a contaminated house. It will generate very large amounts of waste, which must be managed safely. It is expensive, particularly as it implies re-housing, and can generate severe social problems.

2.4.4. Example of schematic overview

Figure 4 shows an example of a schematic overview of the different options described by Brown et al. (2007) for a surface category – here for roads and other paved areas. Shielding countermeasures are identified with grey shading. Such charts can be useful in providing an overview of methods to assist with decision-making. As indicated, generic alternatives to treatment of surfaces are to restrict access, which can in turn introduce severe non-radiological societal repercussions (Brown et al., 2007), and the also not unproblematic option 'Do no recovery'. It should always be kept in mind that it is a valid option to do nothing to actively improve the radiological situation. Indeed, the various other options should always be evaluated against this alternative. However, if it is considered not to introduce countermeasures, it still needs to be established that the choice is justified (see Section 2 of Chapter 10), and a number of considerations should be taken into account, as outlined in Table 1.

2.5. Countermeasure alternatives for different area sizes: an example

Some of the different techniques for different sizes of areas essentially work in the same way, only on different scales. An example of this is the use of the methods of triple digging and skim-and-burial ploughing. First consider triple digging (see Figure 5). This is a simple means for digging in a prescribed way, and requires only the use of a spade. By manually triple digging, the order of three vertical soil layers is changed: as previously summarised, a thin topsoil layer, optimised so that it only just contains the contaminated layer, is buried in the bottom, with the turf facing down, the bottom layer is placed on top of this and the intermediate layer, which must not be inverted, is placed at the very top. Thereby substantial shielding is achieved with a minimised impact on soil fertility. In terms of dose reduction, this is a more effective solution than alternatives like normal garden digging or rotovating, which would merely homogenise the contamination over a topsoil layer. Figure 6 shows an example of the results of a triple digging test carried out in 1995 in Novye Bobovichi, near Novozybkov, Russia, which had been contaminated by the Chernobyl accident (Roed et al., 1999).

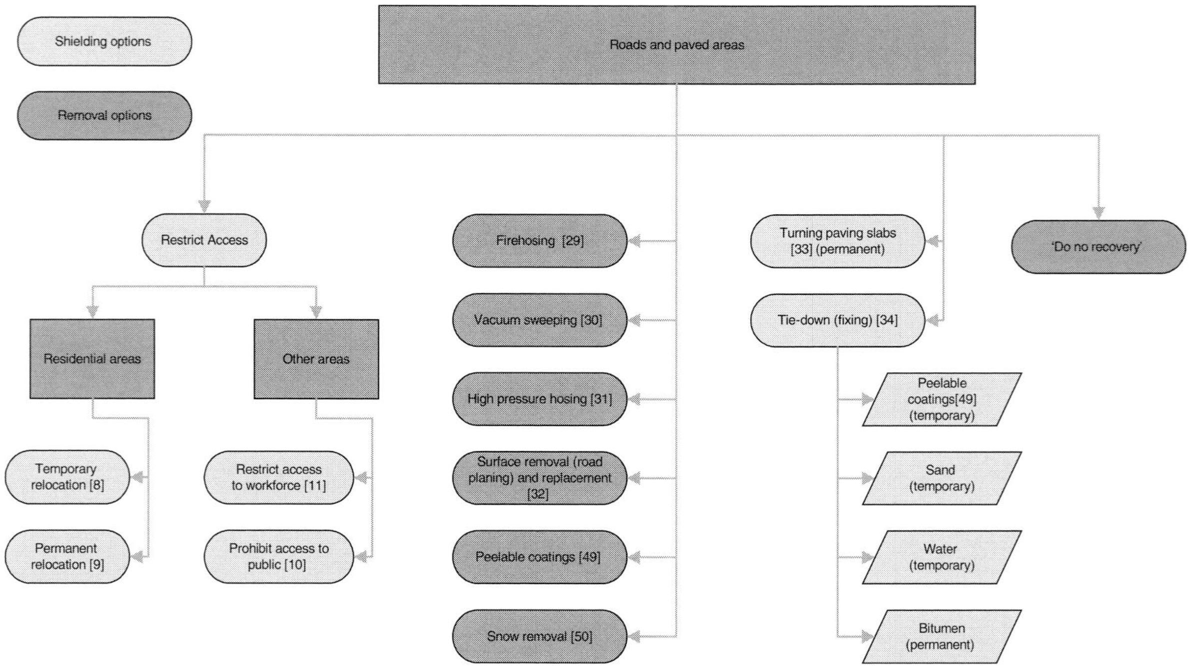

Figure 4 Recovery countermeasure options for roads and other paved areas (from Brown et al., 2007).

Table 1 Issues to consider when assessing the 'Do no recovery' option.

Potential advantages	Potential disadvantages
Implementing countermeasures may be perceived as indicating that there is a problem, even if doses are so low that they are being undertaken to provide reassurance.	Requires very good communication with the community in order to convince people of the decision not to implement countermeasures, that is demonstrating that risks are low.
Perception of affected area from outside may be better, that is incident is not perceived as a real problem; people are living normally. Economic blight may be less.	The implementation of countermeasures is visible and may provide reassurance to people inside and outside the contaminated area.
Sends out a clear message that risks are low and so not implementing countermeasures is justified. Builds public confidence in decision-makers. Saying that the risk is low and still undertaking countermeasures may give out a mixed message.	Needs linking with a very rigorous monitoring strategy. This might not be time or resource effective compared to implementing countermeasures.
No clean-up waste is produced. Some countermeasures that may be undertaken for public reassurance can create a lot of contaminated waste, for example grass cutting.	Not implementing any countermeasures may send out a message that the response organisations and other organisations don't care enough about the community.
If countermeasures are implemented the public may be reluctant to return to their homes.	Need to define the boundary of the area in which recovery options are not implemented. (However, no different to defining outer area beyond which no countermeasures are taken.)
Promotes return to 'normal living' in the area, that is to a situation where people can live without the radiological emergency/contamination being at the forefront of their mind.	If restrictions have been placed on food consumption, it needs to be carefully explained of why these are required but in inhabited areas, no action is taken.

Source: From Brown et al. (2007).

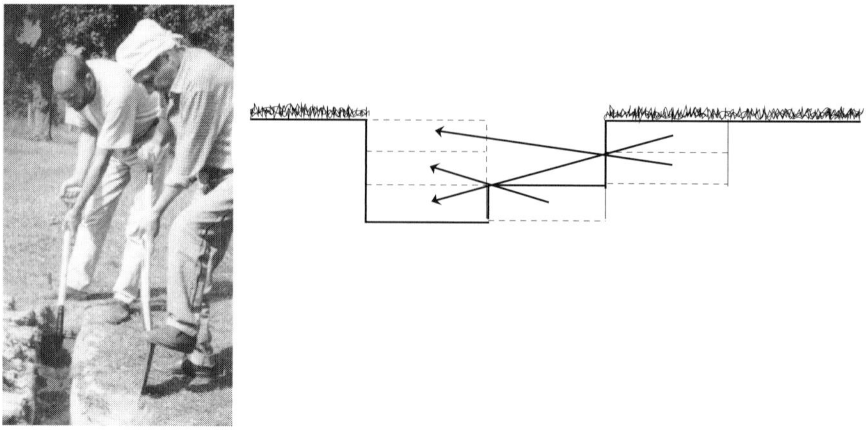

Figure 5 Diagrammatic sketch showing the principle of the triple digging countermeasure (right) and triple digging in the field in a contaminated garden in Russia (left).

In this area, where triple digging was carried out to a depth of about 40 cm, some 80% of the contamination was measured to have been placed at a depth greater than 20 cm, whereby a substantial reduction in external dose rate from the Chernobyl radiocaesium in the soil (by more than a factor of 10) was obtained. Reassessments of the vertical contamination distribution some years later showed that the downward contaminant migration (towards the groundwater level) was negligible (Ramzaev et al., 2006), although the mobility of radiocaesium in the rather peaty soils of this area had been reported to be three to five times higher than for instance that in typical soils in the 30-km zone at Chernobyl (Ivanov et al., 1997).

As the photo above captures, triple digging is time consuming and hard physical work, and can realistically only be carried out on small land plots. An alternative means for burying contamination in large open land areas (e.g. parks) is ploughing. An advantage of ploughing procedures as large-scale dose-reducing measures is that they can be carried out by local agricultural workers, although careful instructions on the special objective of the task need to be given. In terms of cost and availability, normal ploughing is an attractive countermeasure, but is of limited effectiveness since the normal mouldboard plough operates down to a depth of only some 25 cm. Also, subsequent ordinary digging of the area could bring much of the contamination back to the surface. By deep ploughing to 45 cm instead, the dose-reductive effect would be significantly better as the contamination would then be placed under a much thicker shielding soil layer. Another important benefit of the increased ploughing depth is that the risk of subsequently bringing contamination back to the surface would be greatly reduced (Roed et al., 1996a; Roed, 1982; Menzel et al., 1968).

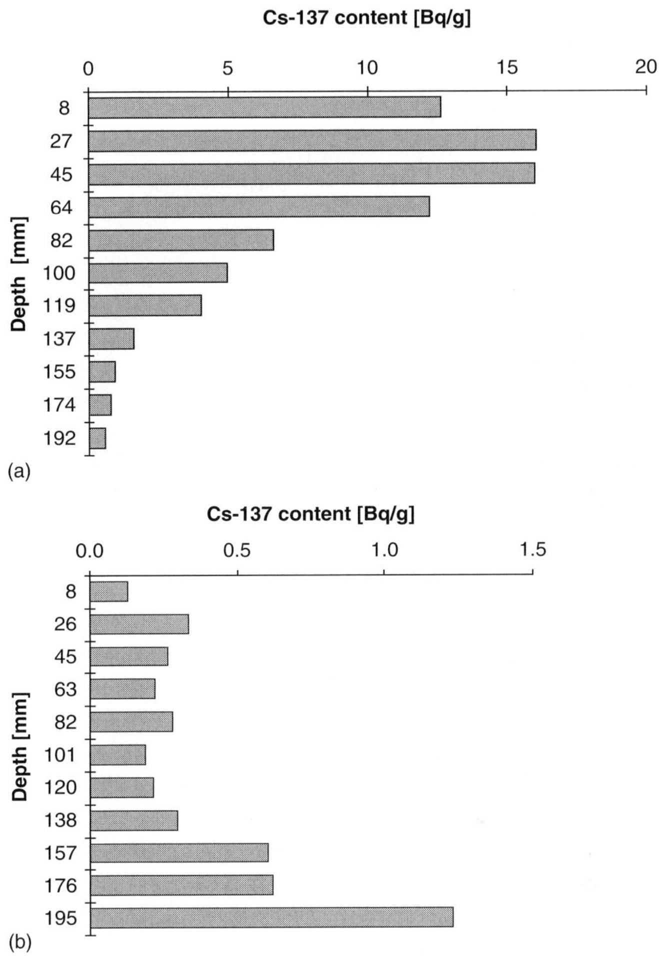

Figure 6 Typical vertical distribution in 1995 of Chernobyl radiocaesium in a soil in Novye Bobovichi, Russia, before (a) and after (b) triple digging.

The main disadvantage of deep ploughing is that poorer quality soil is brought to the surface. In, for instance, the shallow soils of the Chernobyl-contaminated areas of Russia, Belarus and Ukraine, this could literally create a desert. In order to overcome the problems associated with normal ploughing procedures, a special plough type has been developed and successfully tested in contaminated areas of Russia (Roed et al., 1996a, 1998). This plough (see Figure 7) is equipped with one ploughshare (denoted '2' in the picture), which skims off a topsoil layer of adjustable thickness (perhaps 5 cm) and buries it in the bottom of a trench made with another ploughshare (denoted '1' in the picture), at a depth of some

Figure 7 The skim-and-burial plough.

40–50 cm, without inverting the lifted bottom soil layer – hence the name 'skim-and-burial plough'. The displacement of only a thin (adjustable) top layer just containing the deposited contamination without inverting the subsoil minimises the adverse effect on soil fertility – exactly as is the case with triple digging. Essentially the only differences between triple digging and skim-and-burial ploughing lie in the use of different equipment, and implications for, for example speed, manoeuvrability, scale and implementation costs.

It needs to be stressed that, contrary to other ploughs, the skim-and-burial plough is not readily available in quantities that would enable immediate use in the event of a contaminating incident. However, as the natural downward migration in soil of particularly some of the key potential contaminants is exceedingly slow (see Chapter 5), much dose could still be averted even if it took a few years to produce the required ploughs.

Similarly, removal of snow or topsoil layers may in small open areas be accomplished by hand, but even for limited areas around buildings, for instance a 'bobcat' mini-bulldozer could be used, since it does not require much space to manoeuvre, and it could do the job much faster.

3. Systematic Countermeasure Descriptions

In selecting and optimising the implementation of a countermeasure, it is important to have a good overview of the different factors that determine its positive and negative sides. The concept of systematically describing countermeasure features in a datasheet format, taking into account, where possible, 'hands-on' experience, was introduced in the mid-1990s (Roed et al., 1995; Andersson, 1996). Standardised datasheets capturing the key information required to evaluate options are useful in providing a clear and consistent overview of a wide range of options, aiding the decision-making process and ensuring that potentially important issues are not overlooked. They can also outline the requirements (e.g. equipment, consumables, number of skilled workers) that are needed to carry out the countermeasures. Consequently, for the most practicable options within a given area, the required resources can be secured prior to any contaminating incident and other issues considered, such as waste disposal.

Over the years, various 'generations' of countermeasure datasheets have emerged, and the level of detail in descriptions has increased steadily, for example to recognise the need for data representing more diverse conditions and to look at a wider range of radiological hazards. Justification and optimisation of countermeasure strategies cannot be based solely on a balancing of technical features like direct intervention costs against the averted dose, and other, not easily quantifiable, factors must also enter the decision-matrix (see Chapter 9). According to the ICRP (1999), potential advantages of intervention include 'the consequent reassurance gained by the population and the decrease in anxiety created by the situation', whereas 'disadvantages include costs, harm and social disruption associated with it' (see Section 2 of Chapter 10). Such considerations were incorporated in the later generations of countermeasure datasheet descriptions (Howard et al., 2005; Andersson et al., 2003; Eged et al., 2003), introducing viewpoints from experts on social, ethical and economical sciences, and also considering observed reactions of the public and other participants in connection with countermeasure trials in the Former Soviet Union. Indeed, as described in detail in Chapter 9, highly cost-effective countermeasures may be implemented which do not at all reduce dose.

In its latest version, the countermeasure database has been further developed and the format modified, primarily to make it more suitable for describing countermeasure features of specific relevance to inhabited areas (Brown et al., 2007). Also countermeasures for the pre-release and early emergency phase of an incident had not previously been included. Further, whereas previous databases for inhabited areas focused almost solely on radiocaesium contamination, as this would be likely to govern the long-term doses after a major nuclear power plant accident, the current state-of-the-art countermeasure compendium includes an initial effort to take into account a wider range of radionuclides and related hazards that may be important also in connection with other types of contaminating incidents.

However, it should be noted that, since a very large part of the available practical information relates to Chernobyl radiocaesium, the information base for other contaminants is currently limited. For instance, a 'dirty bomb' scenario may produce comparatively large, not readily soluble contaminant airborne particles (Andersson, 2005; Harper et al., 2007), which would be removed much more easily from impermeable surfaces in the environment than would contaminants associated with smaller and soluble particles. This is illustrated by the very high decontamination factor of 50 obtained by Clark and Cobbin (1964) when hosing water on a street contaminated by insoluble contaminant tracer particles in the 44–100 μm range. When hosing a street contaminated by the small, soluble Chernobyl radiocaesium particles, a decontamination factor of 2 was difficult to obtain (Brown et al., 2007). The countermeasure descriptions of Brown et al. (2007) are an integrated part of a European decision-aiding handbook for contaminated inhabited areas, giving advice to decision makers on the development of optimised countermeasure strategies for specific situations, for example through series of flow-charts, tables and check-lists (see Chapter 10).

To illustrate the format used in the countermeasure database of Brown et al. (2007), the following few pages show the datasheet for the countermeasure 'Roof cleaning with pressurised hot water'. The applied template includes a number of criteria that decision makers might wish to consider when evaluating different countermeasures. These include:

- the objectives of the option
- a short description of the option
- constraints on its implementation
- effectiveness
- requirements
- waste generated
- doses received by those implementing the option
- costs
- side-effects
- practical experience.

Values for all data quantities presented in the data sheets should be treated as indicative, as the actual values will be very dependent on the specific circumstances. The main purpose of including values of quantities in the data sheets is to provide the user with a rough image of the implications of implementing a countermeasure in terms of cost, resources, waste generated, doses to workers, etc. This also enables a more direct comparison across different options. Practicability aspects and potential pitfalls are also very important, and these are addressed in the individual data sheets.

ID: 17	Buildings – external surfaces
	Roof cleaning with pressurised hot water
Objective	**To reduce external gamma and beta doses from contamination on roofs of buildings within inhabited areas, and reduce inhalation dose from material resuspended from those surfaces.**
Other benefits	Will remove contamination from roof surfaces.
Countermeasure description	Rotating nozzles are driven by hot water (ca. 65 °C) at high pressure (typically 150 bar). Cleaning is performed in a closed (shielded) 'box' system. The device is mounted on a trolley that can be drawn across the roof. It is operated from the top of the roof, lowered down the roof using the pressure water hose. Note that the use of hotter water (ca. 80 °C) and detergent can considerably increase the effectiveness of the procedure. Care must be taken not to block drains with moss, etc. The wastewater can be easily collected via downpipes. However, water may be allowed to pass into drains or to soakaways via gutters and drainpipes. Cleaning of these should be considered after implementation. The implementation of options to the surrounding ground surfaces should also be considered following roof cleaning if contaminated water drained onto the ground surrounding the buildings. If the implementation of any other options to the surrounding ground surfaces is planned, roof cleaning should be implemented first. Recontamination of surface by resuspended contaminants will be insignificant, so repeated application is not required. Dust creation during implementation is unlikely to be a problem and so methods are not required to reduce the resuspension hazard to workers.
Target surface or population	Contaminated roofs of buildings, both inhabited and industrial.
Target radionuclides	All long-lived radionuclides. Unlikely to be used for removal of short-lived radionuclides.

(*Continued*)

ID: 17	**Buildings – external surfaces**
	Roof cleaning with pressurised hot water
Objective	**To reduce external gamma and beta doses from contamination on roofs of buildings within inhabited areas, and reduce inhalation dose from material resuspended from those surfaces.**
Scale of application	Any size. Suitable for small roofs (e.g. houses) and large roofs (e.g. industrial buildings, schools, etc.). Use depends on availability of equipment.
Timing of implementation	Maximum benefit if carried out soon after deposition when maximum contamination is still on the surfaces. However, roof cleaning can be effective up to 10 years after deposition depending on the roof material and removable debris/growth.
Constraints on implementation	
Legal	Ownership and access to property. Disposal of contaminated water via the public sewer system, if required. Use on listed or other historical buildings.
Environmental/ technical	If the weather is so cold that water might freeze to ice in equipment and form ice on surfaces rather than running off them, the method can only be implemented if the water is heated (to at least 60 °C at the point of application). This option cannot be used if roof surfaces are covered with snow/ice. Roof construction must resist water at high pressure.
Effectiveness	
Reduction in contamination on the surface	A decontamination factor (DF) of between 2 and 7 can be achieved if this option is implemented soon after deposition. Using hot water and detergent improves the effectiveness considerably.

	In the short term, the quoted DF can be considered to be same for all radionuclides, with the exception of elemental iodine and tritium, for which thorough washing of impermeable surfaces will lead to virtually full removal. Even after 10 years, a DF of 2–4 can be achieved. The DF will be lowest for slate, clay and concrete roofs, and highest for silicon-treated slate, and possibly even higher for aluminium/iron. If a surface layer of moss/algae covers the roof at the time of deposition, almost all the contamination may be removable.
Reduction in surface dose rates	External gamma and beta dose rates above decontaminated roofs of buildings will be reduced by approximately the value of the DF.
Reduction in resuspension	Resuspended air concentrations above the roof surface will be reduced by the value of the DF. However, it should be noted that resuspension from roofs is unlikely to be significant.
Averted doses	Reductions in external gamma dose rate shortly after decontamination of the roof surface received by a member of the public living in an inhabited area could be expected to be around 7–8%. This is an illustrative value and should only be used to provide an indication of the likely effectiveness of this option and to compare across options. *Factors influencing dose reduction*: Consistency in effective implementation of option over a large area. Care taken to wash contamination to the roof gutter and not just translocate it onto other parts of the roof. Special care must be taken to clean roof gutters and drainpipes thoroughly after implementation. Whether the ground surfaces below the roof (onto which run-off may have occurred) have been decontaminated after treating the roof (especially if there is no gutter and wastewater is not collected). Number of buildings in the area, that is environment type/land use. The time spent by individuals close to buildings. Time of implementation. The impact of cleaning roofs on the overall doses will be reduced with time as there will be less contamination on the surfaces due to natural weathering. Industrial buildings often have light (not well shielding) roof constructions, and the often shallow slope of the roof may result in relatively high contamination levels and resulting dose rates.

(Continued)

ID: 17	Buildings – external surfaces
	Roof cleaning with pressurised hot water
Objective	**To reduce external gamma and beta doses from contamination on roofs of buildings within inhabited areas, and reduce inhalation dose from material resuspended from those surfaces.**
Additional doses	*Exposure pathways workers could be exposed to are:* External exposure from environment and contaminated equipment. Inhalation of radioactive material resuspended from the ground and other surfaces (may be enhanced over normal levels). *Inadvertent ingestion of dust from workers' hands.* Contributions from pathways in italics will not be significant and using personal protective equipment (PPE) can control doses from these pathways. Exposure routes from transport and disposal of waste are not included. *Beta/gamma hazard* For radionuclides that present a beta/gamma hazard, external dose to workers from contamination in the environment will be a few times higher than public doses over the period of implementation. Even under very dusty conditions, the inhalation dose from resuspended material will only make a small contribution to the total worker dose. *Alpha hazard* For radionuclides that present an alpha hazard, inhalation dose to workers from resuspended material will typically be a few times higher than public doses over the period of implementation. External dose from contamination in the environment can be ignored.

Factors influencing effectiveness of procedure (technical)	Consistent application of cleaning over the contaminated area (i.e. operator skill). Material from which roof is constructed (see above). Evenness, condition of the surface. Amount of moss/debris on roof. Time of operation (the longer the time between deposition and implementation of the option, the less effective it will be due to the contamination becoming fixed to the surface). Water pressure, amount of water, water temperature (hotter water is more effective), use of detergent.
Factors influencing effectiveness of procedure (social)	–
Requirements	
Required specific equipment	Roof cleaning trolley (about €500) and high-pressure hot water generator (about €37,500). Scaffolds or mobile lifts for operation from the roof. Transport vehicles for equipment (and waste).
Required utilities and infrastructure	Water supply could be a problem in periods of drought (water may be pumped from a lake or river if not available from the domestic water supply or a hydrant). Public sewer system. Roads for transporting equipment (and waste). Power supply (petrol-driven mobile generator may be required if power is not available).
Required consumables	Fuel and parts for generators if required ($8\,l\,h^{-1}$) and transport vehicles. Water (about $30\,l\,m^{-2}$).
Required skills	Can be carried out with little instruction – one person on the rooftop and one on the ground administrating supplies. Professionals are required for work on the roof.
Required safety precautions	For tall buildings: lifeline and safety helmet. Water resistant clothing should be recommended, particularly in highly contaminated areas.

(Continued)

ID: 17	Buildings – external surfaces
	Roof cleaning with pressurised hot water
Objective	**To reduce external gamma and beta doses from contamination on roofs of buildings within inhabited areas, and reduce inhalation dose from material resuspended from those surfaces.**
	As the cleaning is carried out in a closed wet environment, the dust (inhalation) hazard is negligible, so respiratory protective equipment (RPE) is unlikely to be required. Precautions are needed to ensure that people making connections to mains water supplies do not inadvertently contaminate the water supply, for example by back-flow from vessels containing radioactivity or other contaminants, or operate hydrants in a way that disturbs settled deposits within the water main system.
Waste	
Amount and type	Generates some $30\,l\,m^{-2}$ of liquid waste, with ca. $0.2\,kg\,m^{-2}$ of solid waste containing nearly all the contamination. Waste may be toxic (asbestos). Water can be collected via the downpipes and filtered using a simple filter prior to disposal via the drains or can be recycled.
Intervention costs (see Appendix D)	
Equipment	€$2\,m^{-2}$. Excludes cost for scaffolding and lifts, if required.
Consumables	$30\,l\,m^{-2}$ of water, at current prices. $8\,l$ petrol per hour for mobile compressor if required, at current prices.
Labour	€$2 \times 10^1\,m^{-2}$.

Operator time	3 m^2 per team hour (team size: two people). Work rate excludes setting up scaffolding.
Factors influencing costs	*The following factors influence the time taken to implement the option and hence labour costs:* Weather Building size: size of scaffolds/mobile lifts Roof gradient and amount of debris on roof Access Proximity of water supplies Operator skill.
Side-effects/impact	
Environmental impact	Wastewater from run-off from buildings in inhabited areas will result following rainfall over a period. Roof cleaning will also create contaminated wastewater. However, it may be easier to control the environmental impact of this water and to ensure appropriate monitoring in the sewer infrastructure. The disposal or storage of waste arising from the implementation of this option may have an environmental impact. However, this should be minimised through the control of any disposal route and relevant authorisations. If wastewater is not collected, some of it will run onto other surfaces (roads, soil, grass, etc). These may require subsequent clean-up, generating more waste.
Social impact	Acceptability of active disposal of contaminated wastewater into the public sewer system. However, note point above. Cleaning roofs will make buildings look cleaner; implementation may give public reassurance. Repair work on roof, etc. may be required but this is unlikely.
Practical experience	Tested on realistic scale on selected roofs of different types in the CIS after the Chernobyl accident.
Key references	Andersson et al. (2003); Andersson et al. (2001); IAEA (2001).

4. Management of Waste Generated by Countermeasures

A number of recovery options that might be considered would generate contaminated waste (liquid, solid or mixed). The impact on normal waste disposal practices and the workforce managing these can be assessed and controlled. Any decision to undertake clean-up which generates contaminated waste should be supported by an assessment of the impact that the generated waste will have on the public, workers and the environment (Brown et al., 2007). This section gives a series of descriptions of possible routes of transport, treatment and storage/disposal of the wastes that may be generated by countermeasures for contaminated inhabited areas.

4.1. Management of clean-up waste prior to disposal

The management of waste may include a number of the following processes prior to the final storage.

4.1.1. Loading and transportation

If possible, it would be advantageous to dispose of most types of waste in the local area, where the countermeasure was implemented. This would greatly reduce transportation costs and doses to transport workers. It might well also be considered most reasonable by the population, if the radioactive waste problems, if any, are shared by the whole affected population rather than massively imposed on a specific selected part of the inhabitants living near a large, centralised repository. It should, however, be recognised that for instance some advantageous techniques to reduce waste volumes prior to final disposal may not be possible to carry out *in situ*. It needs to be secured in advance of an incident that the required transportation vehicles, equipment, operators, etc. would be available for potentially attractive waste management routes.

4.1.2. Waste storage

If the dose rate is dominated by contributions from short-lived radionuclides, and the waste for instance requires the use of various treatment processes prior to final disposal, it may be beneficial first to store the waste for a while in a temporary repository (a vessel of suitable material) to save worker doses. Interim storage may also be necessary, while solutions for permanent disposal are being prepared.

4.1.3. Filtration of solid particles out of wastewater

A number of countermeasures that may be implemented in a contaminated inhabited environment involve the use of water, often applied at high pressure. Such techniques – for instance applied on a contaminated roof – will wash off particles consisting of roofing materials, sludge (e.g. from roof gutters, which would also be decontaminated), algae and moss. Such materials will normally retain the contamination (particularly caesium) well. The washed-off particles can be collected with the wash-water from the drainpipe. Simple filtration through an inexpensive polymer fibre textile with a pore size of about 0.1 mm has in areas contaminated by the Chernobyl accident been found to be highly effective in isolating solid particles. These contained virtually all caesium contamination, from the water, which could safely be led to the sewer or even re-applied on the roof (Fogh et al., 1999).

4.1.4. Treatments for contaminants in liquid waste

Some countermeasures involve the use of detergent solutions. Some of these detergents will be dilute and non-aggressive, whereas others may be highly acidic or alkaline. The solution acidity will to a great extent determine the degree of contaminant association with particles. For the removal of contaminant ions from the waste solution, if required prior to disposal, several methods may be envisaged.

Technically, it would appear that one of the more simple methods would be to concentrate the contamination in a solid residue, by evaporating the solution to dryness. This, however, requires very large amounts of energy ($>1{,}000\,\mathrm{kWh\,m^{-3}}$) (Turner et al., 1994), and may not be easy to handle in reality with strong, reactive solutions. Further, the presence of volatile contaminants, such as ruthenium, would be problematic in connection with this process.

An alternative might be to precipitate contaminants by adding a 'flocculant' agent and adjusting the solution pH (neutralisation by adding, e.g. lime). However, the relatively costly neutralisation process would generally be expected to lead to the generation of comparatively large amounts of precipitate (IAEA, 1993), as not only the targeted contaminant ions would precipitate. Also, the typical decontaminating effect of gravitational settling by neutralisation has been reported to be limited (maximum DF of about 10) (Turner et al., 1994). In connection with both evaporation and precipitation, very large, specialised handling facilities would be required.

A further, potentially attractive, alternative to the above would be to remove the contaminants from the solution by ion exchange (IX). This has been reported to be a highly efficient method (decontamination factor of 10–100, depending on, e.g. flow rate), and also, the required size of the handling facility would be much less than that of an evaporation or gravity settling plant (Turner et al., 1994). In principle, the process could be carried out using only a couple of containers and an ion exchange column. The absorber material must, however generally be disposed of together with the removed contaminants, and the waste volume reduction factor is consequently not quite as great as that of the alternative techniques described (it is about 500, compared with more than 1,000 for gravitational settling). The costs for ion exchange material vary somewhat according to material type, but as a rough estimate, it can be assumed that about 20 m^3 of solution can be decontaminated satisfactorily using 0.5 kg of resin, at a resin cost of ca. 10–100 Euro per kg.

For treatment of relatively large amounts of contaminated liquid, membrane filters based on the reverse osmosis principle may be highly attractive. The cost for a portable device that can process ca. 120 $l\,h^{-1}$ is about 10,000 Euro, but it can process very large volumes of liquid with only small maintenance expenses (Buch and Holm, 2005, personal communication). Membrane filters are reported to be highly efficient in reducing the liquid radioactive waste concentration (DF of several hundreds per cycle) (Zakrzewska-Trznadel et al., 2001).

Finally, a perhaps socially not very attractive (and possibly illegal) solution is to dilute the concentration of the radioactive matter to a sufficiently low concentration to allow disposal as ordinary waste liquid. Stirring systems to ensure the homogeneity of radionuclide concentrations in liquid waste have been developed for this purpose (Ogata and Nishizawa, 1999). However, dilution must be sufficiently effective with respect to both toxicity, acidity and radionuclide content.

When the contamination is in a wet medium this will provide significant attenuation of the radiation from the contamination. Monitoring throughout the handling process will show the need for special worker protection from the contamination (e.g. by shielding or increasing distance through remote operation).

4.1.5. Reduction of volume of organic waste

Organic waste from an inhabited area may be grass or turf removed from a lawn or trees and shrubs removed from, for example gardens and park areas. In particular, the specific activity of grass may be high if the grass is cut early after a dry contamination has occurred. Also leaves on a tree or shrub may have high specific activity right after contamination.

A number of methods may be considered for treating contaminated biomass, depending on the contamination level. For example, aerobic degradation (composting) will produce material that may be useful for soil fertilising, whereas anaerobic degradation produces gas that may be used in energy production. Core wood from contaminated trees may, particularly early after an accident when the contamination will largely be confined to the outer surface, be applied in industry, for example for making furniture. The IAEA has prepared a report, which provides estimates of the conversion factors between biomass (wood) contamination levels and annual doses that would be received due to the contamination, assuming conditions that are believed to adequately reflect 'typical' situations (Balonov et al., 2003).

The wood pulping process in connection with paper manufacturing may significantly reduce the contamination in the paper product. A special wood pulping treatment has been described by Roed et al. (1995) giving a decontamination factor of as much as 50–100. A problem in application of contaminated wood in a pulp mill recycling system is, however, that the radiocaesium is accumulated in the digesting liquor. Ravila and Holm (1994) estimated that, even in Sweden, where contamination levels are not very high, projected 50 y doses to wood pulp factory workers may reach 50 mSv.

An option for comparatively strongly contaminated wood, wood waste and other biomass (e.g. shrubs) is to chip it and combust it in safely designed power plants, which provides adequate protection of workers as well as of the environment (Andersson et al., 1999). This method generates energy and also reduces the mass of the waste by a factor of 10–100. With modern bag filter constructions, the stack release problem can be eliminated (Junker et al., 1998).

According to the recommendations of Junker et al. (1998), the ash from combustion can be disposed of in simple ground repositories in thick plastic 'bin bags' with typical volumes of ca. 2 m^3. Without combustion, the biomass repositories would need to be 10–100 times bigger, and the wood would still need to be chipped.

Also spreading of ash for fertilising fields has been suggested. The fertiliser may in some soils significantly reduce contaminant uptake to plants, and the total effect could thus reduce the received dose, depending on the ash contamination level. The legality and acceptability of this (or any other) solution should of course first be assessed.

4.1.6. Stabilisation of solid waste to avoid migration of contaminants

Some types of collected solid waste arising from countermeasure implementation (e.g. street dust, ash from combustion of contaminated biomass) can contain particularly high concentrations of radionuclides. In constructing ground repositories for strongly contaminated solid waste, it may be desirable to introduce special measures to prevent migration of contaminants to the groundwater. Thick plastic bags/membranes around the contaminated material will generally provide good protection together with clay barriers and draining layers of gravel, and would be recommended for any ground repository for solid waste.

Cementation of highly contaminated wastes could be considered to further stabilise the waste, particularly if the contaminants would otherwise have high environmental mobility. For instance, fly ash from combustion would be a 'natural' ingredient in a cement mixture. However, conventional cementation processes would not be possible for all materials, as the presence of some materials, for example humic materials, retard or prevent solidification. For such cases, alternative additives have been suggested (Brodersen, 1993).

4.2. Waste disposal options

Waste disposal schemes for solid contaminated waste must be selected with care. For coping with an incident and its implications, it must be possible to plan for waste management, including creation of the needed repositories, over a short period of time. This means that the responsible authorities should secure in advance that the required materials, transport vehicles, skilled workers, infrastructure, etc. are available to enable the construction of the type(s) of repositories that would be considered most suitable. It also means that concrete trenches and other complicated and time-consuming engineering projects, which might be considered for wastes originating from controlled practices, could not realistically be finalised within the required time frame. In any case, countermeasures would be useless in an area where the contamination level was so high that the dose rate could not be brought down to an acceptable level (see Section 2 of Chapter 10). This means that there would be limits as to how contaminated the wastes might be, and generally the more expensive and complicated repository options would not be likely to be called for. It should be kept in mind that the waste disposal strategy should be adapted for the task. In other words, the construction of repositories should provide adequate protection of workers and members of the public, but at as low cost as possible.

In selecting the disposal site, a number of factors need to be considered, including:

- Local repositories would generally be less expensive and give less total worker dose than centralised/distant repositories.

- The hydrogeological characteristics of the site should be considered. From the bottom of the constructed repository to the ground water level, there should be at least 3 m. If trench repositories cannot be constructed to meet this requirement, it may be possible to construct elevated surface mound repositories on the ground. Old gravel pits should not be exploited for this purpose, as they are likely to provide too little distance to the ground water. It should also be ensured that the area is not prone to flooding, for example by a nearby river.
- The area should not be prone to earthquakes.
- The future land use needs to be considered. Edible crops should for instance not be grown on top of a waste repository, both to completely rule out the possibility of uptake by plant roots and to reassure the local inhabitants.

In constructing the repository, a number of other factors need to be considered:

- The shielding effect of the repository construction should be sufficient that the contamination in the repository does not contribute significantly to the external dose rate in the area. As the shielding of the radiation provided by the contaminated soil itself ('self-attenuation') is great, this problem can largely be overcome even with very simple repository designs. An example of this is the formation of simple, uncovered waste pile 'hills' in connection with a decontamination exercise in the Chernobyl-contaminated Novozybkov area in Russia in 1995 (Roed et al., 1996b). The primary radionuclide of concern here was ^{137}Cs. It was found that the dose rate to a person standing on top of one of these hills containing contaminated topsoil removed from a vast area was only 15% higher than that in the surrounding contaminated area. By covering the contamination with, for example a layer of uncontaminated soil excavated from deeper soil layers of the same area, the dose rate from the contaminated topsoil can be greatly reduced. Further, the formation of a 'hill' or bank of earth in the area will shield well against radiation from contamination far away.
- The ecological characteristics in the area should be assessed. The construction of the repository should, as far as reasonably possible, prevent intrusion by humans, animals and the roots of plants. In practice, this means that the waste should be stored under some suitable protective layers, for example soil and plastic membranes.
- Migration of radionuclides from the waste with rainwater must be prevented. This can be done by draining the rainwater off in a gravel layer on top of the waste layer. Also flow barriers made of clay, which

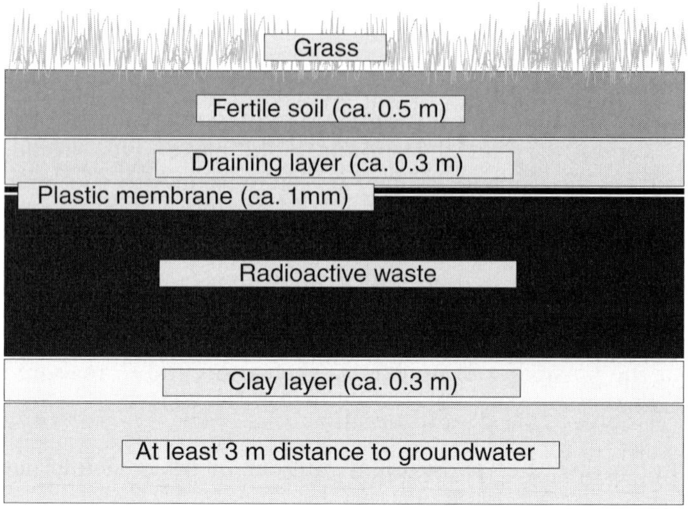

Figure 8 Example of a repository for radioactively contaminated waste (e.g. soil), based on the recommendations of Junker et al. (1998). Other variants of surface mound/trench repositories are described by Salbu et al. (1994).

effectively retains many contaminant ions (especially caesium – see Chapter 5), and thick plastic are useful. A ditch should be dug around the repository to collect the drained-off rainwater. If the repository is arched at the top, the rainwater will conveniently run off the repository side and end up in the ditch.

- The repository should be protected against erosion. This can, for instance, be accomplished by growing grass in the top soil layer of the repository.
- The acceptability of constructing repositories, possibly on private property, needs to be assessed. Communication issues are of great importance, in carefully explaining the implications and possible alternatives. Information leaflets may be produced in advance of an incident.
- The repository may be equipped with a drainpipe at the bottom for inspection of the radionuclide contents in any water passing through the construction.
- The design should be simple, inexpensive and possible to construct over a short time. It has been estimated that the construction of a simple surface mound or trench repository (based on the principles shown in Figure 8) for ca. 500 m^3 solid waste would cost some 3,000 Euro (Andersson et al., 2003).

Alternative disposal strategies may be considered for special cases:

- If the contamination is present in a material which retains it well, contaminant migration would not be an issue that calls for special action. This may for instance be the case if a contaminated roof is replaced. The removed roof tiles (and most other common building materials) would be particularly effective in retaining deposited caesium ions. Indeed, it may take many years for wind and weather to halve the caesium contamination level (see Chapter 5). This means that roof tiles contaminated with caesium may be stored in piles in a restricted area without much risk of contaminant migration. It should be recognised that for instance burial of asbestos products could be problematic and illegal, due to the toxicity of this material. Also the downward migration of contaminants (particularly caesium) in soils would be very limited, primarily due to the presence of essentially the same clay minerals as are present in clay, concrete and slate building materials.
- Granulated asphalt waste generated from a road surface by road planing may be mixed in new hot asphalt and re-applied on a road. The dilution together with the shielding in the asphalt material will mean that the radiation from a road paved with the mix will be much less than that from the planed road, where virtually all contamination will be on top of the surface. The technology known as 'mixing in place' (mixing *on site* and applying just after planing) can, however, not be recommended as it should be carefully assessed whether the volumes of new asphalt would dilute the contamination sufficiently.
- Removal of snow in an urban or industrial area may lead to extremely large amounts of waste (Qvenild and Tveten, 1984), which cannot realistically be melted and decontaminated (e.g. by ion exchange and/or filtration). The snow masses may be dumped in the oceans, where the impact on the ecosystem would be considered to be limited. It should be ensured that the snow is not disposed of in enclosed water bodies, for example lakes, where the waste may give rise to significant sediment contamination problems or lead to contamination of drinking water. As the snow may thus need to be transported over large distances, the transport expenses would often be high.

Figure 9 shows an overview diagram of the most important waste management decisions that might need to be made in connection with introduction of countermeasures in an inhabited area.

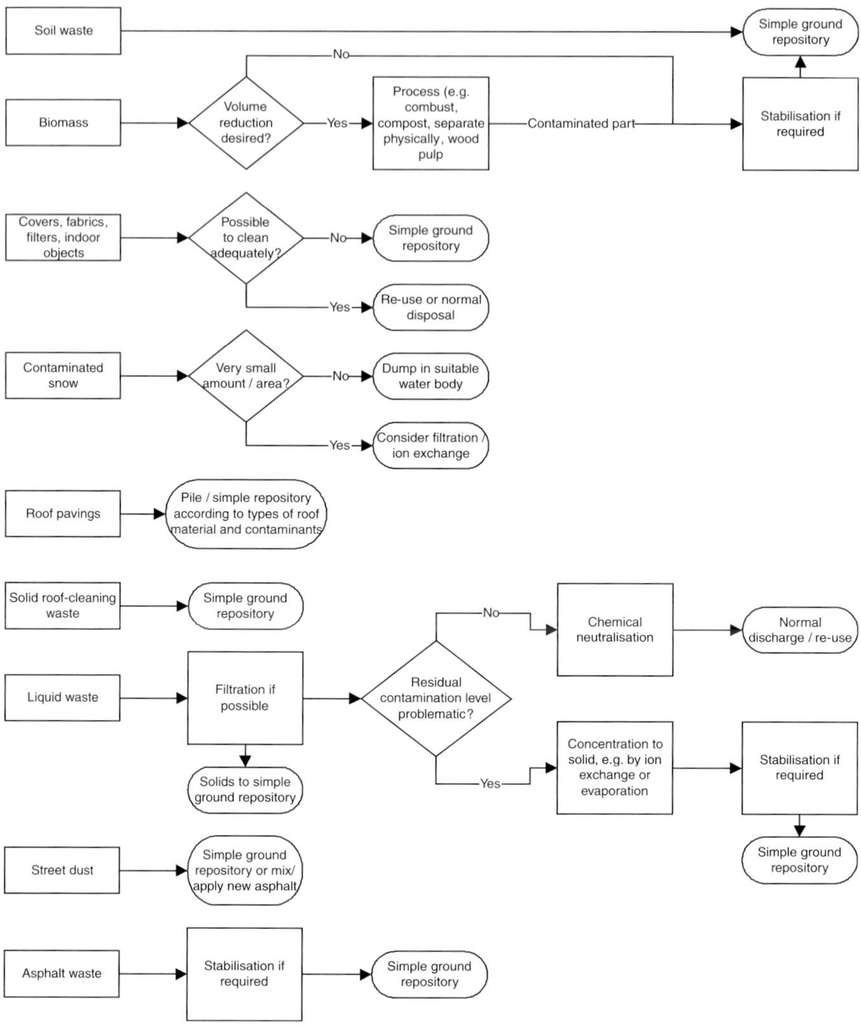

Figure 9 Suggested management schemes for different countermeasure wastes. It is assumed that the various types of wastes contain high levels of long-lived contaminants and thus require special management (from Brown et al., 2007).

ACKNOWLEDGEMENTS

Valuable commenting and advice from Dr. Joanne Brown, Health Protection Agency, UK, during the drafting of this chapter is gratefully acknowledged. The photographs shown in this chapter were taken by Henrik Prip, Risø National Laboratory for Sustainable Energy, Technical University of Denmark, and are reproduced with kind permission.

REFERENCES

Andersson, K. G. (1996). *Evaluation of Early Phase Nuclear Accident Clean-Up Procedures for Nordic Residential Areas.* NKS Report NKS/EKO-5(96)18, ISBN 87-550-2250-2.

Andersson, K. G. (2005). Dose contributions from malicious radionuclide dispersion in an urban area. In: *Radiological Protection in Transition, Proceedings of the XIV Regular Meeting of the Nordic Society for Radiation Protection*, Rättvik, Sweden, 27–31 August 2005, SSI Report 2005:15, Swedish Radiation Protection Authority, Stockholm, Sweden, ISSN 0282-4434, pp. 59–62.

Andersson, K. G., J. Brown, K. Mortimer, J. A. Jones, T. Charnock, S. Thykier-Nielsen, J. C. Kaiser, G. Proehl, and S. P. Nielsen. (2008a). New developments to support decision-making in contaminated inhabited areas following incidents involving a release of radioactivity to the environment. *Journal of Environmental Radioactivity*, 99, 439–454.

Andersson, K. G., C. L. Fogh, and J. Roed. (1999). Occupational exposure at a contemplated Belarussian power plant fired with contaminated bio-mass. *Radiation Protection Dosimetry*, 83(4), 339–344.

Andersson, K. G., T. Mikkelsen, P. Astrup, S. Thykier-Nielsen, L. H. Jacobsen, S. C. Hoe, and S. P. Nielsen. (2008b). 'Dirty bomb' attack in a city area: Requirements to estimate consequences of contaminant dispersion. *Proceedings of the International Conference on Radioecology & Environmental Radioactivity* (Eds P. Strand, J. Brown and T. Jølle), 15–20 June 2008, Bergen, Norway, Oral and Oral Poster Presentations, Part 1, Norwegian Radiation Protection Authority, Østerås, Norway, ISBN 978-82-90362-25-1, pp. 44–47.

Andersson, K. G., J. Roed, M. A. Byrne, H. Hession, P. Clark, E. Elahi, A. Byskov, X. L. Hou, H. Prip, S. K. Olsen, and T. Roed. (2004). *Airborne Contamination in the Indoor Environment and its Implications for Dose.* Risø-R-1462(EN). Risø National Laboratory, Roskilde, Denmark, ISBN 87-550-3317-2.

Andersson, K. G., J. Roed, K. Eged, Z. Kis, G. Voigt, R. Meckbach, D. H. Oughton, J. Hunt, R. Lee, N. A. Beresford, and F. J. Sandalls. (2003). *Physical Countermeasures to Sustain Acceptable Living and Working Conditions in Radioactively Contaminated Residential Areas.* Risø-R-1396(EN). Risø National Laboratory, Roskilde, Denmark, ISBN 87-550-3190-0.

Anisimova, L. I., A. V. Chesnokov, A. P. Govorun, O. P. Invanov, B. N. Potapov, S. B. Shcherbak, L. I. Urutskoev, V. I. Kovalenko, C. V. Krivonosov, A. V. Ponomarev, V. P. Ramzaev, and V. A. Trusov. (1994). *Analysis of Some Countermeasures Efficiency Based on Radioecological Data, Deposit Measurements and Models for External Dose Formation. Case Study of Two Russian Settlements Yalovka and Zaborie.* Annual Progress Report of the ECP4 Project supported by the European Commission, EMERCOM, Moscow, Russia.

Balonov, M., E. Holm, F. Gera, A. V. Panfilov, G. Shaw, C. Torres Vidal, E. Uspenskaya, and A. Venter. (2003). *Methodology for Defining Intervention Exemption Levels of Radionuclides in Timber and Wood Products.* IAEA TECDOC Draft Report. International Atomic Energy Agency, Vienna.

Brodersen, K. (1993). *Cement Solidification of Soil and Interactions Between Cement and Radioactive Soil.* Risø-I-721(EN). Risø National Laboratory, Roskilde, Denmark.

Brown, J., K. Mortimer, K. Andersson, T. Duranova, A. Mrskova, R. Hänninen, T. Ikäheimonen, G. Kirchner, V. Bertsch, F. Gallay, and N. Reales. (2007). *Generic Handbook for Assisting in the Management of Contaminated Inhabited Areas in Europe Following a Radiological Emergency.* Parts I–V. Final Report of the EC-EURANOS Project Activities CAT1RTD02 and CAT1RTD04, EURANOS project web site, www.euranos.fzk.de

Clark, D. E., and W. C. Cobbin. (1964). *Removal of Simulated Fallout from Pavements by Conventional Street Flushers*. US Naval Radiological Defense Lab, San Francisco, CA, USNRDL-TR-797.

Dorrian, M. D. (1997). Particle size distributions of radioactive aerosols in the environment. *Radiation Protection Dosimetry*, 69(2), 117–132.

Eged, K., Z. Kis, G. Voigt, K. G. Andersson, J. Roed, and K. Varga. (2003). *Guidelines for Planning Interventions against External Exposure in Industrial Area after a Nuclear Accident – Part I: A Holistic Approach of Countermeasure Application*. GSF-Bericht 01/03, GSF-Forschungszentrum für Umwelt und Gesundheit GmbH, Neuherberg, Germany, ISSN 0721-1694, 67pp.

Fogh, C. L., K. G. Andersson, A. N. Barkovsky, A. S. Mishine, A. V. Ponamarjov, V. P. Ramzaev, and J. Roed. (1999). Decontamination in a Russian settlement. *Health Physics*, 76(4), 421–430.

Garland, J. A., and I. R. Pomeroy. (1994). Resuspension of fall-out material following the Chernobyl accident. *Journal of Aerosol Science*, 25(5), 793–806.

Harper, F. T., S. V. Musolino, and W. B. Wente. (2007). Realistic radiological dispersal device hazard boundaries and ramifications for early consequence management decisions. *Health Physics*, 93(1), 1–16.

Hession, H., M. A. Byrne, S. Cleary, K. G. Andersson, and J. Roed. (2006). Measurement of contaminant removal from skin using a portable fluorescence scanning system. *Journal of Environmental Radioactivity*, 85(2–3), 196–204.

Howard, B. J., N. A. Beresford, A. Nisbet, G. Cox, D. H. Oughton, J. Hunt, B. Alvarez, K. G. Andersson, A. Liland, and G. Voigt. (2005). The STRATEGY project: Decision tools to aid sustainable restoration and long-term management of contaminated agricultural ecosystems. *Journal of Environmental Radioactivity*, 83, 275–295.

Hubert, P., L. Annisomova, G. Antsipov, V. Ramsaev, and V. Sobotovitch. (1996). *Strategies of Decontamination*. Experimental Collaboration Project 4, European Commission. EUR 16530 EN, ISBN 92-827-5195-3.

IAEA. (1993). *Feasibility of Separation and Utilization of Caesium and Strontium from High Level Liquid Waste*. Technical Report Series No. 356. International Atomic Energy Agency, Vienna.

ICRP. (1999). Protection of the public in situations of prolonged radiation exposure. The application of the Commission's system of radiological protection to controllable radiation exposure due to natural sources and long-lived radioactive residues, ICRP Publication 82. *Annals of the ICRP*, Vienna, Austria, ISSN 0146-6453.

ICRP. (2007). The 2007 recommendations of the International Commission on Radiological Protection, ICRP Publication 103. *Annals of the ICRP*, Vienna, Austria, ISSN 0146-6453.

Ivanov, Y. A., N. Lewyckyj, S. E. Levchuk, B. S. Prister, S. K. Firsakova, N. P. Arkhipov, A. N. Arkhipov, S. V. Kruglov, R. M. Alexakhin, J. Sandalls, and S. Askbrant. (1997). Migration of ^{137}Cs and ^{90}Sr from Chernobyl fallout in Ukrainian, Belarussian and Russian soils. *Journal of Environmental Radioactivity*, 35(1), 1–21.

Junker, H., J. M. Jensen, H. B. Jørgensen, J. Roed, K. G. Andersson, P. D. Kofman, E. Bøllehuus, L. Baxter, and A. Grebenkov. (1998). *Chernobyl Bioenergy Project*. Final Report, Phase 1, ELSAMPROJEKT Report No. EP 8204-98-n600hju.

Menzel, R. G., H. V. Eck, P. E. James, and D. E. Wilkins. (1968). Reduction of strontium-85 uptake in field crops by deep-ploughing and sodium carbonate applications. *Agronomy Journal*, 60, 499–502.

Nisbet, A. F., J. A. Mercer, N. Hesketh, A. Liland, H. Thørring, T. Bergan, N. A. Beresford, B. J. Howard, J. Hunt, and D. H. Oughton. (2004). *Datasheets on Countermeasures and Waste Disposal Options for the Management of Food Production Systems*

Contaminated following a Nuclear Accident. NRPB-W58. National Radiological Protection Board, Chilton, UK, ISBN 0 85951 539 7.

Ogata, Y., and K. Nishizawa. (1999). Stirring system for radioactive waste water storage tank. *Health Physics*, 77, 89–96.

Qvenild, C., and U. Tveten. (1984). *Decontamination and Winter Conditions*. Institute for Energy Technology, Kjeller, Norway, ISBN 82-7017-067-4.

Ramzaev, V., K. G. Andersson, A. Barkovsky, C. L. Fogh, A. Mishine, and J. Roed. (2006). Long-term stability of decontamination effect in recreational areas of the town Novozybkov, Bryansk Region, Russia. *Journal of Environmental Radioactivity*, 85(2–3), 280–298.

Ravila, A., and E. Holm. (1994). Radioactive elements in the forest industry. *Science of the Total Environment*, 157, 339–356.

Reineking, A., K. H. Becker, J. Porstendörfer, and A. Wicke. (1987). Air activity concentrations and particle size distributions of the Chernobyl aerosol. *Radiation Protection Dosimetry*, 19(3), 159–163.

Roed, J. (1982). *Reduktion af dosis ved nedpløjning af gamma-aktive isotoper (in Danish)*. Risø-M-2275. Risø National Laboratory, Roskilde, Denmark, ISBN 87-550-0916-6.

Roed, J., and K. G. Andersson. (1996). Clean-up of urban areas in the CIS Countries contaminated by Chernobyl fallout. *Journal of Environmental Radioactivity*, 33(2), 107–116.

Roed, J., K. G. Andersson, A. N. Barkovsky, C. L. Fogh, A. S. Mishine, S. K. Olsen, A. V. Ponomarjov, H. Prip, V. P. Ramzaev, and B. F. Vorobiev. (1998). *Mechanical Decontamination Tests in Areas Affected by the Chernobyl Accident*. Risø-R-1029. Risø National Laboratory, Roskilde, Denmark, ISBN 87-550-2361-4.

Roed, J., K. G. Andersson, A. N. Barkovsky, C. L. Fogh, A. S. Mishine, A. V. Ponamarjov, and V. P. Ramzaev. (2006). Reduction of external dose in a wet-contaminated housing area in the Bryansk Region, Russia. *Journal of Environmental Radioactivity*, 85(2–3), 265–279.

Roed, J., K. G. Andersson, C. L. Fogh, A. N. Barkovski, B. F. Vorobiev, V. N. Potapov, and A. V. Chesnokov. (1999). Triple digging – A simple method for restoration of radioactively contaminated urban soil areas. *Journal of Environmental Radioactivity*, 45(2), 173–183.

Roed, J., K. G. Andersson, and H. Prip (Eds) (1995). *Practical Means for Decontamination 9 Years after a Nuclear Accident*. Risø-R-828(EN). Risø National Laboratory, Roskilde, Denmark, ISBN 87-550-2080-1, ISSN 0106-2840.

Roed, J., K. G. Andersson, and H. Prip. (1996a). The Skim and burial plough: A new implement for reclamation of radioactively contaminated land. *Journal of Environmental Radioactivity*, 33(2), 117–128.

Roed, J., C. Lange, K. G. Andersson, H. Prip, S. Olsen, V. P. Ramzaev, A. V. Ponomarjov, A. N. Barkovsky, A. S. Mishine, B. F. Vorobiev, A. V. Chesnokov, V. N. Potapov, and S. B. Shcherbak. (1996b). *Decontamination in a Russian Settlement*. Risø-R-870. Risø National Laboratory, Roskilde, Denmark, ISBN 87-550-2152-2.

Rulik, P., I. Bucina, and I. Malatova. (1989). Aerosol particle size distribution in dependence on the type of radionuclide after the Chernobyl accident and in the NPP effluents. *Proceeding of the XVth Regional Congress of IRPA in Visby*, Sweden, September 1989.

Salbu, B., B. Braskerud, G. Østby, and D. Oughton. (1994). Experimental studies on the removal and disposal of radioactive contaminated soil from agricultural areas. *Cleanup of Large Radioactive-Contaminated Areas and Disposal of Generated Waste* (Ed. J. Lehto). Final Report of the KAN2 Project, TemaNord 1994:567, ISBN 92 9120 488 9.

Turner, A. D., N. J. Bridger, C. P. Jones, J. S. Pottinger, A. R. Junkison, P. A. Fletcher, M. D. Neville, P. M. Allen, R. I. Taylor, W. T. A. Fox, and P. G. Griffiths. (1994). *Electrochemical Ion-Exchange for Active Liquid Waste Treatment.* European Commission, Report EUR 14997 EN, ISBN 92-826-7372-3.

Zakrzewska-Trznadel, G., M. Harasimowicz, and A. G. Chmielewski. (2001). Membrane processes in nuclear technology-application for liquid radioactive waste treatment. *Separation and Purification Technology*, 22–23, 617–625.

CHAPTER 9

Non-Radiological Perspectives: Holistic Value Assessment of Countermeasure Strategies

Deborah H. Oughton[1,*] *and* Ellen-Marie Forsberg[2]

Contents

*Corresponding author. Tel.: +47 64965544; Fax: +47 64 96 70 01
E-mail address: deborah.oughton@umb.no

[1] Norwegian University of Life Sciences, N-1432 Aas, Norway
[2] Work Research Institute, St Olavs plass, N-0130 Oslo, Norway

Radioactivity in the Environment, Volume 15
ISSN 1569-4860, DOI 10.1016/S1569-4860(09)00409-4

1. Introduction

Over the past decade, people involved in radiation protection policy have shown an increased awareness that risk management needs to take account of social and ethical issues. As mentioned in Chapter 8, it is now generally accepted that the costs of remediation can be greater than the financial investment required to carry out countermeasures, and that the benefits of some actions may extend beyond dose reduction. It is recognised that these additional costs and benefits may influence the social and ethical acceptability of countermeasures (e.g. Oughton, 1996; Morrey and Allen, 1996; Hériard Dubreuil et al., 1999; Kelly, 2001; Nisbet and Mondon, 2001). The social and ethical aspects should, therefore, be taken into account when deciding on appropriate countermeasure strategies and social and ethical criteria should be included in countermeasure evaluation. However, this inclusion is not without difficulties. For example, problems may arise due to the many dimensions to the criteria, complexity (many countermeasures have both positive and negative social and ethical consequences), various 'trade-offs' that may be required, a possible lack of agreement within society on what is practical or acceptable, and the lack of established procedures, and experience, in systematically incorporating these dimensions in decision-making.

If decisions about countermeasure selection are going to be ethically and rationally defendable, decision-makers require advice on what criteria are important to consider and why, and also a methodology to ensure a systematic and transparent procedure for balancing these criteria. This chapter will discuss these criteria and suggest a methodology for evaluating such criteria (i.e. an ethical matrix). Also, the chapter will introduce examples of 'social countermeasures' where the prime aim is not dose reduction.

2. Holistic Assessment of Countermeasures

In radiation protection, the primary objective of countermeasure implementation is usually dose reduction. 1991 ICRP recommendations state that, for an action to be justified, the benefits from dose reduction or averted dose should outweigh the costs of implementing the countermeasure. More recent recommendations acknowledge that the costs of countermeasures may be both social and economic, and that there may be benefits other than dose reduction (ICRP, 2000). Moreover, these ethical, legal and social factors will bear on the acceptability and effectiveness of countermeasures. Therefore, they are issues that should be taken into account in the overall balancing and judgement of the negative and positive attributes of a countermeasure. In other instances, they are constraints that veto the action outright, irrespective of its potential cost-effectiveness.

It should be noted that investing money to reduce exposure to radiation is a trade-off in itself, and yet this relatively simple trade-off is not without controversy in radiation protection. Even based on the two criteria of monetary cost and dose, a decision on how to go about reducing exposure to radiation will be an ethical judgement: we are making choices about which health risks to reduce and at what cost. It is clear that countermeasures may impact on social values in a number of ways (Hunt and Wynne, 2002). Moreover, they may have different impact on different stakeholder groups. Accordingly, deciding on a remediation strategy based on a multi-criteria evaluation is going to require a whole suite of trade-offs and value judgements.

Stakeholder evaluation of countermeasure templates in the STRATEGY project (Sustainable Restoration and Long-Term Management of Contaminated Rural, Urban and Industrial Ecosystems)[1] suggested that many countermeasures (especially in the UK) were as likely to be rejected on socio-ethical grounds as technical and economic grounds (Nisbet, 2002). Socio-ethical issues will, therefore, influence the potential efficiency of implementing the countermeasures. Accordingly, it is important to undertake an evaluation of the socio-ethical aspects of implementation of the countermeasure in the specific local context in which they may be used.

[1] Financed under the EU 5th Framework Program. See www.strategy-eu.org.uk

It is important to design a thorough countermeasure evaluation process in order to create robust countermeasure strategies that are not vulnerable to public resistance. This process in itself needs to be holistic and able to survive public scrutiny.

2.1. Assessing countermeasure strategies

Countermeasures are not used in isolation, but combined into a holistic strategy. Hence, the strategy's *overall* impact on a region's population and environment also needs to be evaluated. Whilst individual countermeasures may be satisfactory (from both an efficiency viewpoint and from a social and ethical point of view), their combined use within a restoration strategy may lead to an unfair distribution of costs, doses and benefits across the population. On the other hand, countermeasures can be combined to ensure a more even distribution, or adjusted to ensure that minority populations have either consented to, and/or been compensated for, any prior inequity. Whether or not a countermeasure is acceptable in practice will also depend on the available alternatives. A viable countermeasure may become inappropriate if there is a less ethically problematic or socially intrusive alternative. The STRATEGY project constructed a framework to assess countermeasure *strategies* on a number of evaluation criteria, including technical and economic factors, as well as practicality and acceptability, socio-ethical aspects, environmental consequences and indirect side-effect costs (Howard et al., 2002). As mentioned in Chapter 8, this multi-criteria assessment has been carried forward in the 6th Framework EURANOS project and handbook (EURANOS, 2006). In summary, the concept of holism, in the STRATEGY context, can be understood as pointing both to assessing a selection of countermeasures as a whole and compared to alternative strategies, as well as assessing countermeasures by a wider set of evaluation criteria. Ethical evaluation of a holistic remediation strategy will need to build on the assessment of individual countermeasures and their interactions. As a procedure for ensuring transparent and systematic consideration of ethical issues, we suggest the use of an ethical matrix. This method will be presented below. First, however, we would like to shed more light on some of the most important ethical issues with regard to remediation of contaminated areas.

3. General Ethical Issues

A holistic assessment of a countermeasure strategy involves taking a wide variety of ethical concerns into account. Many of the issues will be relevant for any risk assessment, including other areas of radiation protection (Shrader-Frechette, 1991; Oughton, 1996, 2000; Shrader-Frechette and Persson, 1997; Oughton et al., 2004). Taking ethical considerations into account does not necessarily mean to introduce concerns previously thought as external to countermeasure assessment. For instance, dose reduction is in itself a very fundamental ethical concern, as is economic costs (that may have large implications for the welfare of societies).[2] When concerns here are labelled *ethical*, it is a way to bring out explicitly the value dimensions that are inherent to choices of countermeasures and countermeasure strategies. Some important issues are discussed below and summarised in the box.

> *Important ethical issues to consider when deciding on countermeasure strategies:*
>
> - Disruption of everyday life and self-help
> - Free informed consent of workers (to risks of radiation exposure and/or chemical exposure) and consent of private owners for access to property
> - Distribution of dose, costs and benefits
> - Liability and/or compensation for unforeseen health or property effects
> - Animal welfare issues
> - Change in public perception or use of an amenity
> - Uncertainty
> - Environmental risk from ecosystem changes, groundwater contamination, etc.
> - Environmental consequences of waste generation and treatment (chemical and radioactive).

3.1. Disruption of everyday life and self-help

Countermeasures vary enormously in the amount of disruption they cause, ranging from total social upheaval (such as imposed relocation) to minimal changes (e.g. dietary advice). The degree of 'self-help' considers the extent to which the affected persons themselves can implement countermeasures, and their degree of control or choice over the situation. Voluntary countermeasures that are carried out by the public or affected individuals themselves, or that increase personal understanding or control over the situation, are usually deemed positive as the action respects the fundamental ethical values of autonomy (i.e. respect for freedom of choice of individuals), liberty and dignity. Concrete examples include the provision of counting equipment, as well as dietary and behavioural advice. On the

[2] The widespread disruption, economic burdens and social breakdown attributable to the Chernobyl accident in Belarus, Ukraine and Russia are poignant examples. In 1991, Belarus was spending 20% of its gross national product on remediation and compensation, and Ukraine lost more than of 30% of its agricultural land, leading to widespread cultural upheaval, particularly in vulnerable rural communities (UNDP, 2002; Bay and Oughton, 2005).

contrary, imposed countermeasures that are highly disruptive, infringe upon liberty or restrict normal practices are usually judged to be negative. Examples would include relocation, bans on amenity use or other radical changes in everyday routines.

3.2. Free informed consent of workers (to risks of radiation exposure and/or chemical exposure) and consent of private owners for access to property

The issue of consent is strongly linked to the fundamental ethical value of autonomy. Employers have a duty to obtain the informed consent of any worker who may be exposed to chemical and/or radiation risk. This is particularly important if lower paid workers (e.g. cleaners for industrial countermeasures) are employed to carry out the measure, as it has been suggested that the necessary conditions for free informed consent are often violated for these groups (Bullard, 1990; Shrader-Frechette, 2001). The increased risk may justify some form of compensation via higher wage premiums, but the opportunity for certain sub-groups of the population to make a profit at the expense of others can have negative social consequences (e.g. increased inequity – see below). Furthermore, compensation itself can raise questions of whether or not this may coerce people into taking risks they would otherwise not have (Bullard, 1990; Rawles, 2002).

3.3. Distribution of dose, costs and benefits

Although a countermeasure may be expected to provide a net benefit to society, it can also influence the distribution of costs, risks and benefits. The way in which a countermeasure may influence this distribution has significance due to the fundamental ethical values of equity, justice and fairness. Costs, benefits and risks may vary over both space and time, and between different members of a community. Dose distribution is obviously a main consideration for radiation protection, and many countermeasures that reduce collective dose (man Sv) may also change the distribution of dose, for example, from inhabitants in residential areas, children using playgrounds and football fields and other users of public amenities to liquidators, workers or populations around waste facilities.

The question of who is paying the monetary and social costs of the countermeasure and who will receive the benefits must also be addressed. Some countermeasures, and sets of countermeasures, result in an equitable distribution of cost and dose reduction, such as investment by tax payers to reduce activity concentrations in a common food product; others are less equitable, for example, when a reduction of dose to the majority is only possible at the expense of a higher dose, cost or welfare burden, on a minority (like the burden that has to be borne by populations around waste facilities). Another question is whether the countermeasure has implications for vulnerable or already disadvantaged members of society (children, ethnic or cultural minorities). Finally, some members of the population may profit significantly from the countermeasure implementation, such as the owners or hirers of heavy machinery. In small communities, this can be particularly problematic, as was seen in the case of foot-and-mouth disease in the UK (DEFRA, 2002).

3.4. Liability and/or compensation for unforeseen health or property effects

Employers usually hold legal and ethical responsibilities over their employees, and contractors or industries may be held legally or financially liable for any damage they may cause to public or private property. The matter of who bears liability is relevant both from the perspective of responsibility (moral and legal) and because of links to equity issues. Liability can become particularly important if outside contractors are paid to carry out the countermeasure, both for the contractor themselves – Will I be sued if the countermeasure causes unforeseen damage? – and the workers/property owners who may risk injury – Will I be compensated if the countermeasure causes me damage?

3.5. Animal welfare issues

Animal welfare concerns the amount of suffering the countermeasure may inflict on non-human sentient animals. In the context of countermeasures for inhabited areas, this will concern pets mainly.[3] For example, a ban on use of recreational areas may have implications for dogs. There is a long-standing philosophical debate around the question of why one should value non-human living beings, and whether or not they have moral standing (Regan and Singer, 1989; Oughton, 2003). Nevertheless, in many countries, animal welfare issues are covered by law, and may result in both legal and ethical constraints on some countermeasures.

3.6. Change in public perception or use of an amenity

If a countermeasure has some effect on the public's use of a particular amenity (such as a park), then this will have an influence on the acceptability of that countermeasure. But such effects can have deeper relevance than whether or not people are able to use the amenity. Perceptions can include, for example that something has changed from being 'natural' to 'unnatural' or from 'clean' to 'damaged'. This is an issue

[3] Although, it will also apply to other animals, not least birds like swans, sparrows, pigeons, etc.

with which the public has a strong tendency to attach moral relevance (Thompson, 1997), and certainly impacts upon people's sense of their quality of life. Some countermeasures may threaten botanical gardens that may contain unique flora and may be an important part of the cultural heritage and identity of a city. Similarly, implementing invasive countermeasures in botanical gardens or parks may be a very visible sign of disruption and destruction of social identity.

3.7. Uncertainty

Uncertainty in this context can be taken to refer to an evaluation of the risk (environmental, technical or social) associated with the countermeasure, and relate to the question of what the actual consequences of the countermeasure implementation will be and the probability that that outcome will occur. Uncertainties can arise due to variability (e.g. environmental factors influencing ecosystem transfer of radionuclides, or differences in individual susceptibility to disease) and statistical error. In some cases, uncertainties can be reduced by further research, thus knowledge from previous experience of countermeasures will be important, but there may also be consequences that are inevitably indeterminate. How one deals with uncertainty is an ethical issue. People's acceptance of measures with uncertain consequences vary. The principle of respect for autonomy, as well as the precautionary principle, require that measures with uncertain consequences should be discussed with those that may be affected.[4] In evaluating countermeasures one needs to consider what the main uncertainties associated with the countermeasure are, as well as what action might be taken to avoid or reduce these uncertainties and what the consequences of being wrong are.

3.8. Environmental risk from ecosystem changes, groundwater contamination, etc.

Countermeasures that change or interfere with ecosystems (e.g. removing or adding topsoil) may produce negative environmental consequences. In addition to the obvious questions of uncertainty, environmental risk raises a variety of ethical issues including consequences for future generations, sustainability, cross-boundary pollution and balancing harms to the environment/animals against benefits to humans. The ethical acceptability of the countermeasure will be highly dependent on the ecological status of the area and the degree to which the action diverges from usual practice. In most cases, environmental legislation must be considered.

[4] For a clear statement of the Precautionary Principle, please confer UNESCO-COMEST 2005: http://unesdoc.unesco.org/images/0013/001395/139578e.pdf

3.9. Environmental consequences of waste generation and treatment (chemical and radioactive)

Countermeasures that produce waste will raise both equity and environmental risk issues. Waste disposal can lead to a 'redistribution' of radiation exposures, and the environmental impact of disposal sites will need to address both ethical and legal implications. For example, balancing human (present and future generations) and non-human (animal, environment) interests. Hence, any countermeasure involving the generation of waste and/or its treatment will have ethical relevance, and controversy, in itself. Treatment of waste *in situ* can be positive as it avoids problems arising from redistribution of exposures to persons living close to disposal site. But, *in situ* treatment may also have negative side-effects due to complicating future waste removal or 'causing' contamination of soil. In the STRATEGY project, the handling of waste was one issue that showed large cultural variability, particularly the values attributed to 'dilute and disperse' measures such as land spreading of contaminated milk or discharging to sea (Nisbet, 2002).

4. The Ethical Matrix as a Case Specific Tool for Mapping Ethical Concerns

As the previous sections have shown, there are a number of ethically relevant issues that need to be taken into account. However, in most complex decision-making situations all concerns cannot be respected to the same degree and values may come into conflict. Nevertheless, these considerations can be understood as *prima facie*, that is they are important concerns that one is obliged to take into consideration and that require explicit justification if they are violated. For example, the serious social and ethical disruption caused by forced evacuation can only be justified if the dose reduction to populations is large enough. While the overarching concept is familiar from the principle of justification in radiation protection, balancing values is rarely straightforward, particularly if the final decision is to be transparent and ethically justifiable.

Good practical ethical decision-making is built upon three conditions: high-quality and relevant information (facts), ethical argument informed by relevant ethical theories (values) and moral judgement. In pluralistic societies, views on both facts and values typically differ, which makes it difficult to ascertain that all relevant information has been collected and that there is no bias towards particular ethical approaches or particular scientific viewpoints. An ethical matrix is a tool to ensure the systematic consideration of all affected values and to indicate what facts are needed for making the decision (Kaiser and Forsberg, 2001).

The matrix derives from principle-based ethics. Rather than relying on an overarching ethical theory to evaluate actions (usually resorting to an appeal to either utilitarian or deontological doctrines), a principle-based approach holds that some general moral norms are central in moral reasoning, and that these may be constructed as principles or rules. There is a variety of principle-based approaches and, arguably, the most successful in practical ethics are those that adopt a pluralistic view, and start with a selection of principles that are generally acknowledged in society and that can find a broad degree of support from different ethical theories or cultural beliefs. One of the most famous and widely used examples of principle-based ethics is that of Beauchamp and Childress (2001), who identified four basic principles that form the foundation of modern medical ethics.

Basic ethical principles (Beauchamp and Childress, 2001, p. 12):

- Respect for autonomy (a norm of respecting the decision-making capacities of autonomous persons)
- Non-maleficence (a norm of avoiding the causation of harm)
- Beneficence (a group of norms for providing benefits and balancing benefits against risks and costs)
- Justice (a group of norms for distributing benefits, risks and costs fairly)

Inspired by medical ethics, Ben Mepham transferred Beauchamp and Childress' approach to a practical scheme for addressing broader policy-related problems (Mepham, 1995, 1996) (see an example in Table 1). While Beauchamp and Childress addressed issues concerning primarily the relationship between doctor and patient (or the public and the health service), there is a greater complexity in using the approach in societal

Table 1 An example ethical matrix for assessment of dairy technology issues, by Mepham (2004).

	Well-being	Autonomy	Fairness
Dairy farmers	Satisfactory income and working conditions	Managerial freedom of action	Fair trade laws and practices
Consumers	Food safety and acceptability. Quality of life	Democratic, informed choice, for example of food	Availability of affordable food
Dairy cows	Animal welfare	Behavioural freedom	Intrinsic value
The biota	Conservation	Biodiversity	Sustainability

decision-making, which has to cover a much wider area of concern, often including animals and the environment. Mepham suggested that the various principles, values and stakeholders can be best summarised in an ethical matrix. In this way, a bias towards certain kinds of values may be avoided, and the matrix can be used to address conflicts between values in a systematic way, without, necessarily, having to invoke full-fledged theories.

An advantage of the matrix is that it is able to cover most of the relevant values that should be considered in a certain decision area. And although the matrix represents a relatively new methodology, it has been tested within evaluation processes both by Mepham and The Norwegian Research Ethics Committees and has been demonstrated to be a useful tool for use in participatory and deliberative decision processes (e.g. Mepham, 2004; Kaiser and Forsberg, 2001; Kaiser et al., 2007). The conclusion in both Mepham's and NENT's studies was that the ethical matrix works relatively well for the purpose of structuring a discussion. There were, however, certain practical problems connected to specifying the principles and weighting the specifications, agreeing on the facts and undertaking the final evaluation. Also, the ethical matrix is a rather cumbersome tool that takes time to explain to participants. These practical problems are taken up both by Mepham and the NENT project (see the above references). The general conclusion suggests, however, that the advantages in some cases of complex decision-making may outweigh the practical problems related to the process.

The ethical matrix method that is proposed in the STRATEGY project takes its starting point in three fundamental principles, namely:

1. To promote well-being and minimise health risks, welfare burdens and other detriments to affected stakeholders.
2. To respect the dignity of affected stakeholders.
3. To recognise the norm of justice and aim to treat everybody fairly and ensure an equitable distribution of goods among affected stakeholders.

In STRATEGY, a matrix for assessing agricultural countermeasures was developed with detailed specification of these three general principles as derived by a number of co-operating researchers and stakeholders in STRATEGY. It encompassed all the values that were considered relevant. Many of the issues can be traced to those identified as relevant for the ethical and social evaluation of individual countermeasures, and will have several relevant further specifications according to the particular case in question (Oughton et al., 2004).

Rather than seeing the inclusion of ethical and social values as an additional burden on top of what is already a complicated process, our view is that ethical assessment can be a practical instrument to aid decision-makers. We suggest that a systematic consideration of ethical and social issues, including ethical justification of the decision-making procedure itself (with regard to citizen and stakeholder participation), will make countermeasure selection more transparent and less controversial for society. An ethical matrix is a tool to ensure that all relevant concerns are being taken into consideration and to clarify the ethical basis upon which eventual decisions are made. There is no deduction of the correct ethical answer from the matrix: the ethical matrix is not a substitute for ethical judgement, it is a way of doing it.

In practice, an ethical matrix can aid a decision-making group by giving an overall picture of the ethical status of the issue at stake. The consequences of both the accident itself and countermeasures can affect different groups in different ways, and the matrix can be used to help identify the relevant information required for decision-making (i.e. the facts, values and stakeholders affected). However, a method for getting a grip on facts and values will not be sufficient. Even with all the relevant information on the table and with a systematic representation of different values, moral judgement must be exercised. An important question for any decision-making process is, therefore, that of who is to judge? For decisions that concern the whole of society, stakeholder or lay people involvement is central for ensuring a justified and publicly acceptable conclusion. For an ethical matrix evaluation, it is natural to choose stakeholders as the evaluation is structured around values related to stakeholders. However, lay people have also performed ethical matrix evaluations (see Kaiser et al., 2007). When an ethical matrix is combined with a stakeholder process the real value of both the matrix and the participatory approach appears. We, therefore, recommended that an ethical matrix assessment should be performed in a participatory process where the stakeholders themselves are allowed to define what values are at stake and to contribute to the assessment of the countermeasures. In Section 7, we shall return to the justification for doing the ethical assessments in a participatory process.

5. Application to an Inhabited Area Scenario

Choosing a countermeasure strategy for restoring inhabited areas with a holistic perspective requires an ethical framework. We will now take a closer look at how this framework can be used. Table 2 shows a *value matrix* for inhabited area countermeasures adapted from the one constructed in the STRATEGY project. This value matrix contains the different stakeholder categories (with concrete examples of who these might be) and the three principles already mentioned. The cells under the principles describe the most important values at stake when deciding upon countermeasure strategies for inhabited area scenarios. Note that these include both positive and negative aspects of countermeasures.

This value matrix may be considered a map of the value landscape which can then be used to assess concrete countermeasures. It is constructed at a

Table 2 A proposed value matrix for assessment of countermeasures for inhabited areas.

Stakeholder	Examples	Well-being	Dignity	Justice/distribution
		Example: Health and economic welfare	Example: Choice/consent/(legal) rights	Is any sub-group of stakeholders worst-off?
Owners/employers	Government House dweller Land owner Hotel owner Shop owner Business proprietor Factory owner Local authority	Doses to humans Loss/gain in income Loss of property Damage to, or reduction in value of, property Loss of taxes Compensation	Self-help Consent Property rights Being allowed to pay one's duties Contract fulfilment No disruption No insecurity Liberty	Possibility for conflict between different industries or projects Opportunities for entrepreneurs
Workers/employees	Park workers Factory workers Contractors Immigrant workers Other employees	Doses to humans Fear of job loss Gain/loss of income Insecurity Family relationships Compensation	Traditional skills and practices Consent Training	Possibility for disputes and social inequity
Users/community	Neighbours Recreational users Tourists Public amenity (library, town hall, playground, park) users Local community Garden users	Access Aesthetic value Isolation Community values Tourism Compensation	Respect for public heritage Community sense Personal control Self-help Liberty	Potential conflict of age/sex/cultural minorities Availability of alternative amenities

Table 2. (*Continued*)

Stakeholder	Examples	Well-being	Dignity	Justice/distribution
		Example: Health and economic welfare	Example: Choice/consent/(legal) rights	Is any sub-group of stakeholders worst-off?
Consumers	Kitchen gardeners Allotment owners	Doses to humans More expensive goods Loss of jobs Insecurity	Information Choice Self-help	Potential conflict concerning diet and possibility of self gathering
Future generations	Future users of clean air and water Future users of recreational areas, etc.	Loss of opportunities to use areas, resources, common goods, etc.	Respect for the right to keep living according to basic human values	No one future group be sacrificed for the presumed good of other future groups
Environment	Pets Other biota Ecosystems	Dose to biota Other toxic or beneficial effects Compensation	Endangered species Loss of habitat Right to life consistent with their nature	Potential conflict between ecosystems
Waste location stakeholders (if different from accident location)	Including all the above stakeholders connected to the waste site	Uncertainty/risk estimates: Possibilities for monitoring, retrieval and treatment Compensation	Consent Self-help Information, etc.	Potential conflict between stakeholders close to disposal site Opportunities for entrepreneurs

generic level which allows comparison between different proposed countermeasures (roof cleaning, snow removal, turning flagstones, etc.) and comparison with the 'do nothing' scenario. As briefly noted above, this kind of map would ideally be produced in a participatory process where at least some of the most important stakeholder representatives are included. As a minimum one would invite representatives from industry and private enterprise, from public authorities (regional and local), from workers unions and representatives for home owners and for the environmental movement. One should, therefore, consider this as neither the final identification of principles and stakeholders, nor as a final specified matrix, as this is the task of a participatory decision-making process. One may consider it a starting point for discussion. For the sake of simplicity, a proposed value matrix is often prepared by the organisers in advance of the participatory process, but the assessment group will be expected to revise and adjust the value matrix in order for it to incorporate all relevant values. Table 2 illustrates the range of values that might be considered important when making judgements on countermeasures.

With or without the aid of an ethical matrix, any selection of a restoration strategy will have to rank and weight the importance of the values presented above. If we accept the guidance of an ethical matrix framework, we are able to use the value matrix above to assess concrete countermeasures and countermeasure strategies. To illustrate the method we shall start with assessing two individual countermeasures for inhabited areas: lawn mowing and ploughing. This is done by constructing so-called *consequence matrixes* that show how the values are affected by the countermeasure. The completed consequence matrix can give a picture of how benefits and burdens are distributed across the stakeholders and form the basis for opinions about the overall ethical acceptability of the countermeasure.

5.1. Ethical assessment of lawn mowing

In the case of an airborne radiation deposition to an inhabited area, lawns, parks, football fields, etc. will be contaminated. This represents an important source of contamination for all people, but perhaps in particular for children. Pets may also bring contamination from lawns into the homes. Therefore, the removal of contaminated grass by mowing can be a cost-effective method of reducing doses. However, even this straightforward countermeasure is not without problematic issues. The ethical matrix can provide a systematic guide for highlighting possible issues.[5]

[5] A small-scale holistic assessment of countermeasure strategies including an ethical matrix exercise was carried out by the Norwegian preparedness organisation in a meeting in 2004. The scenario for this meeting was exposure to ionising radiation in a northern region in Norway and numbers and figures for the imagined accident was calculated in advance. With regard to 'urban' exposure three countermeasures were considered for this exercise: lawn mowing, dietary advice and local measuring stations. The analysis here will in part build upon reflections from this meeting and from discussion in the STRATEGY project (see Cox et al., 2005).

First, we should consider the countermeasure more closely.[6] While the primary objective is to reduce the external dose rates, a secondary benefit is a possible reduction of ingestion dose from consumption of kitchen garden products. The countermeasure will primarily be applicable in situations where a significant fraction of the radionuclides is deposited to grass cover rather than to soil (as is generally expected in cases of dry deposition to lawns). In these situations, lawn mowing and disposal of the cut grass may prevent much of the contamination from reaching the underlying soil. This countermeasure can be carried out on a large scale where equipment is or can be made available with short notice. As such equipment is part of the general park and garden maintenance machinery, one can expect this usually to be the case. As the natural transfer of contamination from grass to soil has a half-life of only few weeks (depending largely on the amount of rainfall), lawn mowing must be carried out soon after the contamination (and before the first heavy rainfall). Assuming a dry contaminant deposition, the countermeasure would be expected to achieve a reduction in contamination of 50–90%, although this would be influenced by factors like time and amount of precipitation after contamination, length of grass at time of deposition, evenness of ground surface and grass cutting height and effectiveness. The main technical limitation of the countermeasure is whether it can be implemented before it is rendered ineffective by weathering processes, hence doses may be higher due do exposure to short-lived radionuclides. Some of the ethically relevant consequences are mapped in Table 3.

We see that this seems like a countermeasure that efficiently will reduce dose to humans, animals and the environment at the exposure site. However, those performing the lawn mowing may receive up to three times as much exposure as those that do not. Those transporting the grass may receive up to four times as much. Because the countermeasure must be carried out soon after deposition to be effective, individuals may have increased dose from short-lived radionuclides – doses that could be avoided had persons stayed indoors in the immediate time following deposition. Hence for both workers and for private owners using their own lawn mowers, to be acceptable the task must be voluntary and adequate information must be provided so that informed consent can be given. The countermeasure will also produce large amounts of grass for disposal, creating a waste problem. Hence a burden will be moved from one location to another, which may be perceived as unfair. Choosing waste repository locations will always be subject to the NIMBY reaction (Not In My Back Yard) and there is an inherent moral dilemma in letting one group of

[6] This description is taken from the template on lawn mowing produced in the STRATEGY project, available at http://www-infocris.iaea.org/en/w3.exe$EAFull?ID = 101 [July 2007] and http://www.strategy-ec.org.uk/ [July 2007].

Table 3 Consequence matrix for lawn mowing.

Stakeholder	Examples	Well-being	Dignity	Justice/distribution
		Example: Health and economic welfare	Example: Choice/consent/ (legal) rights	Is any sub-group of stakeholders worst-off?
Owners/ employers	Government House owner Land owner Hotel owner Business proprietor Factory owner Local authority	Minimal loss in income Can give extra doses to owners and the public if the lawn mowing has to be carried out before short-lived radionuclides have decayed? (−)	Enhances self-help and control (+) Is consent required – or is mowing imposed? (?) Respects property rights Allows public to contribute (+) Enhances feeling of security (+)	Assumedly not much conflict potential Possible conflicts if the service is provided to only selected owners
Workers/ employees	Park workers Factory workers Contractors Immigrant workers Other employees	Will give extra doses to those mowing the lawn (−) Will they get special compensation for this?	Information and consent required (!)	No equity consequences (?)
Users/community	Neighbours Recreational users Tourists Public amenity (library, town hall, playground, park) users Local community Garden users	Will enhance access (+) Will maintain aesthetic value (+) Will help maintain tourism (+)	Will be respecting public heritage (+) Will give personal control and self-help (+) Does not restrict liberty (+)	Little conflict potential

Table 3. (*Continued*)

Stakeholder	Examples	Well-being	Dignity	Justice/distribution
		Example: Health and economic welfare	Example: Choice/consent/(legal) rights	Is any sub-group of stakeholders worst-off?
Consumers	Kitchen gardeners Allotment owners	May reduce internal doses slightly (+)	Information about effect of mowing on dose in vegetables required (!)	Little conflict potential
Future generations	Future users of clean air and water Future users of recreational areas, etc.	Enhances use of areas (+)	Allows continued living in the area (+)	No negative distributive effects
Environment	Pets Other biota Ecosystems	Dose to biota reduced (+)	No restrictions or lack of respect for integrity (+)	Little potential conflict
Waste location stakeholders (if different from accident location)	Including all the above stakeholders connected to the waste site	Increased dose to humans and biota (−)	Will they be able to consent? Will they receive compensation? (?) Restricts future use of area (−)	May create conflict between stakeholders close to disposal site (−)

Note: '+' indicates that the consequence is supporting the value, '−' indicates that the consequence is in conflict with the value, '?' indicates a need for more knowledge and '!' indicates a requirement.

people take the burden of the whole society, even if it is for the common good. There is also a moral difference between receiving a burden through an accident, and receiving it by being imposed by a community. There is no easy solution to this problem, but it is important that it is discussed in the group of stakeholders.

There are no significant consequences in the justice column, largely because the countermeasure is in line with what is already done on a routine basis. It is, therefore, unlikely to create new conflicts. The exception is the waste location stakeholders, and possibly some private owners if authorities decide to provide the service to selected owners. In the latter case it is important that the selection criteria are acceptable to the community, for example to the elderly and disabled and not only to those who can pay.

Positive consequences for users reflect the fact that lawn mowing is not an invasive countermeasure. Compared to other possible countermeasures like ploughing, removing topsoil or adding topsoil it maintains the aesthetic character of gardens and parks, thus allowing places of heritage and tourism value to remain unchanged. Moreover, it permits people to act on their own, enhancing their own sense of self-help and control over the situation.

5.2. Ethical assessment of deep ploughing

Ploughing will be a potential countermeasure for any long-lived radionuclide that shows a strong fixation in soils. For example, ^{137}Cs deposited on clays or brown earths can remain in the topmost few centimetres for many years. Deep ploughing can effectively reduce the activity in surface soils, leading also to a reduction in activity in vegetation grown on the soil. Techniques range from simple ploughing using standard equipment, to so-called 'skim-and-burial' which removes the surface layer and buries it at depth, thus reducing the extent to which contamination in the surface layer is mixed into the soil. The depth of ploughing also influences the effectiveness of the countermeasure, as well as impacts on soil fertility. The countermeasure can be carried out with either normal equipment or specialist equipment, but, unlike for rural areas, such measures are not routine in inhabited areas. The objective of the countermeasure is the same as for lawn mowing: to reduce the external dose rate in the area, as well as to reduce ingestion dose from consumption of kitchen garden products. The target is soil areas (e.g. grassed or flowers/vegetables) and other open areas around dwellings. Although the countermeasure is much more invasive than lawn mowing, and creates a larger environmental impact, a practical advantage is that the procedure may have nearly the same effect on dose rate after 1 week as after 2 years, hence it could be accomplished over a long period. It is, therefore, not vulnerable to weather conditions. Furthermore, it does not create waste. Other ethical issues are mapped in Table 4. Although waste is not an issue for the countermeasure, the row is kept to illustrate the differences.

This countermeasure has significantly more adverse effects for the users of these gardens, parks, football fields, etc. There may be a removal of flora and fauna and very visible impacts on cultural heritage. It will also require much more effort, and will almost certainly need to be carried out by contractors or workers. Thus opportunities for self-help and control will be minimal. The same issues regarding dose, consent and compensation to workers are relevant here as was relevant for lawn mowing. Likewise, because the measure is likely to be carried out by authorities, there is a stronger potential for conflict over site selection. On the other hand, doses will be averted more permanently than if simply the grass is cut, and without the creation of waste.

5.3. Ethical assessment of a holistic countermeasure strategy

If we compare these two countermeasures and sort them in a diagram where the pluses, minuses and question marks are seen together for each stakeholder group, the differences in distributional effects of the different countermeasures become clear (Table 5).

It appears that lawn mowing would be the most desirable countermeasure if it was not for the waste problem. How this problem will be solved is important for the ethical acceptability of the countermeasure. When planning, one must take into account that the effect of lawn mowing is highly vulnerable to weather conditions. If a heavy rainfall occurs, this countermeasure may be rendered inefficient, and measures targeted to the soil must rather be used. It should also be noted that the countermeasures are not exclusive, and it is possible that both may be applied.

Of course these are only two countermeasures and there are many more that may achieve the same objective, for example topsoil removal and adding clean soil, reducing the external dose rate in the area and reducing ingestion dose from consumption of kitchen garden products, if produced. In addition there are other inhabited area countermeasures designed for achieving different objectives, for example reducing external dose while also reducing contaminant aerosol concentration in the breathing zone. In real life, a countermeasure strategy would consist of much more than three countermeasures. The STRATEGY project operates with 29 'urban' countermeasures (not including the social countermeasures). A complete strategy for an inhabited area would include countermeasures that are designed to reduce external dose in several ways. When such a strategy is generated an ethical assessment can be performed of all the suggested countermeasures individually. Another option is to limit the use of a countermeasure. For instance, lawn mowing may be used in important parks and in parts of ones garden, and ploughing in places that are not so

Table 4 Consequence matrix for deep-ploughing.

Stakeholder	Examples	Well-being	Dignity	Justice/distribution
		Example: Health and economic welfare	Example: Choice/ consent/(legal) rights	Is any sub-group of stakeholders worst-off?
Owners/ employers	Government House owner Land owner Hotel owner Business proprietor Factory owner Local authority	Doses to humans reduced (+) May lead to gain in income from machinery hire (+) May lead to loss, or reduction in value, of property in areas (−) Compensation schemes? (?)	Will it be imposed or with consent? (?) Possibility for owners to contribute (+) Disruptive (−)	Possibility for conflict (−) Site selection criteria will have to be acceptable (!)
Workers/ employees	Park workers Factory workers Contractors Immigrant workers Other employees	Worker doses increase (−) Increased income for contracted workers (+) Compensation for possible dose increase (!) Possible benefits to entrepreneurs (+)	Consent and training required (!)	No major equity consequences (?)
Users/ community	Neighbours Recreational users Tourists Public amenity (library, town hall, playground, park) users Local community Garden users	Access to the areas will be hindered for a while (−) Aesthetic value will be destroyed for a period (−) Tourism will decrease for a period (−) The sight of ploughed areas may be discomforting (−)	Public and private heritage, like botanical parks, etc. will suffer (−) May increase feeling of personal control and self-help (e.g. for allotment users) (+) Liberty will be restricted in these areas (−)	Alternative amenities will likely not be available nearby (−)

Table 4. (*Continued*)

Stakeholder	Examples	Well-being	Dignity	Justice/distribution
		Example: Health and economic welfare	**Example: Choice/ consent/(legal) rights**	**Is any sub-group of stakeholders worst-off?**
Consumers	Kitchen gardeners Allotment owners	Reduction in internal dose (+)	Kitchen gardens and allotments can be used after ploughing (+)	Little conflict potential
Future generations	Future users of clean air and water Future users of recreational areas, etc.	Ploughing will make possible future use of the areas (+)	Allows continued living in the area (+)	Little long-term negative distributive effects
Environment	Pets Other biota Ecosystems	Dose to biota reduced (+) Removal of flora and fauna (−) Risk of groundwater contamination? (?)	Endangered species may suffer (−) Loss of habitat (−) Animals will be restricted from usual environment (−)	Possible conflict between present and future ecosystems (−)
Waste location stakeholders (if different from accident location)	Including all the above stakeholders connected to the waste site	Not applicable	Waste 'dilution' rather than 'containment' (−)	*In situ* treatment of 'waste' (+)

Note: See note as given in Table 3.

Table 5 Summary of consequences.

	Owners/ employers	Workers/ employees	User/ community	Consumers	Future generations	Environment	Waste location workers
Lawn mowing	+++ – ?? 	 – ?? !	++++++	+ !	++	++	 ––– ??
Deep-ploughing	+++ ––– ?? !	++ – ? !!	+ –––––––	++	++	+ ––––– ?	+ –

Note: See note as given in Table 3.

important for people's pleasure and feeling of identity. Finally, one must look at the distributional effects of the strategy as a whole. If it is a general tendency that some stakeholder groups must take all the burdens, while other groups receive all the benefits, this may be a reason to include countermeasures like compensation schemes (a social countermeasure) for the underprivileged groups.

Experts may note that the results of these analyses are not in themselves surprising. The value matrix is not necessary to come to the conclusion that one might like to investigate further ways in which external dose might be reduced, to look at other alternatives or combinations of countermeasures for inhabited areas. However, the decision-makers and consultation group may not be equally acquainted with the countermeasures and their ethical and social side-effects. The value matrix may prove helpful in allowing the decision-makers to identify ethical issues clearly, quickly and objectively, as was indicated by the end-user group in the STRATEGY project (Cox et al., 2004). It will also bring out distributional concerns and facilitate comparison of individual countermeasures and countermeasure strategies. In combination, the optimisation tool and the ethical framework do have a strong potential to be used in a dynamic and interactive decision-making situation. The matrix has a transparent and systematic character that may improve the social legitimacy of countermeasure strategies. It is important that these countermeasure decisions are transparent and well justified (from both a practical and an ethical point of view), and that they are open to revision at a later date.

6. Social Countermeasures

As was mentioned above there may be a social need, or even a moral imperative, to implement countermeasures of a somewhat different kind, so-called 'social countermeasures'. The concept was introduced in the STRATEGY project and includes countermeasures whereby dose reduction is achieved by interventions focused on human activities (e.g. food bans), as well as actions wherein social benefits other than dose reduction are the prime objective (e.g. stakeholder communication and public information strategies). Such countermeasures can enhance people's ability to deal with the deposition situation either by avoiding exposition to radiation or by being given tools to cope with it physically or psychologically. It would be expected that in any emergency situation, authorities will need to consider actions that may negate some of the more disruptive social or ethical impacts of countermeasures, hence measures such as compensation, communication and medical check-up would be addressed for all exposure cases. In both STRATEGY and the follow-up EURANOS project, it was deemed important that authorities also consider the costs, technical requirements and implications of these types of actions, and that they are explicitly considered as part of a restoration strategy. Evacuation, relocation and bans on access to areas are well-established countermeasures wherein the main aim is dose reduction. Some examples where dose reduction is not the main or only consequence are given below. More information can be found in the STRATEGY and EURANOS project templates.[7]

Social countermeasures:

- Evacuation
- Relocation
- Bans on access to areas
- Dietary advice
- Provision of counting/monitoring equipment
- Do nothing
- Raising intervention limits
- Food labelling
- Compensation scheme
- Information/Advice bureau
- Education programme in schools
- Medical check-up
- Stakeholder and public consultation methods.

6.1. Dietary advice

In a radiation accident situation where foodstuffs are affected people need provision of advice on ways to restrict their dietary radionuclide intake. Examples might be advice to reduce consumption of specific foodstuffs (e.g. game, mushrooms, etc.), to not drink water from private supplies or to prepare food in ways that reduce contamination levels (e.g. washing and peeling vegetables and fruit, brining fish, cooking meat, etc.). Advice may vary from suggestions as to which foodstuffs can be eaten without restrictions, which would be acceptable to eat occasionally, and which should be avoided completely. In the early phase, communication would be largely media based (e.g. newspapers, internet) and would need to reflect any uncertainty in the situation and be updated regularly. In later phases, specialised leaflets could also be made available, including material prepared for specific population sub-groups such as for pregnant women and parents.

[7] See http://www.strategy-ec.org.uk/cms/countermsrs.htm and http://www.euranos.fzk.de/index.php?action=euranos&title=products).

The countermeasure might need to be supplemented with monitoring (see local provision of monitoring equipment below). Note that in some cases dietary advice may lead to an *increase* in doses, as people may start to eat food that they had previously avoided completely because of worries of contamination (e.g. vegetables, mushrooms or freshwater fish). Such foods may contain levels of concentration that are beneath constraints, but nevertheless represent the major dose contribution to populations.

6.2. Provision of counting/monitoring equipment

Counting equipment or local measurement stations should be provided so that the general public can check habitats or foodstuffs for radionuclide content (particularly home grown or self-gathered). Facilities could include loan or donation of Geiger counters or dose rate monitors – with suitable training – to enable personal monitoring of external radiation dose levels, or access to independent monitoring services. The measure is particularly important in increasing personal control over the situation, and has proved successful in rural areas in Belarus (Hériard Dubreuil et al., 1999). However, it could also be applied in inhabited areas, for kitchen gardeners or those gathering 'free food'. Other services may include whole body monitoring together with general advice on radiation risks. As is the case for dietary advice, this countermeasure may also lead to an increase in doses.

6.3. Do nothing

This is a decision to not implement active countermeasures to reduce radionuclide levels in the environment or foodstuffs, on the conditions that the direct or side-effect costs of the countermeasure outweigh any benefit from dose reduction, and/or that intervention limits are not exceeded. Intervention limits may need to be monitored and reasons for 'doing nothing' explained and/or approved by affected populations.

6.4. Raising intervention limits

Raising intervention limits in foodstuffs may be considered either in order to protect a particular producer/group or due to revision of dose-risk estimates. In inhabited area scenarios, this is usually most relevant for 'specialist' or 'self-gathered' foodstuffs. The clearest example is perhaps the increase in limits in reindeer meat that was introduced in Norway in 1987, largely to protect the Sami reindeer herders (Bay and Oughton, 2005). The measure is likely to be controversial, so a good communication strategy will be essential.

6.5. Food labelling

Food labelling involves provision of information on activity concentrations, and also possibly on the source of produce, and on any remedial treatment. This information may be provided either by labels on food products or by provision of information at retail outlets. This requires an associated system of analysis/monitoring and provision of information explaining the information provided in the labelling. This information may possibly be instigated by public demand. Consumer awareness is growing and increasingly important in the allocation of market shares (e.g. growth in the demand for organic produce). On the other hand, food labelling is potentially highly contentious and is likely to meet resistance from some production or retailing companies. Therefore, good stakeholder dialogue procedures will be essential.

6.6. Compensation scheme

Introduction of compensation or insurance scheme may be a necessary social countermeasure. For people having higher exposures (including countermeasure operators), this could be in the form of direct monetary compensation for increased risk, guarantee of a lump sum payment should illness arise or some form of general medical insurance. Compensation may also be provided for other negative side-effects of countermeasures, such as economic loss and social or lifestyle disturbance. For whole communities rather than individuals, indirect compensation in the form of community benefits (e.g. schools or communal facilities) can be an alternative to direct monetary payments. Note that in Russia, Belarus and Ukraine, compensation has not been without problems, and it has been suggested that the affected populations' dependence on compensation has delayed the return to normal lifestyle (UNDP, 2002). Since it is often difficult to stop compensation once payment has started, lump sums or a pre-agreed time limit may help to alleviate some of the problems.

6.7. Information/Advice bureau

This social countermeasure involves the provision of information by advice or information centres. This will give an opportunity for specialised and specific provision of information/communication according to individual or local needs and requirements, and for people to have questions answered. Information could be provided through a variety of media channels (leaflets, talks, radio/TV/newspapers, etc., roadshows, public lectures, debates and internet) and social networks (schools, trade unions, local government, market networks, etc.). Attention should be paid to public health workers such as local doctors and midwives, as these are likely to be approached by concerned members of the public. Since the media is likely to be the main source of information in the early phase, rapid and effective communication need to be established, and authorities should have plans for how such a campaign would be managed. The degree to which authorities can influence and 'control' the media is likely to be variable, particularly in the early phase. The media themselves are not likely to see authorities as the only source of information and are likely to seek out other sources. Good emergency preparedness procedures would usually help in setting up successful measures.

6.8. Education programme in schools

Revision of school curriculum to include education on radiological risks, radiation protection and appropriate actions is also important. The aim is to promote understanding of risks (their sources and potential health effects), how to employ self-help strategies and to explain the need for compliance with imposed countermeasures. After an accident, schools are likely to be an important source of information to children.

6.9. Medical check-up

Medical check-up should be introduced for exposed individuals or groups, with the aim of providing better health care in the case of radiation-related illnesses (i.e. early identification and treatment of cancer) and other non-radiation-related diseases. Added advantages include reassurance, the provision of 'after-care' and establishing comprehensive and reliable records of post crisis health effects. Assurance of proper medical follow-up will be particularly important for countermeasure implementers, particularly if there is uncertainty about the doses and exposure levels. There should be specific schemes for pregnant women as experience shows that this group is one of the most likely to seek medical advice (Gould, 1990). This countermeasure may also be seen as a form of compensation for radiation or other (e.g. chemical) risks in sub-groups.

6.10. Stakeholder and public consultation methods

These are various procedures (e.g. Citizen's Jury, Focus Groups, Consensus Conference, or Advisory Committees) that may vary according to participation from experts, laypersons or a mixture of the two. The general objective of using these consultation procedures is to provide informed public and expert input to decisions to ensure that decisions are publicly acceptable. In addition, the measures help promote empowerment, personal control and the legitimacy of decisions. The procedure can be related to both direct radiation-related issues and/or other negative side-effects of countermeasures (e.g. environmental risk). In general, it is beneficial that the main stakeholders are involved in crisis management as early as possible in the event of an accident. During the early phase, experts and main stakeholder representatives (e.g. politicians) are likely to be most prominent. However, the need for broader participation increases with time and good accident management and emergency preparedness will have already identified the relevant representatives for wider participation. As stakeholder involvement also is a central element of the ethical assessment process we shall now look at the justification for this in more detail.

7. Stakeholder Involvement as a Social Management Option

7.1. Arguments for a wider inclusion of citizens and stakeholders in ethical assessments and decision-making

The last decade has seen an increased focus on the importance of stakeholder involvement and public participation in a number of policy areas, particularly those concerned with environmental issues or technology evaluation. Not least, stakeholder involvement has been used quite extensively in radiation management, for instance with regard to waste location decisions. There are a number of arguments for involving lay people or stakeholders in the kind of decision-making processes relevant here. We shall now briefly look at these legislative, sociological and ethical reasons for the focus on participation and involvement of stakeholders and citizens in deciding upon countermeasure strategies.

In environmental legislation this commitment can be found internationally in Principle 10 of the Rio Declaration:

> Environmental issues are best handled with the participation of all concerned citizens, at the relevant level. At the national level, each individual shall have appropriate access to information concerning the environment that is held by public authorities, including information on hazardous materials and activities in their communities, and the opportunity to participate in decision-making processes. States shall facilitate and encourage public awareness and participation by making information widely available. Effective access to judicial and administrative proceedings, including redress and remedy, shall be provided (UNDP, 2002).

Nearly 10 years later, The Aarhus Convention on Access to Information, Public Participation in Decision-making and Access to Justice in Environmental Matters was legally ratified. Since then, the requirement for stakeholder participation in decision-making has been restated within Article 26 of the Johannesburg World Summit on Sustainable

Development (WSSD, 2003) and a number of EC directives, including in connection with biotechnology (EC Directive 2001/18, 2001).

Sociologists have also highlighted a number of reasons to support stakeholder participation, many of which have relevance to issues of justice and equity. The most common include (Martin, 1999):

Empowerment: Affected people need to feel they have some form of control/influence over what people do to their environment and well-being.
Democratisation: The public has a democratic right to take part in decisions affecting their lives.
Efficiency – all stakeholders, including the public, have important knowledge of relevance to the decision-making process and can input a range of issues that might otherwise be overlooked, undervalued or under-utilised.[8]
Success: The public has the capacity to halt many projects, so the policy implementer should acknowledge the needs and desires of these people.
Consensus building: All people involved in a project should be able to work for common goals if there is a process whereby these common goals can be identified and worked towards.

Particularly in a situation like the one discussed in this book, it is important to maintain the public's trust. If decisions on countermeasure strategies are made by experts purporting to be neutral, there is the risk that the public will be sceptical. According to Irwin et al. (1996), 'any effort at disseminating scientific information within a 'loaded' social context' produces a very critical and suspicious reaction from the public. It may be appropriate that scientists should be called upon to elucidate on matters concerned with ethics and values when these have scientific components. And it is, perhaps, not surprising that many scientists and experts rarely express an opinion on the probability of some harm occurring without passing judgement on either its relative size and/or acceptability. However, there is a problem if such input is provided without proper acknowledgement of the impact of social and ethical factors on scientific assessments, and with technical experts blindly assuming that they are taking a purely objective and 'scientific' stance in making statements about the acceptability of risk.

Ethically, the requirement for stakeholder[9] participation within societal decision-making might be said to satisfy the normative principle of *due process* or *fair procedure*. Respect for such principles has been found in just about every human society (Brown, 1991; Bok, 1995), and is deemed prerequisite for both law and ethical theory. The norms of justice and

[8] See also Wynne (1989).
[9] Some definitions make a distinction between stakeholder (directly affected parties) and public (or layperson) participation. However for the rest of this chapter, 'stakeholder' will be used as a general term, covering both laypersons and other groups or individuals either affected by or having an interest in policy; 'public' will be used only when referring specifically to lay and other non-specialist participation procedures.

fairness are often linked to the distribution of goods and the equal treatment of individuals, but in questions of due process one asks not what the final distribution of goods is (i.e. distributive justice), but how that distribution came about (i.e. procedural justice). In other words, what is fair or right depends on the way the decision was made, not what that decision was (Shrader-Frechette, 1991; Rawls, 1971). As a simple analogy, the rule of 'I cut, you choose' is a procedure that leads to a fair division of a cake. More seriously, the verdict of a trial is not the criterion by which we decide whether the trial was a fair one. We would ask whether the defendant was given a fair trial under due process, which means asking whether justice had been administered in accordance with established rules and principles, such as an unbiased jury and with access to all relevant facts and information. With respect to societal decision-making, one might ask whether everyone's interests were taken into account and whether people were offered an opportunity to give consent to the imposition of any risk associated with a particular policy or technology (Shrader-Frechette, 1991). The underlying ethical norm is that affected persons have a moral right to take part in decisions that affect their lives.

With regard to including stakeholder representatives in ethical assessments like the one described above, another strain of reasoning can be applied, namely that of the importance of bottom-up justification of ethical decisions. As we live in pluralist societies and all moral theories are controversial, one cannot simply pick one stance from which to make moral judgements (see argument in Forsberg, 2007, based on Misak, 2000). When justifying a moral decision or advice one needs to consider all morally relevant arguments and, as we have noted above, balance a number of ethical concerns. The more thoroughly a balancing proposal is assessed the more justified it will be. The proposed balancing solution should, therefore, be 'tested' on those people that are affected by the decision in order to assure that all relevant arguments have been taken into account. It is far from certain that one will actually agree on what is the ethically correct thing to do in a situation, but without going through such a process the ethical conclusion will be more easily criticisable and thereby less robust.

All these arguments indicate that participation and dialogue in choices of restoration strategies not only is an important social countermeasure, but should also be part of the framework in which the actual decisions on countermeasure strategies are made.

7.2. A good consultation process is an ethical issue

However, despite the general consensus that public participation is important within policy making, involving stakeholders in decision-making is not without controversy. The debate as to how exactly that participation should be undertaken and by what criteria the various methods for participation should be evaluated continues. It is ethically important that involving citizens or stakeholders in decision-making is carried out in the right way. Discourse ethics (e.g. Habermas, 1989; Apel and Kettner, 1992) has provided guidance as to how ethically justified judgements should be made. There is also a lot of research on how participatory processes should be carried out (e.g. Renn et al., 1993). Decision-making processes have been criticised both for taking too much notice of public opinion – 'it is the job of those in power to take decisions necessary for the sound management of waste' (UK House of Lords Select Committee on Science and Technology, 1999) – or for paying too little attention to stakeholder advice (Webler, 1995; Dienel and Renn, 1995). Procedures have been criticised as being a waste of resources or undermining democracy. For example, in Europe, stakeholder participation has become particularly popular in relation to nuclear issues, leading critics to suggest that this simply is a manoeuvre to exhaust public opposition (LLRC, 2003). One of the most common complaints is that those in charge of the process have misled participants on their impact on the eventual decision and in practice, many participation exercises simply represent a more subtle form of public relations or propaganda. For this reason, a number of criteria for evaluating participation processes have been proposed. These are also intended to help avoid some of the more serious pitfalls and realise the goal of participation as defined in the conventions and directives (e.g. Åarhus, 2001). The evaluation of both procedures and outcomes needs to include consideration of the legitimacy of the procedure, in addition to the more common efficiency evaluation of procedures and outcomes. A number of principles have been proposed by various writers as guidance for participation (e.g. Rowe and Frewer, 2000). In the following discussion, these have been grouped into issues relating to *Representativeness, Transparency and openness* and *Accountability and influence.*

7.2.1. Representativeness

Just about all processes need to address the question of the legitimacy of representation for those persons sitting around the decision-making table. Who do they represent? On what grounds are they there? Were they elected? Do they share the same opinion? Or are they speaking on behalf of those that do not have a voice (e.g. animals or future generations)? Perhaps some people attend simply because they know a lot. Representativeness is perhaps one of the easiest principles to define and the hardest to realise. At a basic level, the criteria reflect that participants should comprise a broadly representative sample of the affected parties. This need not be all voters (as in referenda), but a sample representing a relevant cross-section of a population. However, this raises questions about what a relevant cross-section should contain (i.e. that all opinions are represented or that the representation balances the support for each view in the affected population?); on what grounds an individual can legitimately represent a group; and who decides who participates? It is never possible to represent everyone's opinion, but a good cross-section increases the chance that issues people feel important will have been raised and properly addressed. As it is never possible to obtain a truly representative sample of stakeholders, the procedure should enable all those who wish to participate to do so in some way. This may call for a variety of procedures rather than a single event.

7.2.2. Transparency and openness

Transparency can relate to both the procedure itself and the information used within that procedure. Definitions include that the process should be transparent to enable the public to see what is going on and how decisions are being made. In addition, the reasoning behind actions, deliberations and decisions should be made available. It can often be justified that some part of the decision-making process takes place behind 'closed-doors' – for example, to encourage a freer exchange of information and to allow participants to 'represent themselves rather than their employers' or to 'toe the party line'. Thus, transparency need not require physically open procedures throughout the whole process. Part of the discussion may, for example, take place under the 'Chatham House Rule', which forbids participants to attribute remarks or cite other members opinions (i.e. what was said can be reported but not who said what). However, transparency usually demands that, even where part of the process is to be closed, this is made explicit from the outset, that the discussions also include open procedures and, ultimately, publication of the results and decisions should enable both participants and non-participants to control (or understand) how decisions are made and upon what assumptions.

To achieve transparency, it is fundamental that participants are clear about the overall aims of the process and how the process fits into the broader decision-making context, for example whether the procedure is one of information gathering, provision of advice to policy-makers or actual decision-making.

7.2.3. Accountability and influence

Accountability refers to the fact that those responsible for the process should be responsible or answerable for their actions to all parties. This often includes aspects linked to transparency; for instance they must provide documentation of the reasoning behind decisions (scientific, ethical, political, etc.). It is particularly important in cases where policy-makers did not heed stakeholder advice that clear reasons for this are provided. Accountability requires an interactive process, including giving stakeholders feedback to indicate how their views have or have not been taken into account. Usually, for the process to meet criteria for acceptability and legitimacy, the output of the procedure should have a genuine impact on policy.

As mentioned above, one of the most common complaints against participation procedures in practice is their lack of influence on policy. According to Wallentinus and Paivo (2001), there have been instances where bodies have sat down to listen to stakeholders in a succession of meetings, but no effort was made to adjust proposals to the suggestions of stakeholders. Beder (1999) also cites the lack of accountability and influence as evidence that decisions have already been made and that policy-makers are just going through the motions for the sake of public relations (this objection has been raised with regard to the UK GM Nation debate and nuclear power debates), or to manipulate public acceptance of a pre-chosen proposal, to select from a range of favoured proposals. Superficial public participation efforts serve the decision-makers much more than they serve the stakeholders, to the extent that (as has been proposed in respect to some of the earlier nuclear power consultation) the conclusion of the participants was that they existed 'to legitimise agency decisions, defuse opposition, warn the agency of possible political obstacles and satisfy procedural and legal requirements' (Kraft and Clary, 1993). To date, there is relatively good evidence to show that stakeholder participation can provide a valuable input of information to the decision-making process, but much less evidence of procedures having a direct impact on policy.

It is important that the design of stakeholder consultations in general and participatory ethical matrix processes in particular reflects these basic characteristics of good participatory processes.

REFERENCES

Apel, K.-O., and M. Kettner. (Eds) (1992). *Zur Anwendung der Diskursethik in Politik, Recht und Wissenschaft*. Suhrkamp, Frankfurt, Germany.

Åarhus Convention. (2001). *The Åarhus Convention on Access to Information, Public Participation in Decision-Making and Access to Justice in Environmental Matters*. Adopted on June 1998, ratified on October 2001. Available at: http://www.unece.org/env/pp/welcome.html

Bay, I., and D. H. Oughton. (2005). Social and economic effects. In: *Chernobyl, Catastrophe and Consequences* (Eds J. Smith and N. A. Beresford). Springer-Verlag, Berlin. ISBN 3-540-23866-2, pp. 239–262.

Beauchamp, T., and J. Childress. (2001). *The Principles of Biomedical Ethics*. Oxford University Press, Oxford.

Beder, S. (1999). Public participation or public relations? In: *Technology and Public Participation* (Ed. B. Martin). University of Wollongong Press, Wollongong, Australia.

Bok, S. (1995). *Common Values*. University of Missouri Press, Columbia, MO.

Brown, D. E. (1991). *Human Universals*. McGraw-Hill, New York.

Bullard, R. D. (1990). *Dumping in Dixie: Race, Class and Environmental Quality*. Westview Press, Boulder, CO.

Cox, G., N. Beresford, B. Alvarez, D. Oughton, Z. Kis, K. Egen, K. Andersson, H. Thørring, J. Hunt, S. Wright, C. Barnett, J. Gil, B. Howard, and N. Crout. (2005). Identification of optimal countermeasure strategies for radioactively contaminated regions: Case studies. *Journal of Environmental Radioactivity*, 83, 339–383.

Cox, G. M., N. M. J. Crout, B. J. Howard, N. A. Beresford, S. M. Wright, D. H. Oughton, I. A. Bay, E.-M. Forsberg, B. Alvarez-Fario, and J. M. Gil. (2004). *Case Studies, Deliverable 9 from the STRATEGY Project*. University of Nottingham, Nottingham, UK.

DEFRA. (2002). *Surveys of the Economic Impact of Foot and Mouth Disease in Six Districts*. Department for Environment, Food and Rural Affairs (DEFRA), London. Available at: http://www.defra.gov.uk/animalh/diseases/fmd/2001/index.htm

Dienel, P. C., and O. Renn. (1995). Planning cells: A gate to 'fractal' mediation. In: *Fairness and Competence in Citizen Participation (Technology, Risk and Society)* (Eds O. Renn, T. Webler, and P. Wiedemann). Kluwer Academic Publishers, Dordrecht, The Netherlands, 130pp.

EC. (2001). European Commission Directive 2001/18 of 12 March 2001 on the deliberate release into the environment of GMOs (revision of directive 90/220). Available at: http://www.europa.eu.int

EURANOS. (2006). *Generic Handbook for Assisting in the Management of Contaminated Food Productions Systems in Europe following a Radiological Emergency*. Deliverable from the ERURANOS Project. EURANOS(CAT1)-TN(06)-06. Available at: http://www.euranos.fzk.de/index.php?action=euranos&title=products

Forsberg, E.-M. (2007). Pluralism, the ethical matrix and coming to conclusions. *Journal of Agricultural and Environmental Ethics*, 20(5), 455–468.

Gould, P. (1990). *Fire in the Rain: The Democratic Consequences of Chernobyl*. Polity Press, Cambridge, MA.

Habermas, J. (1989). *Moral Consciousness and Communicative Action*. MIT Press, Cambridge, MA.

Hériard Dubreuil, G. F., J. Lochard, P. Girard, J. F. Guyonnet, G. Le Cardinal, S. Lepicard, P. Livolsi, M. Monroy, H. Ollagon, A. Pena-Vega, V. Pupin, J. Rigby, I. Rolevitch, and T. Schneider. (1999). Chernobyl post-accident management: The ETHOS Project. *Health Physics*, 77, 361–372.

House of Lords Select Committee on Science and Technology. (1999). *Management of Nuclear Waste*. The Stationery Office, London, UK.

Howard, B. J., K. G. Andersson, N. A. Beresford, N. M. J. Crout, J. M. Gil, J. Hunt, A. Liland, A. Nisbet, D. H. Oughton, and G. Voigt. (2002). Sustainable restoration and long-term management of contaminated rural, urban and industrial ecosystems. *Radioprotection – Colloques*, 37(C1), 1067–1072.

Hunt, J., and B. Wynne. (2002). *Social Assumptions in Remediation Strategies, STRATEGY Deliverable 5*. University of Lancaster, Lancaster, UK.

ICRP. (1991). *Principles for Intervention for Protection of the Public in a Radiological Emergency*. Publication 63. *Annals of the ICRP*, Pergamon Press, Oxford.

ICRP. (2000). Protection of the public in situations of prolonged radiation exposure. *ICRP*, Sutton (GB), ISSN 0146-6453, 2000.

Irwin, A., A. Dale, and D. Smith. (1996). Science and hell's kitchen: The local understanding of hazard issues. In: *Misunderstanding Science?: The Public Reconstruction of Science and Technology* (Eds A. Irwin and B. Wynne). Cambridge University Press, Cambridge.

Kaiser, M., and E.-M. Forsberg. (2001). Assessing fisheries – Using an ethical matrix in a participatory process. *Journal of Agricultural and Environmental Ethics*, 14, 191–200.

Kaiser, M., K. Millar, E. Thorstensen, and S. Tomkins. (2007). Developing the ethical matrix as a decision support framework: GM fish as a case study. *Journal of Agricultural and Environmental Ethics*, 20(1), 65–80.

Kelly, N. (2001). Post accident management. In: *Radioactive Pollutants: Impact on the Environment* (Eds B. J. Howard and F. Bréchignac). EDP Sciences, France, pp. 75–87.
Kraft, M. E., and B. B. Clary. (1993). Public testimony in nuclear waste repository hearings: A content analysis. In: *Public Reactions to Nuclear Waste* (Eds R. E. Dunlap, et al.). Duke University Press, London, UK.
LLRC. (2003). Low level radiation campaign. www.llrc.org
Martin, B., (Ed.) (1999). *Technology and Public Participation*. University of Wollongong Printery, Wollongong, Australia.
Mepham, T. B. (1995). Ethical impacts of biotechnology in dairying. In: *Progress in Dairy Science* (Ed. C. J. C. Phillips). CAB International, Wallingford, pp. 375–395.
Mepham, T. B. (1996). Ethical analysis of food biotechnologies: An evaluative framework. In: *Food Ethics* (Ed. T. B. Mepham). Routledge, London, UK, pp. 101–119.
Mepham, T. B. (2004). A decade of the Ethical Matrix: A response to criticisms. *Proceedings of the Fourth Congress of the European Society for Agricultural and Food Ethics (EurSafe)*, 2–4 September 2004, Leuven, Belgium, pp. 271–274.
Misak, C. (2000). *Truth, Politics, Morality. Pragmatism and Deliberation*. Routledge, London and New York.
Morrey, M., and P. Allen. (1996). The role of social and psychological factors in radiation protection after accidents. *Radiation Protection Dosimetry*, 68, 267–271.
Nisbet, A. F. (2002). Management options for food production systems contaminated as a result of a nuclear accident. *Radioprotection – Colloques*, 37(C1), 115–120.
Nisbet, A. F., and K. J. Mondon. (2001). *Development of Strategies for Responding to Environmental Contamination Incidents Involving Radioactivity*. The UK Agriculture and Food Countermeasures Working Group 1997-2000. NRPB-R331.
Oughton, D. H. (1996). Ethical values in radiological protection. *Radiation Protection Dosimetry*, 68, 203–208.
Oughton, D. H. (2000). Ethical aspects of ICRP recommendations for radiation protection. In: *Ethical Issues in Radiation Protection* (Ed. L. Persson). SSI, Stockholm, Sweden, pp. 30–32.
Oughton, D. H. (2003). Protection of the environment from ionising radiation: Ethical issues. *Journal of Environmental Radioactivity*, 66, 3–18.
Oughton, D., E.-M. Forsberg, I. Bay, M. Kaiser, and B. Howard. (2004). An ethical dimension to sustainable restoration and long-term management of contaminated areas. *Journal of Environmental Radioactivity*, 74, 171–183.
Rawles, K. (2002). *Compensation in Radioactive Waste Management: Ethical Issues in the Treatment of Host Communities – A Report for Nirex*. Nirex, Harwell, UK.
Rawls, J. (1971). *A Theory of Justice*. The Belknap Press of Harvard University Press, Cambridge, MA (Revised in 1999).
Regan, T., and P. Singer. (1989). *Animal Rights and Human Obligations*. Prentice-Hall, Englewood Cliffs, NJ.
Renn, O., T. Webler, H. Rakel, H. Dienel, and B. Johnston. (1993). Public participation in decision making: A three step procedure. *Policy Sciences*, 26, 191–214.
Rowe, G., and L. J. Frewer. (2000). Public participation methods: A framework for evaluation. *Science, Technology and Human Values*, 25(1), 3–29.
Shrader-Frechette, K. (1991). *Risk and Rationality*. University of California Press, Berkeley, CA.
Shrader-Frechette, K. (2001). Risky business: Nuclear workers, ethics and the market efficiency argument. *Ethics and the Environment*, 7, 1–19.
Shrader-Frechette, K., and L. Persson. (1997). Ethical issues in radiation protection. *Health Physics*, 73, 378–382.
Thompson, P. (1997). *Food Biotechnology in Ethical Perspectives*. Blackie, London, UK.

UNDP. (2002). Rio Declaration. United Nations Environment Programme, www.un.org/documents/ga/conf151/aconf15126-1annex1.htm

UNESCO – COMEST. (2005). *The Precautionary Principle*. Report, UNESCO, Paris, http://unesdoc.unesco.org/images/0013/001395/139578e.pdf

Wallentinus, H.- G., and J. Paivo. (2001). The Hallandsas railway tunnel project in Sweden, In: *The Role of Environmental Impact assessment in the Planning and Decision Process of Large Development Projects in Nordic Countries* (Ed. T. Hilding-Rydevik). Norderegio Report, Stockholm.

Webler, T. (1995). 'Right' discourse in citizen participation: An evaluative Yardstick. In: *Fairness and Competence in Citizen Participation (Technology, Risk and Society)* (Eds O. Renn, T. Webler, and P. Wiedemann). Kluwer Academic Publishers, Dordrecht, The Netherlands, pp. 35–86.

WSSD. (2003). *The Johannesburg Declaration on Sustainable Development: World Summit on Sustainable Development*. UNEP. Available at: http://www.unep.fr/scp/csd/wssd/postjoburg/wssdoutcomes.htm

Wynne, B. (1989). Sheep farming after Chernobyl: A case study in communicating scientific information. *Environment*, 31, 10–15.

CHAPTER 10

Strategies for Restoration of Contaminated Inhabited Areas

Kasper G. Andersson*

Contents

1. Introduction

This chapter is aimed at providing an overview of the requirements to construct and implement the right countermeasure strategy for restoration of a specific contaminated inhabited area. A countermeasure strategy may consist of one or several of the different types of countermeasures described

*Corresponding author. Tel.: +45 46 77 41 73; Fax: +45 4677 5330
E-mail address: kasper.andersson@risoe.dk

Risoe National Laboratory for Sustainable Energy, Technical University of Denmark, DK-4000 Roskilde, Denmark

Radioactivity in the Environment, Volume 15
ISSN 1569-4860, DOI 10.1016/S1569-4860(09)00410-0

in Chapters 8 and 9. There are thus many different possibilities for intervention, but although individual countermeasures can have very different immediate objectives (e.g. to reduce doses to the population, to reduce anxiety and social disruption), the overall goal of a strategy is in all cases to spend resources as wisely and reasonably as at all possible, and maximise the benefit. The International Commission on Radiological Protection (ICRP) has issued generic recommendations as to how this can best be achieved, and the implications of this are discussed in this chapter. The chapter further deals with the influences on restoration strategy formation of the type and scale of the incident leading to the contamination of an inhabited area. For instance, terror attacks would be expected to lead to contamination of smaller areas and different psychological impacts than large accidents at nuclear power plants. A few practical examples of clean-up work are given, highlighting some important issues that can be decisive for the success of a strategy in reducing doses in an inhabited area. Finally, the difficult and complex task of decision-making is discussed, describing some of the different types of supporting tools that are currently available to enhance the overview for decision-makers and their advisors.

2. Overall Purposes and Criteria for Restoration Strategies

As mentioned in Section 2 of Chapter 8, a radiological emergency leading to airborne contamination of inhabited areas would, seen from a radiation protection perspective, be followed by different, more or less distinct time phases. In the earliest phase, the primary concern would be to adequately protect potentially affected populations against possibly comparatively high risks that might be expected to occur or start in the short term. The implementation of such countermeasures would thus need to occur rapidly, and the responsible decision-makers would at that stage typically have insufficient knowledge of the situation to allow very thorough evaluation of specific advantages and disadvantages of intervention and insufficient time to apply dialogue-based decision instruments. The countermeasures that could be implemented in this early phase range from issuing simple instructions to the public via various communication channels on protective measures (sheltering, closing air ducts, etc.) to the logistically and in practice highly challenging measure of evacuating the population of a perhaps large area. Due to the urgent nature of such countermeasures, it is highly valuable to have set, in advance, intervention criteria that can be applied swiftly (see also Section 8 Chapter 7). For instance, the ICRP (1993) recommends temporary relocation of people if the individual dose over a month is likely to exceed 10 mSv.

Later on, when the acute emergency phase is over, and due assessments have been made of the contamination situation, decision-makers and their advisors would have obtained sufficient knowledge to issue more reliable estimates of the actual health risk implications, including primary dose pathways, and the planning for a more detailed strategy to sustain acceptable living conditions in affected areas can begin. The case has, according to ICRP terminology, then gone from being an 'emergency exposure situation' to an 'existing exposure situation' (ICRP, 2007). It needs to be stressed very strongly that a wide range of types of risks other than radiological are generally at play here, and that radiological risks can in some cases even be insignificant compared with the disruptive effects imposed on society by the implementation of a given countermeasure strategy.

The basic principle that should always be fulfilled with respect to *justification* of any countermeasure strategy is that the advantages of intervening offset the disadvantages (ICRP, 1991). Among the justified options, the *optimum* protection option is the action which results in the maximum net benefit. Consequently, the optimum protection option is not necessarily the option that results in the lowest residual individual or collective dose (ICRP, 1999). It is thus not possible to give rigid *generic* recommendations for contamination or dose threshold levels above which countermeasures should be implemented, or below which countermeasures should not be recommended. The exception is that individual doses exceeding about 100 mSv either acutely or within a year should always be avoided, as they imply an increased likelihood of deterministic effects and a significant risk of cancer (ICRP, 2007). In agreement with the ICRP system of protection, it can be beneficial to reduce even a low dose if the associated *total* costs are outweighed by the *total* benefits of the action. However, rough indications of a classification system with intervention criteria based on dose levels have been suggested by, for instance, the IAEA (1997) (see Table 1), which can be compared with the ICRP recommendation that an existing annual dose approaching about 10 mSv would often

Table 1 IAEA (1997) dose criteria for clean-up[a] of contaminated land.

Band number	Range of annual doses attributable to the contamination ($mSv\ y^{-1}$)	Is clean-up[a] needed?
6	>100	Always
5	10–100	Almost always
4	1–10	Usually
3	0.1–1	Sometimes
2	0.01–0.1	Rarely
1	<0.01	Almost never

[a]Clean-up used by IAEA to mean decontamination or other recovery options.

not be likely to justify intervention for prolonged exposure situations, whereas an existing annual dose approaching 100 mSv will practically always justify intervention. Any distinct threshold values for intervention are nevertheless *politically* determined and not the result of an *optimisation* process, which would be case specific, and might also need to consider multiple (e.g. radiological, chemical and physical) hazards occurring either simultaneously or at time intervals. However, as mentioned, unacceptably high exposure may occur, and the ICRP (2007) defines the term 'reference level' to represent the level of dose or risk above which it is deemed inappropriate to plan to allow exposures to occur, and for which protective options would therefore in general need to be planned and optimised. The ICRP judges that reference levels for the highest planned residual dose from a radiological emergency would be in the range between 20 and 100 mSv (acute or annual dose). On the use of reference levels, the view of the ICRP (1999) is 'The use of predetermined specific reference levels can facilitate timely decisions on interventions and the effective deployment of resources; however, an improper use may lead to inconsistencies with the principles of justification and optimisation'. It is thus clear that their application requires great caution. It should also be noted that naturally occurring radionuclides may contribute significantly to the annual dose in some areas, but not in others, which means that different degrees of intervention may be called for in similar areas with the same existing annual dose to reach a given residual dose rate.

Selecting the optimal countermeasure strategy for a specific emergency situation is by no means an easy task, as countermeasure implementation can impact on society in a wide range of more or less foreseeable ways. Some aspects can relatively unproblematically be quantified in for instance monetary terms, thereby facilitating inter-comparison between methods. Such aspects include the use of machinery, consumables, transport and worker wages (see Section 4 of Chapter 8), although due attention must be paid to local and temporal variation. However, as stated by the ICRP in their latest recommendations (ICRP, 1999, 2007), problems like social disruption, loss of property value and loss of income due to the contamination situation as well as due to countermeasure implementation should also be taken into account (see Chapter 9). Depending on the specific scenario, there may be a wealth of such 'indirect' factors, the importance of which is in general very difficult to describe. When evaluating the possible implications of a countermeasure strategy, it is important always to measure these against the implications of doing nothing (see Section 3 of Chapter 8). If nothing is done to reduce the exposure problems there will be an equally long (quite possibly longer) list of different types of adverse effects on society, both radiological and non-radiological.

In addition to ensuring that any strategy for restoration is justified and optimised considering a wide range of aspects, it is of course essential to ensure that the strategy achieves the ultimate goal of intervention: that the health consequences and other adverse impacts of the contamination are reduced sufficiently to allow the affected population and society as a whole to resume their lives and functions in the area. In extreme cases, contamination levels might be so high that even the most effective existing countermeasures would be insufficient to reduce the problems to an acceptable level that would permit a population to remain (or return) and function in the area, and any clean-up effort would thus then be in vain. A clean-up strategy that would not reduce residual doses to less than an agreed reference level should never be carried out (ICRP, 2007). However, practically any remedial means for mitigation would generally be justified to avoid permanent removal (or desertion) of (parts of) a population, as this can have immense societal repercussions.

Collective dose to a population is a term used in optimisation, for comparing radiological technologies and protection procedures. However, aspects like the age and gender of exposed persons, the magnitudes of individual doses and the time periods over which they are received also need careful consideration. To identify sub-groups of members of the public receiving comparatively high doses, the ICRP (2007) now recommends the use of the 'Representative Person' (ICRP, 2006) instead of the previously used 'Critical Group' approach. In terms of factors like breathing rate, location and use of resources, the 'Representative Person' should reflect the typical habits of a small number of people who receive the highest level of exposure, rather than the extreme habits of a single person. As mentioned in Section 3 of Chapter 9, the ethical principle of equity is also important to take into account, to ensure that it is not one part of the population that receives practically all the benefits of an intervention, whereas the costs and disadvantages are primarily inflicted on another population group. In connection with the European STRATEGY project (Howard et al., 2005), it was considered to introduce a quantity in the decision matrix which accommodated for the injustice in the dose distribution profile resulting from countermeasures. This was based on an integral over different residual dose levels of the products of the number of people receiving the given dose times the difference between the dose they receive and the average dose. This value might then be weighted by a politically determined conversion factor to convert to monetary or other ranking units facilitating comparison with other optimisation elements. It would seem reasonable to somehow enable decision-makers to take into account, for example, the inequity of higher dose contributions received by inhabitants on the top floor in a contaminated multi-storey building environment. These people, who are living directly under the roof, can receive a considerable dose contribution from the contamination on the roof. However, this dose contribution virtually disappears in a calculation of collective doses to people living on all perhaps 10 floors of a building, and cleaning of the roof would thus on the basis of collective dose alone not be prioritised.

As mentioned in Section 6 of Chapter 9, the overall purpose of including a countermeasure in a strategy is not necessarily to reduce the radiological health consequences posed by the contamination, but perhaps to provide information, advice and reassurance to the public and answer the many questions that a contamination scenario would prompt. Following the Chernobyl accident, the information level from the authorities, the press and other sources to the public in the worst-affected areas was over many years extremely low, and myths arose, spreading undue anxiety and misunderstandings that hindered proper response to the actual health risks. A European financed investigation in the affected areas of Russia, Belarus and Ukraine (ENVREG 9602, 2000) concluded that the state of knowledge of relevant issues was very poor among those persons in the local communities from whom people would often seek information and advice (e.g. teachers at various levels and general practitioners). As a consequence, a series of education pilot projects were initiated to improve the state of knowledge. Also things like medical check-ups of exposed persons and compensation for increased risk might prove to be advantageous – though not uncomplicated – countermeasures under some circumstances.

3. Influences of Type and Scale of the Contaminating Incident

Depending on the type of the contaminating incident, large areas of land including inhabited areas may become contaminated. In such cases it can be very useful to divide a large area into sub-areas with different required urgency of actions to reduce doses. This would require rapid assessment of contamination levels or dose rates at relatively low spatial resolution, for which airborne measurements using aeroplanes or helicopters (Figure 1) would be ideal (see Section 5.2 of Chapter 7). On this background, early-phase countermeasures may be implemented that might save comparatively high doses to inhabitants. If the affected area is very large, as the Chernobyl accident demonstrated could be the case following a major nuclear power plant accident, the resources available might be insufficient to immediately start implementation of countermeasures everywhere to deal with the long-term contamination problems, and a prioritisation of sub-areas may need to take place. Here it is also important to take into account whether people have been evacuated or still remain in their contaminated living area. The prioritisation would in practice not be an easy task to carry out, as it would be difficult to explain to affected persons why countermeasures are at one point implemented in one sub-area and not in another also significantly

Figure 1 A helicopter measuring the gamma spectrum from contaminants deposited on a defined area of the ground. The normal measuring height is ca. 100 m. On the right, measurement points are plotted on a map, from an airborne gamma spectrometric survey over Risø National Laboratory in Denmark, which has nuclear installations under decommissioning (Reproduced with kind permission from the ARGOS consortium).

contaminated sub-area close by. This would clearly require careful dialogue with the public to avoid dispute.

Optimised long-term (recovery) restoration strategies derived for inhabited sub-areas with different levels of contamination, area features and population densities/behaviour can be very different. A person present in a contaminated inhabited area will be exposed to radiation from contamination deposited on a complex of different types of surfaces. Both the nature of the released contaminants (Chapter 1), the initial contamination levels on the different surfaces (Chapters 2, 3 and 4), the subsequent contaminant migration (Chapter 5) and the sizes, locations and orientations of contaminated surfaces and shielding elements in relation to the time periods spent by inhabitants in different locations (Chapter 6) can greatly influence the magnitude of the doses received over time. For instance, in a suburban area with detached houses with relatively thin walls and large grassed areas, the location-averaged individual dose over the first year from a contamination with ^{137}Cs in readily soluble form may easily be a factor of 5–10 higher than that received by an individual in a highly urbanised area with the same type and initial level of contamination. This is largely because the latter type of area often provides much better shielding and has a lower fraction of open soil areas that usually efficiently retain ^{137}Cs near the surface. This in turn means that remediation strategies focusing on different types of surfaces are needed for areas with different building features. Further, since the total doses that would without intervention be received by individuals over the first year would differ between such areas also the remediation effort that would be needed to comply with the ICRP recommendations regarding, for example reference levels can be quite different.

As outlined in Chapter 8, some countermeasures that may, under specific circumstances, be particularly highly advantageous in achieving a great dose reduction at a low cost need to be implemented over a short period of time following the contamination. Early cutting and removal of contaminated grass is an example of such a countermeasure. This needs to be done before the first heavy rainshower, which will wash the contamination down on and into the soil, where it will be much more complicated to remove. Depending on the size of the contaminated area, it may well not be possible for the usually rather limited number of available professional workers to carry out this countermeasure within the time limits, but a simple method like this, for which the required equipment would normally be available in great numbers, and no complex skills are required, could be carried out by the property owners themselves. This is a typical example of a countermeasure that could be useful to implement on a 'self-help' basis. Other than facilitating rapid intervention over large areas, the use of 'self-help' countermeasures is also advantageous in the sense that it involves individuals in contaminated areas more personally and actively in an effort to improve their own living conditions. When they have been directly involved in the work, they will know exactly what has been carried out, and how well it has been done. This will help to prevent myths and misunderstandings from arising among the inhabitants, and they will get a better feeling of control of the situation. Further the knowledge gained will provide an improved understanding of the relative importance of different exposure routes. This will in turn enable inhabitants to judge what they can do to reduce their own personal exposure. In terms of ethics (see Chapter 9), voluntarily implemented 'self-help' countermeasures are normally considered positive, as they respect the fundamental ethical values of autonomy, liberty and dignity. 'Self-help' can also be comparatively attractive in terms of labour costs. On the negative side, there will always be a risk of misunderstanding or misinterpretation of instructions given, which might in some cases lead to an irreversible worsening of the situation. It is therefore of great importance to provide carefully worded and detailed instructions to the people that participate.

It should be noted that, although it is often an advantage if a countermeasure is based on a method that is carried out routinely, it may also be a problem. For instance, ploughing is suggested as an option for large contaminated soil areas (e.g. parks). When the so-called skim-and-burial plough was tested as a remedial measure in a Chernobyl-contaminated area in the Bryansk region in Russia, the local plough operator was carefully instructed on how to operate the plough to achieve the highest dose-reducing effect. However, he continuously slipped rapidly into conducting 'business as usual', and was more concerned about keeping in a straight line and doing the job quickly than about burying a soil layer of the right thickness. Supervision of the implementation of countermeasure strategies may therefore also be useful.

Other types of contaminating incidents than nuclear power plant accidents could also lead to highly significant airborne contamination, but over much smaller areas, and imply very different requirements for contamination assessment as well as for intervention. For instance, as mentioned in Chapter 8, a malicious contamination incident could involve the atmospheric dispersion of a single pure alpha emitter. Due to the short range of alpha particles in air, this would severely complicate monitoring efforts to determine the resultant contaminant distribution on the ground and on surfaces in an inhabited area. This could literally take ages to carry out even over the comparatively limited area that could become contaminated following a relatively low-altitude release that would for instance arise after the detonation of a 'dirty bomb'. In such cases, computerised decision support systems, using, for example recorded weather and plume rise data, to estimate the plume propagation over a complex inhabited area, may be valuable in quickly identifying the areas that are most likely to have become significantly contaminated (see Section 5.2). As mentioned in Chapter 8, a strategy for restoration of areas contaminated by a pure alpha emitter would be aimed at reducing internal doses (from inhalation and ingestion of contaminants), and thus consist of a very different suite of countermeasures than those which would be considered in connection with, for example a terror attack involving a beta/gamma emitter or a major release from a nuclear installation accident. However, it should be noted that pure alpha emitters deposited in the environment may also lead to internal doses, for example after resuspension caused by wind or traffic or after contact transfer to humans.

An example from the past of an incident leading to contamination of a relatively small inhabited area is the Goiânia accident that occurred in September 1987 in Brazil (Leao and Oberhofer, 1988). This accident involved the spreading of parts of a strong (50.9 TBq) ^{137}CsCl teletherapy source over different areas of Goiânia city. The source was in readily soluble powder form and was rapidly spread out through a multitude of different transport pathways. It has, however, been stated that the main cause of contamination of buildings in Goiânia was atmospheric dispersion of the fine contaminant powder (Da Silva et al., 1991). The decontamination strategy applied was not based on a carefully optimised plan, but reflected a strong popular demand to eliminate the problem, and an emergency response that, on the foundation of very limited practical experience, implemented countermeasures on a 'hit and miss' basis. Roofs with contamination levels ranging up to 700 kBq m^{-2} were for instance decontaminated using pressurised water-jets, but the result of this effort was disappointingly a removal of only about 20% of the deposited contamination, and some roofs were actually subsequently replaced.

The restoration strategy also comprised a considerable effort to decontaminate indoor surfaces. However, even though a rather good effect was generally obtained with simple techniques, the local inhabitants refused to accept any measurable residual contamination, and for instance cooking utensils were discarded if they showed any detectable level of contamination at all (Da Silva et al., 1991). The fact that the concept of acceptable levels of contamination was in this case rejected by the population demonstrated that in some cases where the anxiety is very great, as in this case reflected by symptoms like elevated blood pressure, gastrointestinal problems and cardiovascular symptoms (Steinhäusler, 2005), methods that have been proven sufficiently effective in eliminating radiation hazards may still not solve the societal problems satisfactorily.

The Goiânia accident was also interesting in the sense that the affected area was probably similar in size to that which might become affected by the explosion of a 'dirty bomb' (Andersson et al., 2008c). In the Goiânia case, the psychological and social impact was immense, and on public demand, practically all possible efforts were made to clean the affected areas as thoroughly as at all possible, in spite of the poor economic situation of the area in general. This would not have been possible if vast areas had been contaminated, as in the case of the Chernobyl accident. In the event of a malicious action, which is by definition aimed at provoking severe psychological reactions in the public, a very high degree of anxiety and disruption would be expected, and authorities would be likely to be pressed to go very far to remove even small traces of any substances that were placed in the environment to intentionally cause harm – also because it would be likely to be the public perception that the authorities had failed in providing an adequate level of security. This could mean that effective countermeasure variants which would, in connection with other types of contaminating incidents, be deemed too expensive and thus not sufficiently cost-effective might possibly become attractive parts of a strategy for restoration of sites contaminated by a malicious dispersion event.

4. Practical Examples of Implementation of Dose Reduction Strategies for Inhabited Areas

An optimised strategy to reduce external dose from airborne contaminants deposited in an inhabited area would often consist of several countermeasures applied in sequence on different types of surfaces. In such cases, it is important to consider the order in which the different operations of the strategy are carried out. Due to the risk of translocation (e.g. by resuspension) of contaminants, the most heavily contaminated surfaces should as a rule of thumb generally be treated first. Also the orientation of the surfaces should be taken into account. A procedure like water hosing of walls should for instance be implemented before the paved or soil surfaces immediately under the wall are treated, even though the latter surfaces would often be much more strongly contaminated. This is because the horizontal surfaces would inevitably receive some of the radioactive material that is washed off the wall. This type of simple advice may seem logical, but is in practice not always followed, since it may not comply with practical work schedules and may require some prior assessment work.

4.1. Clean-up attempts by the Soviet army in 1989

It can in general be highly valuable to assess both the relative levels of contamination on the different types of surfaces and the depth of penetration of contaminants into permeable surfaces, to ensure that a countermeasure strategy is carried out in an optimised way. This has been demonstrated in practice through the results of the attempts made in 1989 by Soviet army units in Russia, Belarus and Ukraine to decontaminate inhabited areas that had become contaminated with radiocaesium from the Chernobyl accident release. This was a major operation that alone in the Bryansk region of Russia comprised 93 settlements with a total population of about 90,000. The decontamination strategy carried out in all the areas was the same. Essentially, there were only two countermeasures that were applied:

- removal of a topsoil layer, which would ideally contain practically all the contamination in the soil;
- subsequent application of a layer of uncontaminated sand or gravel in the decontaminated areas.

The operation in general focused exclusively on areas in gardens of private houses, around public buildings and in the Bryansk region along roads in settlements with a total length of 190 km. The dose-reducing effect recorded was generally disappointing. The external dose rate was typically only reduced by a factor of 1.1–1.5, and in some areas it even rose (Anisimova et al., 1994; Roed et al., 2006). The direct implementation costs of the work in the strongest contaminated areas of the Bryansk region were in 1989 found to amount to about 15,000 roubles per averted man-Sv, and since a man-Sv had at that time been assigned the value of 5,000 roubles by the Soviet authorities, the operation was clearly not cost-effective. The general conclusion made by the authorities was that, from that point on, decontamination of inhabited areas was not deemed to be a worthwhile option. During that same year, the so-called safe living concept had been implemented by the Soviet National Committee on Radiological Protection, defining a threshold dose, excluding contributions from naturally occurring radioactivity, of 350 mSv over a lifetime, above which intervention was deemed necessary. This level of dose would correspond to a ^{137}Cs contamination on reference soil surfaces by some 1.5 MBq m^{-2} (Fogh et al., 1999; Balonov, 2000). Relocation of the population was therefore continuously carried out over large areas, at very high societal and personal costs.

Practically nothing has been published about how the countermeasure strategy was actually carried out, but the authorities are actually known to have respected the ethical value (Section 3 of Chapter 9) of seeking the consent of owners before any decontamination operations were carried out on private property. Such rural areas in Russia often consist of large open soil areas (e.g. kitchen gardens), divided into limited individual ground lots with small wooden houses, which do not offer much shielding against radiation. This means that a considerable part of the average dose rate to persons living in the area could well come from contamination on their neighbours' ground lots. Roed et al. (2006) estimated that the implementation of the topsoil removal procedure over a very large area could in some cases lead to five times greater reduction in the external dose rate from the ground than would application of the method over an area of only $10 \times 10\,m^2$. So if people do not allow decontamination of their ground areas, it sets limits as to how much their neighbours' exposure can be reduced.

Another and quite possibly more severe problem in connection with the decontamination work carried out by the Soviet army in 1989 related to the vertical distribution of the contamination in the soil, which was at the time of the implementation described as rather inhomogeneous (Anisimova et al., 1994). Judging from the maps of the radiological effect of the work, it was not carried out consistently, even over small soil areas, and the thickness of the removed soil layer does not seem to have been optimised in relation to the actual vertical contaminant distribution in the soil in the given locality.

Figure 2 shows an example of a radiocaesium soil profile recorded in the Bryansk area a few years later. It is by no means a typical radiocaesium profile in the region, but nevertheless one that was recorded in some areas. In this case, the soil had a thick organic horizon, which the contaminants rather rapidly penetrated through. It is clear that, if for instance a 5–7 cm thick topsoil layer were removed here, the dose rate would increase considerably, as effectively a practically uncontaminated layer of shielding soil would then have been taken off. Instead, if the soil removal procedure had relied on measurements of the local contaminant distribution, a removal of the top 15 cm of soil could in this case in principle have reduced the dose rate from the ground by a factor of as much as 50–100 (however, due to inadvertent mixing of soil layers the effect would in practice always be less). Also the subsequent application of uncontaminated sand or gravel seems to have had little effect in reducing dose rate. This must also pertain to the way that it was applied, but the details are unclear from the available descriptions.

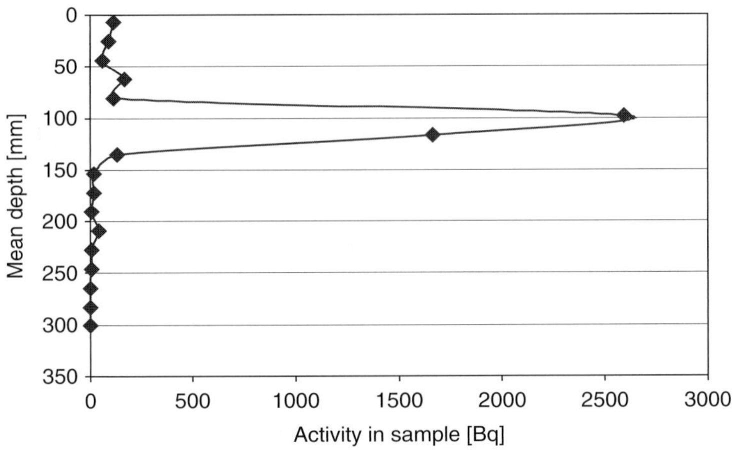

Figure 2 An example of a vertical ^{137}Cs contamination profile recorded in 1995 in a soil with a thick organic horizon in the Bryansk region (data from Roed et al., 1996).

4.2. Clean-up tests in the Bryansk area in 1997

A more focused effort to determine the effect of applying practically the same restoration strategy in another area of the Bryansk region was made in the late summer of 1997 by a combined Russian and Danish team (Roed et al., 1998, 2006). It should be noted that the work was here purely experimental, and was not preceded by any selection procedure justifying or optimising the intervention for the particular area. The objective was simply to assess how much the dose rate in such villages could be reduced at the time using a simple strategy, and to record the direct implementation costs. By 1997, it would be expected to have become more difficult to achieve a good dose rate reduction, since the contaminants would have become more firmly fixed in many different types of construction materials (see Chapter 5), and the contamination might also have migrated deeper into the soil, necessitating removal of thicker soil layers. The small settlement in Guta-Muravinka, where the work was done, is located some 20 km south-east of Novozybkov, in a clearing of a forested area, and had not been permanently inhabited since the Chernobyl accident. The houses practically all had a ground area of about 120 m^2, were of two stories and were made of wood, with asbestos sheets mounted on the roofs and on the outer wall surfaces. Measurements revealed that about 95–98% of the contamination in soil in this area still lay in the top 5 cm layer. Using a 'Bobcat' mini-bulldozer, a topsoil layer was removed (Figure 3) over an area of about 2,000 m^2 around the buildings. The removal depth was carefully optimised in relation to assessments of the local vertical radiocaesium profile in the soil (ca. 7 cm in thickness). Since the fertile

Figure 3 Topsoil removal in Guta-Muravinka, Russia, using a 'Bobcat' mini-bulldozer. The removal depth was optimised according to the results of measurements in a portable NaI(Tl) detector-based spectrometer system, of vertical 2–5 cm sections of excavated soil cores.

soil layer in this area was thin, it was important not to remove a thicker layer than necessary, and this would also minimise the amount of waste produced.

A layer of uncontaminated sand or soil was then taken from a ground excavation and applied over the garden areas around the buildings, where a topsoil layer had been removed (Figure 4), partially to restore the ground level around the building foundations, and partially to shield against any contaminants that might have penetrated deeper into the soil than the removed top layer.

The third countermeasure included in this strategy was a replacement of the asbestos roof sheets on one of the houses (Figure 5). This had been included since measurements had shown that roof materials in the area were relatively strongly contaminated.

The dose reduction achieved through the various steps of the strategy can be seen in Figure 6, inside and outside the house, where the roof was replaced. The overall result of the work was a reduction in the dose rate outdoors by nearly a factor of 6, and indoors by a factor of about 3. If larger soil areas had been treated and more roofs replaced, the reduction in dose rate would have been even greater. A comparison with the results of the Russian army's efforts in the area in 1989 highlights the importance of optimisation of the implementation of each countermeasure in relation to assessments of the contaminant distribution. It has later been established that re-contamination of the decontaminated areas over nearly a decade through resuspension of contaminants from adjacent areas is insignificant (Ramzaev et al., 2006).

Figure 4 Application of uncontaminated sand around buildings in the Guta-Muravinka settlement, after topsoil removal.

Figure 5 Replacement of an asbestos roof in the Guta-Muravinka settlement. As can be seen, the area was deserted and some of the buildings no longer had doors or windows.

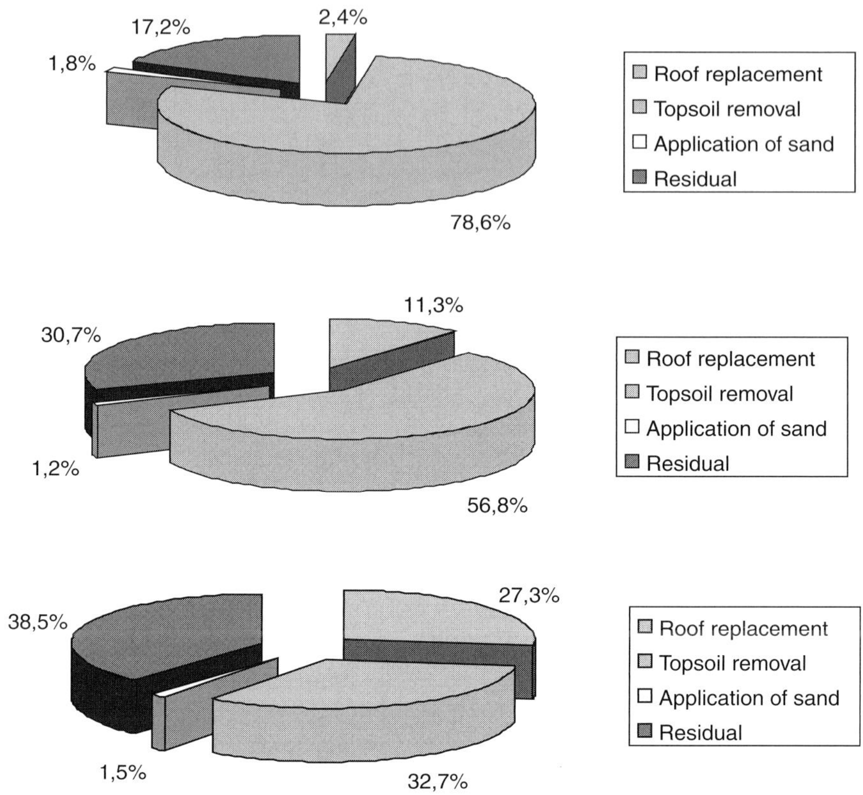

Figure 6 Dose rate reduction recorded by implementation of each of the three countermeasures in the strategy used in the Guta-Muravinka settlement. The top pie chart shows the dose rate reductions outdoors in an area between two buildings. The middle pie chart shows the corresponding dose rate reductions as measured on the ground floor of the building, where the roof was replaced. The bottom pie chart shows the dose rate reductions in the same building, but on the first floor (data from Roed et al., 2006).

In Table 2 the direct implementation costs of the work in 1997 in a single house area are compared with estimates of the dose that would have been averted over a lifetime (70 years), assuming that 6 people live in a house. For comparison, the ICRP in 1993 recommended a minimum value of 3,000 US$ per averted man-Sv for developing countries, and a value of 100,000 US$ per averted man-Sv for rich developed countries. Although the value estimated for this case is lower than both these figures, it should be noted that a full cost–benefit analysis to identify whether the intervention strategy was justified would require consideration of a much wider range of aspects, including social and ethical concerns as well as economic factors in a broader context. Also, it might be that the necessary equipment,

Table 2 Estimates of direct implementation costs of the restoration strategy carried out in one house and garden lot in the Guta-Muravinka area, and estimates of averted integrated collective dose assuming that six people live in the house.

Cost element	Cost estimate (US$)
Initial radiological survey	40
Topsoil removal	400
Excavation of sand to be applied around houses	50
Manual soil removal very near obstacles	100
Roof replacement	414[a]
Total costs	1,004
Averted collective dose over 70 years	0.45 man-Sv
Implementation costs per averted dose unit	2,231 US$ per man-Sv

Source: Data from Roed et al. (2006)
[a]Actual cost encountered in the specific case.

consumables and skilled personnel would not be locally available in an affected area, and transportation costs would then need to be added.

5. Decision Support Tools

5.1. Decision support handbooks

Over the years a number of tools for decision support in connection with airborne radioactive contamination incidents have emerged and continuously been refined and extended. For instance, handbooks have in recent years been produced with the aim of clarifying the requirements and facilitating preparedness planning prior to a contamination from an airborne release of contaminants, and also assisting in the identification and implementation of optimised countermeasure strategies in the event of a contaminating incident. The target groups for such handbooks are generally the persons and organisations that have responsibilities for selecting, prioritising, organising and supervising restoration strategies. These key persons need to ensure that selected parts of the material are made available, explained and discussed with workers and representatives of all groups of the population that might become affected by the decisions made during different time phases following an emergency. Important objectives may here include the strengthening of the basis for public dialogue, and enabling countermeasures to be carried out in a prescribed way. The organisations involved will be highly dependent on the way radiological preparedness is organised in a particular country or region. In many member states of the

European Union, the responsibility for decision-making, including decisions related to implementation of countermeasures, rests within organisations at a national level. However, for instance in Sweden, such decisions are taken on a regional basis by district councils, which may seek guidance and advice from central government bodies. As the background knowledge of the user group may vary considerably, handbooks should be equipped with separate text sections explaining general radiation protection principles and criteria, to clarify objectives.

Such decision support handbooks for contaminated inhabited areas have recently been developed on British national (HPA-RPD, 2008), Nordic regional (Andersson et al., 2008a) and European (Brown et al., 2007) initiatives, with slightly different scopes, but mainly differing in the extent to which they are targeted for the potential needs of a specific country or region. The application of a common approach for decision support on, for instance, the European level can be useful in harmonising the decision-making in Europe, thus reducing the risk of making 'contrasting' decisions (particularly in border areas), which could be perceived as illogical and might possibly have a disruptive effect on society. Nevertheless, prior to its use in any specific community, the generic European handbook will need to undergo a 'customisation' process, to extract and adapt the information of use to the particular country/area, as for instance not all countermeasures and waste management options will locally be relevant, practicable, acceptable and legal. Also some optimisation elements may well need modification for local contextualisation. An annex to the European handbook is aimed at providing the information and guidance that is needed in connection with 'customisation' of the material for a specific country or region.

The European handbook contains the following elements:

- Information of general relevance in establishing a common understanding of radiological problems and their direct implications, thereby providing a common platform for discussion between decision-makers, experts and the public. This can be regarded as a 'primer' to knowledge on radioactive contamination problems in inhabited areas.
- Presentation of a decision framework, which can help decision-makers to quickly identify those countermeasures that might be of relevance in the given situation. The selection framework consists of a series of texts, flowcharts, check lists and tables. For instance, a framework of decision trees has been created to help the user through some of the different aspects of selecting a countermeasure strategy for a specific purpose. An example of a decision tree from the handbook is shown in Figure 7, for elimination of countermeasure options that are under different conditions not relevant for treatment of building surfaces.
- Information on protection of workers carrying out countermeasures or related tasks, including waste management. Individual worker risks can

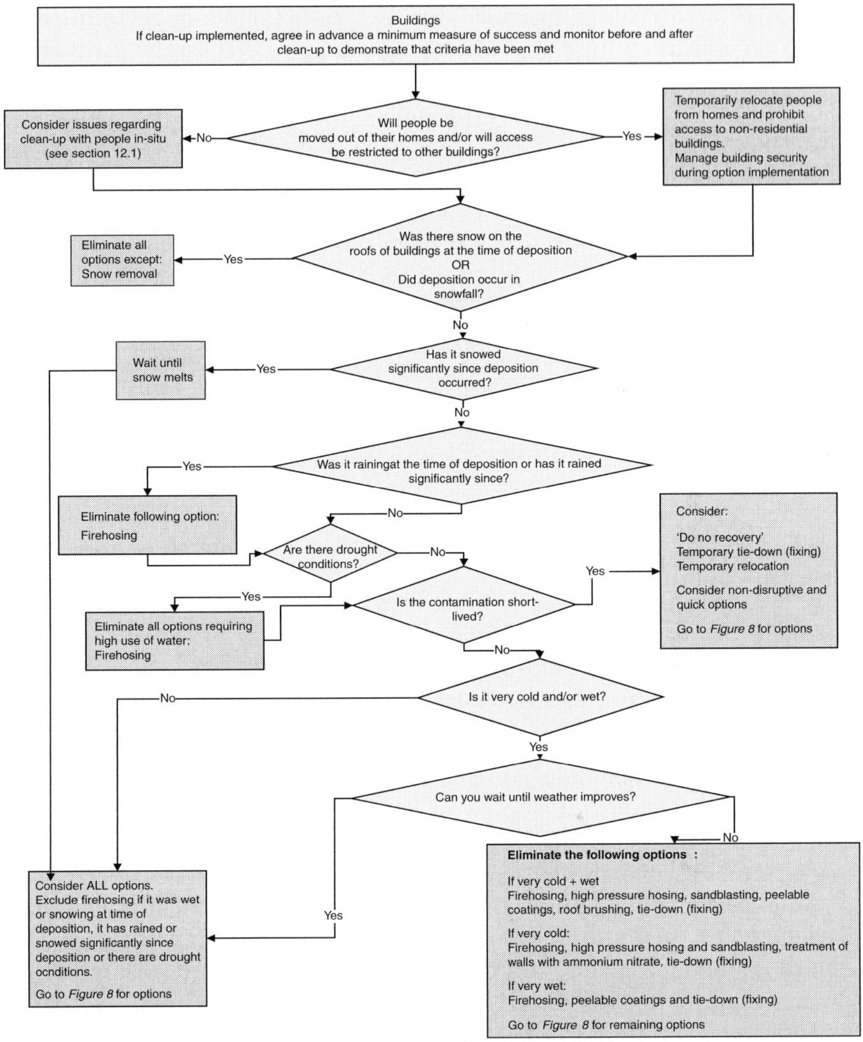

Figure 7 Branch of decision tree to aid option elimination for buildings (from Brown et al., 2007).

for instance be addressed through work time restrictions, delaying operations, shielding, ventilation, fixation of contaminants to avoid resuspension in air, respiratory protection, protective clothing (including helmets and eye protection) and protection against falling from heights. The European handbook also gives an overview of current international labour protection standards of relevance, and includes information that can be applied to easily estimate doses that would be received by workers in connection with specific intervention situations.

- Information on waste management methods, including recommendations on construction of waste handling strategies that can be included in cost-effective countermeasure strategies.
- Description of some of the social and ethical factors that may influence the choice of countermeasures.
- A suite of different countermeasure descriptions in standardised format, as described in Chapter 8.
- A few examples of how it is envisaged that the handbook could be of value in developing a countermeasure strategy under specific circumstances with different types of airborne contaminant releases.

The European handbook was evaluated in 'stakeholder' networks consisting of key players in the decision-making process in Finland, France, Germany and the Slovak Republic, to determine its suitability from different perspectives in Europe, with respect to scope, format and content. On the basis of the received comments and suggestions, it has been revised in an interactive process between the originators and the 'stakeholder' networks. Finally, the handbook has been tested by decision fora in a number of European countries in connection with discussions of specific cases.

5.2. Computerised decision support systems for consequence assessment

Common to the different decision support handbooks for contaminated inhabited areas is that they contain a simplified methodology ('look-up' tables) for calculation of some of the dose aspects in relation to a contaminating incident. However, as described in Chapter 6, a much more detailed and flexible model, ERMIN, for calculating doses received in contaminated inhabited areas contaminated by a large nuclear power plant accident, with or without introduction of countermeasure strategies, has been developed in parallel, and has been implemented in the computerised decision support systems ARGOS (2008) and RODOS (2008), which are currently used by preparedness organisations in a wide range of countries inside as well as outside Europe. In general, such decision support systems apply assumptions regarding source terms together with atmospheric dispersion models and data libraries of deposition data to estimate the contamination pattern at a given locality.

Some systems enable the use of manually entered or online real-time wind field data from meteorological towers, and radar images for wet deposition estimates (Figure 8), and are able to generate forecast data that can be applied in consequence prognoses (ARGOS, 2008). As mentioned in Chapter 2, different types of dispersion models are needed for different dispersion scales (long, meso, street scale), with different resolutions and degrees of simplification of, for instance, the features of the area above

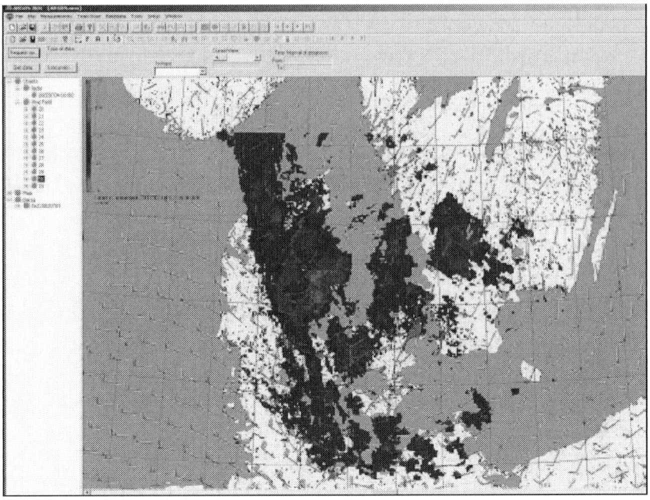

Figure 8 An example of an ARGOS screen showing wind fields and rain densities (Reproduced with kind permission from the ARGOS consortium).

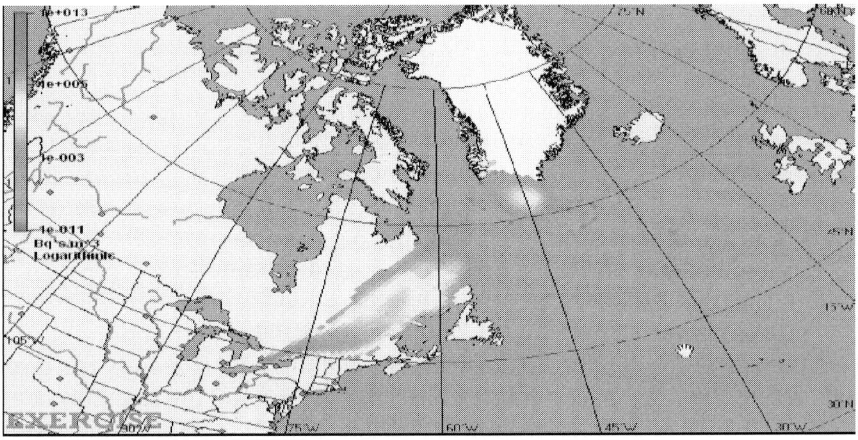

Figure 9 An example of an ARGOS screen showing estimates of time-integrated contaminant air concentrations over large scale from a hypothetical release (Reproduced with kind permission from the ARGOS consortium).

which the contaminated plume is dispersed. Figure 9 shows an example of how ARGOS can present time-integrated contaminant air concentrations from large-scale dispersion as overlays on a map. Data from, for example airborne gamma monitoring systems, car-borne measurement units or other types and formats of measurement data can be integrated in some decision support systems, facilitating the generation of improved prognoses through application of data assimilation modules (Kaiser et al., 2008). Data from numerous practical investigations (Andersson et al., 2008b; Jones et al., 2005) have been applied to estimate the post-deposition behaviour of the contaminants on/in the various types of outdoor and indoor surfaces in the inhabited environment (Chapter 5). Resultant doses can then be calculated using methodologies as described in Chapter 6. A decision support system like ARGOS enables use of mapped information on population densities and other area features to determine representative building characteristics for the purpose of highly detailed dose calculations. This is important, as average doses received by inhabitants in different types of dwelling areas can, as mentioned above, deviate considerably. Also, very different types of surfaces can contribute significantly to time-integrated doses in such areas, and it is thus important to obtain a reasonably detailed picture of the area's features to ensure that resources allocated for intervention are spent on treatment of the right surfaces.

As mentioned in Chapter 1, the approach of planning for a radiological incident alone must be considered to be outmoded in view of the multiple threats that populations in inhabited areas are facing today, for instance from terror attacks, which might pose combined radiological, chemical, biological and/or physical hazards. This is acknowledged in the features of for instance the American HPAC decision support system (HPAC, 2008), and similar expansions are under development to enhance the capabilities of the European-based ARGOS system, with highly detailed modelling of contamination problems related to all airborne hazard groups in inhabited areas. ARGOS and RODOS were originally developed exclusively to support decisions in relation to accidents at nuclear installations, and a scenario like a radiological terror attack would set new requirements to the features and application of a decision support system.

In modelling the dispersion in an inhabited complex of aerosolised radionuclides following a 'dirty bomb' attack, a requirement is to determine the initial plume dimensions and elevation height distribution of the contaminant aerosols generated by the blast. Plume rise obviously has great influence on the size of the affected area, and determines the significance of plume interaction with environmental structures (e.g. buildings and trees). The parameterisation of plume rise from a 'dirty bomb' may be based on measurements from blast studies. Figure 10 shows a 'qualitative' example of curves describing the initial plume top height and cap diameter as a function of the TNT equivalent explosive mass, on the basis of data from a US blast test series conducted to investigate the effect of accidental conventional explosions spreading radioactive material from a nuclear device.

For modelling the subsequent airborne dispersion of the contaminants over the environment, it is important to apply methods that adequately account for the mechanisms governing the flow and deposition in relation to the given dispersion altitude and scale (see Chapter 2). Here a considerable degree of simplification is traditionally applied in decision support

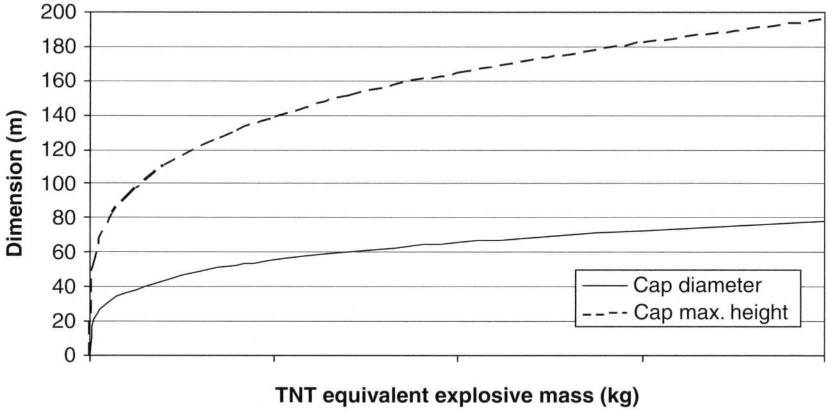

Figure 10 Plume top height and cap diameter following the detonation of a 'dirty bomb', as a function of TNT equivalent explosive mass. 'Qualitative' curves based on US test data.

models, which are designed to describe the long-range transport of contaminants from high-altitude releases following large nuclear power plant accidents. Since the plume here generally passes well over the various environmental obstacles, the influences on the plume propagation in inhabited areas can be reasonably simulated through the use of different overall roughness and deposition rate parameters compared with, for example rural areas (Päsler-Sauer, 1997; Mikkelsen et al., 1984). However, recent investigations have shown such simplifications to be problematic, when the contaminant dispersion partially takes place at street level, as could be the case following a 'dirty bomb' explosion. This is particularly true for scenarios involving shifting wind directions (Astrup et al., 2005). Here, higher-resolution models are required to address the issues of plume interaction with, and flow along, obstacles in the inhabited environment. Such systems include the UDM code developed by the UK Defence Science and Technology Laboratory (Hall et al., 2002) and the URD code developed at Risø National Laboratory (Astrup et al., 2005).

For instance, the URD code incorporates plume interaction with environmental obstacles in three different ways. The obstacles form barriers limiting the magnitude of horizontal eddies in the atmosphere, which in turn reduces the large-scale horizontal dispersion. At the same time, the interaction will increase the small-scale turbulence over a city, which will lead to greater small-scale dispersion. Finally, obstacles like downstream building walls constitute barriers that will to some extent delay the further dispersion of the contaminants. URD calculations for a test scenario involving an airborne release in central Copenhagen at a height of 2 m above ground level have illustrated that the plume interaction effect over short distances, which influences the larger-scale dispersion, is particularly important (Andersson et al., 2008b). Overall, a much more diffuse dispersion pattern was observed in this case when the modelling was carried out alone with ARGOS' longer-range dispersion module RIMPUFF (Mikkelsen et al., 1984), and time-integrated downwind air concentrations were underestimated within the nearest few hundred metres of the release point, whereas the dispersion pattern at greater distances was distorted and could misguide decision-making.

5.3. Multi-criteria analysis tools for optimisation of countermeasure strategies

Optimised decision-making for intervention is an extremely complex process, which should take into consideration the full range of benefits and costs inflicted on the population as a whole as well as on population subgroups. As mentioned in Chapter 9, individual countermeasures as well as strategies consisting of several countermeasures may have both positive and negative social and ethical implications. Also making an investment to reduce doses should in itself be seen as a social/ethical trade-off, as the resources could alternatively be applied to solve other societal problems, which might then not be dealt with adequately. A suggested tool to be applied in a decision process, in addition to an actual optimisation tool, is an ethical matrix, as described in Section 5 of Chapter 9, to ensure that ethical issues of each element of a countermeasure strategy are dealt with in a holistic and transparent way. As described in Section 7 of Chapter 9, involvement of citizens and stakeholders in the decision process for long-term restoration strategies is an important instrument in reaching generally acceptable and robust decisions that would not be prone to public resistance. During that process, a number of not easily comparable issues need to be balanced against each other, in a decision matrix.

Acknowledging the recommendations of ICRP (2007), a number of politically driven factors that cannot be quantified on a generic scale will need to enter the decision matrix for intervention in the event of any contaminating incident. Such issues are often highly site and case specific, and can generally only be addressed to some limited extent in advance of a contaminating incident. Some examples of factors that need to be included in a holistic justification/optimisation of intervention in the event of contamination of inhabited areas, but require quantification on the basis of political decisions, are:

- The value of each unit of averted dose. To express the benefit of averting dose in units that can be compared with other benefits and costs, it may be convenient to transpose the total averted dose to monetary units. Various approaches have been suggested to equate averted doses with a monetary value (e.g. Guenther and Thein, 1997; Eged et al., 2001;

Hedemann Jensen and Yatsalo, 1998), and it is likely that conversion factors would vary across Europe, depending on the resources available in the given area.

- The value of equity across the population in the different outcomes of a strategy and of avoidance of dispute.
- The value of public reassurance and psychological well-being, as opposed to social disruption, panic and possibly mass desertion.
- The value of maintaining societal functions. For instance, to prevent the local society from falling apart, it is important to some extent to stabilise economic conditions, prioritise commercial businesses, shops, roads, railways, hospitals, etc., to secure acceptable living conditions for those people that may remain in (or return to) the area.
- The value of lost income, for example if businesses in the area cannot be maintained, or if marketing possibilities are lost due to stigmatisation of the area, which could for instance affect export and tourism.
- The value of lost or damaged personal property.
- The societal value of, for example a scenic park area or a preserved building, which might need to be destroyed or damaged to reduce dose in the area.
- The value of avoiding environmental risks, for example from groundwater pollution or changes to an area's ecosystem.

In the European project STRATEGY (Howard et al., 2005), a series of choice experiments was conducted to assess how groups of affected populations would rank various adverse effects of a countermeasure implementation. The experiments were based on a person's willingness to allocate money from a limited budget in a way he/she felt would maximise utility. Although in these experiments, which took place in Spain and the UK, the preservation of a landscape was for instance generally found to rank somewhat higher than disruption, it was not possible to draw any firm conclusions that would be considered valid in a generic sense. A problem with such questionnaires made in 'peace times' is in general that it is difficult to imagine how far people would in reality go to eliminate a given threat or problem that may possibly receive the undivided attention of the media. The balancing factors would be highly case specific, and, for example differ according to location, as investigations have shown that risk perception varies rather widely between countries (Salt et al., 1999; Sjöberg et al., 2000). Risk perception also varies between population groups within the same country with, for instance, different levels of education (Salt et al., 1999).

An example of a tool that is used to facilitate the actual decision process is the Web-HIPRE (hierarchical preference analysis on the web) software (Hämäläinen and Mustajoki, 2008) for decision analytic problem structuring, multi-criteria evaluation and prioritisation (Belton and Stewart, 2002).

Similar developments have been made by other workers, for example in connection with the European RESTRAT project (Jackson et al., 1999; Zeevaert et al., 2001). Web-HIPRE, which is integrated in the RODOS decision support system (Raskob et al., 2005), applies the approach of multi-attribute value theory, providing decision analysis to support the evaluation of multiple alternatives on multiple criteria (Keeney and Raiffa, 1976; Hämäläinen et al., 2000). Basically, each countermeasure strategy alternative (including the option of not introducing countermeasures) is rated on each of a number of identified attributes of importance. The consequences of these attributes are then weighted according to an assigned relative importance. Provided that attributes are mutually preferentially independent, an overall value of each countermeasure strategy option in relation to a specific problem can be estimated according to the formula

$$v(\boldsymbol{x}) = \sum_{i=1}^{n} w_i v_i(x_i)$$

(Mustajoki et al., 2007), where $\boldsymbol{x}$ is a vector describing the consequence of the option with respect to attribute i, $v_i(x_i)$ is its rating on a scale between 0 and 1, and w_i is the weight fraction assigned to attribute i. The actual implementation of the tool in a decision-making process incorporating multiple participants representing different stakeholder groups (these may be physically present or can take part through Internet-based presentations) requires the collective formation of a case-specific decision tree, outlining the different types of attributes of concern. An example is given in Figure 11, based on the

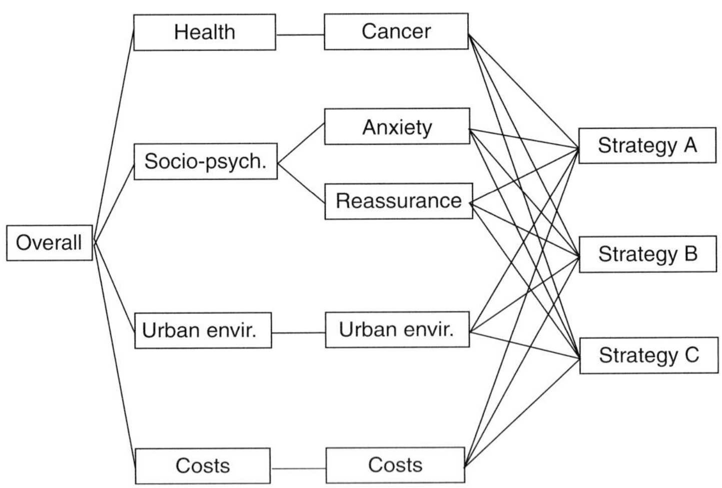

Figure 11 Example of a simple decision tree to structure and systematise decisions to be made in relation to a situation involving contamination in an inhabited (urban) area.

suggestions of Mustajoki et al. (2007). In this case, four different attribute groups have been identified, each consisting of one or several attributes, which can be related to each of a number of alternative countermeasure strategies.

The use of such a system in a participatory forum can, with the right facilitation and advisory support, form the basis for a case-specific ranking, which may seem reasonable and transparent to all involved, thus providing a useful platform for reaching agreement on the seemingly 'best' solution. The weakness in the use of this approach lies in the valuing of rating and weighting factors, and the summation over often many attributes to reach an estimate of the overall value of each option might, to some extent, remove the focus from issues that demand undivided attention. However, tools for sensitivity analysis to estimate the robustness of a decision and for identification of particularly important factors have been developed (e.g. Papamichail and French, 2005; Papamichail, 2000; Geldermann et al., 2005).

ACKNOWLEDGEMENTS

Valuable advice from Dr. Joanne Brown, Health Protection Agency, UK, during the drafting of this chapter is gratefully acknowledged. The photographs shown in Figures 3–5 were taken by Henrik Prip, Risø National Laboratory for Sustainable Energy, Technical University of Denmark, and are reproduced with kind permission.

REFERENCES

Andersson, K. G., M. Ammann, S. Backe, and K. Rosén. (2008a). URBHAND-*Decision Support Handbook for Recovery of Contaminated Inhabited Areas.* NKS Report NKS-175. Nordic Nuclear Safety Research, Roskilde, Denmark, ISBN 978-87-7893-241-9.

Andersson, K. G., J. Brown, K. Mortimer, J. A. Jones, T. Charnock, S. Thykier-Nielsen, J. C. Kaiser, G. Proehl, and S. P. Nielsen. (2008b). New developments to support decision-making in contaminated inhabited areas following incidents involving a release of radioactivity to the environment. *Journal of Environmental Radioactivity*, 99, 439–454.

Andersson, K. G., T. Mikkelsen, P. Astrup, S. Thykier-Nielsen, L. H. Jacobsen, L. Schou-Jensen, S. C. Hoe, and S. P. Nielsen. (2008c). Estimation of health hazards resulting from a radiological terrorist attack in a city. *Radiation Protection Dosimetry*, 131(3), 297–307.

Anisimova, L. I., A. V. Chesnokov, A. P. Govorun, O. P. Invanov, B. N. Potapov, S. B. Shcherbak, L. I. Urutskoev, V. I. Kovalenko, C. V. Krivonosov, A. V. Ponomarev, V. P. Ramzaev, and V. A. Trusov. (1994). *Analysis of Some Countermeasures Efficiency Based on Radioecological Data, Deposit Measurements and Models for External Dose Formation. Case Study of Two Russian Settlements Yalovka and Zaborie.* Annual Progress Report of the ECP4 Project supported by the European Commission, EMERCOM, Moscow, Russia.

ARGOS. (2008). ARGOS CBRN Whitepaper. Decision support for emergency management, www.pdc.dk/argos

Astrup, P., S. Thykier-Nielsen, and T. Mikkelsen. (2005). *In-Town Dispersion Calculations with RIMPUFF and UDM.* Risø-R-1539(EN). Risø National Laboratory, Roskilde, Denmark, ISBN 87-550-3485-3.

Balonov, M. I. (2000). The legacy of the Chernobyl accident in the Bryansk region: Radiological assessment, population protection and area rehabilitation. *Proceedings of the International Symposium in Arlington, Virginia, 1999, on Restoration of Environments with Radioactive Residues*, IAEA, Vienna, pp. 407–417, ISBN 92-0-102600-5.

Belton, V., and T. Stewart. (2002). *Multiple Criteria Decision Analysis: An Integrated Approach.* Kluwer Academic Publishers, Boston, MA, ISBN 0-7923-7505-X.

Brown, J., K. Mortimer, K. Andersson, T. Duranova, A. Mrskova, R. Hänninen, T. Ikäheimonen, G. Kirchner, V. Bertsch, F. Gallay, and N. Reales. (2007). *Generic Handbook for Assisting in the Management of Contaminated Inhabited Areas in Europe Following a Radiological Emergency, Parts I–V.* Final Report of the EC-EURANOS Project Activities CAT1RTD02 and CAT1RTD04. Radiation Protection Division, Health Protection Agency, UK.

Da Silva, C. J., J. U. Delgado, M. T. B. Luiz, P. G. Cunha, and P. D. de Barros. (1991). Considerations related to the decontamination of houses in Goiânia: Limitations and implications. *Health Physics*, 60(1), 87–90.

Eged, K., B. Kanyár, Z. Kis, T. Tatay, Á. Avády, and G. Volent. (2001). Determination and use of the monetary values of the averted person-sievert for use in radiation protection decisions in Hungary. *Health Physics*, 80(2), 137–141.

ENVREG 9602. (2000). Project EC/TACIS ENVREG 9602 "Addressing the Clean-up and Secondary Medical Effects of the Chernobyl disaster", www.chernobyl.info

Fogh, C. L., K. G. Andersson, A. N. Barkovsky, A. S. Mishine, A. V. Ponamarjov, V. P. Ramzaev, and J. Roed. (1999). Decontamination in a Russian Settlement. *Health Physics*, 76(4), 421–430.

Geldermann, J., M. Treitz, V. Bertsch, and O. Rentz. (2005). Moderated decision support and countermeasures planning for off-site emergency management. In: *Energy and Environment: Modelling and Analysis* (Eds R. Loulou, J.-P. Waaub, and G. Zaccour). Kluwer Academic Publishers, Dordrecht, The Netherlands. ISBN: 978-0-387-25351-0.

Guenther, C. F., and C. Thein. (1997). Estimated cost of person-Sv exposure. *Health Physics*, 72(2), 204–221.

Hall, D. J., A. M. Spanton, I. H. Griffiths, M. Hargarve, and S. Walker. (2002). The Urban Dispersion Model (UDM): Version 2.2 Technical Documentation. DSTL/TR04774, Defence Science and Technology Laboratory, Porton Down, UK.

Hämäläinen, R. P., M. R. K. Lindstedt, and K. Sinkko. (2000). Multiattribute risk analysis in nuclear emergency management. *Risk Analysis*, 20(4), 455–467.

Hämäläinen, R. P., and J. Mustajoki. (2008). Web-HIPRE – Global decision support, computer software, Systems Analysis Laboratory, Helsinki University of Technology, Finland, www.hipre.hut.fi

Hedemann Jensen, P., and B. I. Yatsalo. (1998). Methodology for determining action levels for clean-up of contaminated urban and agricultural environments. *Health Physics*, 75(2), 120–129.

Howard, B. J., N. A. Beresford, A. Nisbet, G. Cox, D. H. Oughton, J. Hunt, B. Alvarez, K. G. Andersson, A. Liland, and G. Voigt. (2005). The STRATEGY project: decision tools to aid sustainable restoration and long-term management of contaminated agricultural ecosystems. *Journal of Environmental Radioactivity*, 83, 275–295.

HPAC. (2008). Hazard prediction and assessment capability, version 4.04, software available from US Defense and Threat Reduction Agency, USA, www.dtra.mil

HPA-RPD. (2008). *UK Recovery Handbook for Radiation Incidents*. HPA-RPD-002, Health Protection Agency – Radiological Protection Division, Chilton, UK.

IAEA. (1997). *Application of Radiation Protection Principles to the Cleanup of Contaminated Areas* IAEA-TECDOC-987. IAEA, Vienna, Austria.

ICRP (1991). *1990 Recommendations of the International Commission on Radiological Protection: Adopted by the Commission in November 1990.* ICRP Publication 60. *Annals of the ICRP*, Pergamon Press, Oxford, ISBN 0-08-041144-4.

ICRP. (1993). *Protection from Potential Exposure: A Conceptual Framework*. ICRP Publication 64. *Annals of the ICRP*, Pergamon Press, Oxford, ISBN 0-08-042205-5.

ICRP. (1999). *Protection of the Public in Situations of Prolonged Radiation Exposure. The Application of the Commission's System of Radiological Protection to Controllable Radiation Exposure due to Natural Sources and Long-Lived Radioactive Residues*. ICRP Publication 82. *Annals of the ICRP*, Vienna, Austria, ISSN 0146-6453.

ICRP. (2006). *Assessing Dose of the Representative Person for the Purpose of Radiation Protection of the Public and The Optimisation of Radiological Protection: Broadening the Process*. ICRP Publication 101. *Annals of the ICRP*, Vienna, Austria, ISSN 0146-6453.

ICRP. (2007). *The 2007 Recommendations of the International Commission on Radiological Protection*. ICRP Publication 103. *Annals of the ICRP*, Vienna, Austria, ISSN 0146-6453.

Jackson, D., S. Wragg, A. Bousher, Th. Zeevaert, T. Stiglund, V. Brendler, P. Hedemann Jensen, and S. Nordlinder. (1999). Establishing a method for assessing and ranking restoration strategies for radioactively contaminated sites and their immediate surroundings. *Nuclear Energy*, 38(4), 223–231.

Jones, J. A., T. W. Charnock, L. Singer, J. Roed, K. Andersson, S. Thykier Nielsen, T. Mikkelsen, P. Astrup, J. C. Kaiser, H. Müller, G. Pröhl, W. Raskob, S. Hoe, L. H. Jacobsen, L. Schou-Jensen, and F. Gering. (2005). *Description of Inhabited Area Model*. Final Report of the EC-EURANOS Project Activity CAT2RTD02, EURANOS(CAT2)-TN(05)-04. Health Protection Agency, Centre for Radiation, Chemical and Environmental Hazards, Chilton, UK.

Kaiser, J. C., F. Gering, K. Andersson, T. Charnock, S. Hoe, and L. H. Jacobsen. (2008). Correction of model predictions for the radioactive contamination of the environment with data assimilation. In: *Proceedings of the International Conference on Radioecology & Environmental Radioactivity* (Eds P. Strand, J. Brown and T. Jølle), 15–20 June, 2008, Bergen, Norway, Oral & Oral Poster Presentations, Part 1, Norwegian Radiation Protection Authority, Østerås, Norway, ISBN 978-82-90362-25-1, pp. 130–133.

Keeney, R. L., and H. Raiffa. (1976). *Decisions with Multiple Objectives*. Wiley, New York, ISBN-13: 9780521441858.

Leao, J. L. B., and M. Oberhofer. (1988). The radiological accident at Goiânia. *Proceedings of the International Conference on Radiation Protection in Nuclear Energy* (Vol. 2). International Atomic Energy Agency, Vienna, Austria.

Mikkelsen, T., S. E. Larsen, and S. Thykier-Nielsen. (1984). Description of the Risø Puff Diffusion Model. *Nuclear Technology*, 67, 56–65.

Mustajoki, J., R. P. Hämäläinen, and K. Sinkko. (2007). Interactive computer support in decision conferencing: Two cases on off-site nuclear emergency management. *Decision Support Systems*, 42, 2247–2260.

Papamichail, K. N. (2000). Intelligent decision support for nuclear emergencies. PhD Thesis, Department of Computer Science, University of Manchester, UK.

Papamichail, K. N., and S. French. (2005). Design and evaluation of an intelligent decision support system for nuclear emergencies. *Decision Support Systems*, 41, 84–111.

Päsler-Sauer, J. (1997). Description of the atmospheric dispersion model ATSTEP. RODOS (WG2)-TN(97)01, Progress report of RODOS, Draft version, November 1997, FZK, Karlsruhe, Germany, www.rodos.fzk.de

Ramzaev, V., K. G. Andersson, A. Barkovsky, C. L. Fogh, A. Mishine, and J. Roed. (2006). Long-term stability of decontamination effect in recreational areas of the town Novozybkov, Bryansk Region, Russia. *Journal of Environmental Radioactivity*, 85(2–3), 280–298.

Raskob, W., V. Bertsch, J. Geldermann, S. Baig, and F. Gering. (2005). Demands to and experience with the decision support system RODOS for off-site emergency management in the decision making process in Germany. In: *Proceedings of the 2nd International ISCRAM Conference* (Eds B. van de Walle and B. Carlé), Brussels, Belgium, April 2005, pp. 269–278.

RODOS. (2008). Homepage: Realtime online decision support system for nuclear emergency management, www.rodos.fzk.de

Roed, J., K. G. Andersson, A. N. Barkovsky, C. L. Fogh, A. S. Mishine, S. K. Olsen, A. V. Ponomarjov, H. Prip, V. P. Ramzaev, and B. F. Vorobiev. (1998). *Mechanical Decontamination Tests in Areas Affected by the Chernobyl Accident*. Risø-R-1029. Risø National Laboratory, Roskilde, Denmark, ISBN 87-550-2361-4.

Roed, J., K. G. Andersson, A. N. Barkovsky, C. L. Fogh, A. S. Mishine, A. V. Ponamarjov, and V. P. Ramzaev. (2006). Reduction of external dose in a wet-contaminated housing area in the Bryansk Region, Russia. *Journal of Environmental Radioactivity*, 85(2–3), 265–279.

Roed, J., C. Lange, K. G. Andersson, H. Prip, S. Olsen, V. P. Ramzaev, A. V. Ponomarjov, A. N. Barkovsky, A. S. Mishine, B. F. Vorobiev, A. V. Chesnokov, V. N. Potapov, and S. B. Shcherbak. (1996). *Decontamination in a Russian Settlement*. Risø-R-870. Risø National Laboratory, Roskilde, Denmark, ISBN 87-550-2152-2.

Salt, C. A., J. Grande, N. Hanley, H. Solheim Hansen, H. Lettner, G. Kirchner, S. Rekolainen, I. Bärlund, R. Baumgartner, M. Berreck, M. Culligan Dunsmore, H. Ehlers, S. Ehlken, M. Gastberger, K. Haselwandter, V. Hormann, F. Hosner, T. Peer, M. Pintaric, S. Tattari, and M. Wilson. (1999). *Countermeasures: Environmental and Socio-Economic Responses*. Final Report of the Project CESER, co-funded by the Nuclear Safety Programme of the European Commission, Contract FI 4P-CT95-0021, University of Stirling, UK.

Sjöberg, L., B. Jansson, and M. Viklund. (2000). *Risk Perception by the Swedish Public 10 years after Chernobyl*. Project Report from the RISKPERCOM Project (Contract FI4PCT950016).

Steinhäusler, F. (2005). Chernobyl and Goiânia lessons for responding to radiological terrorism. *Health Physics*, 89(5), 566–574.

Zeevaert, Th., A. Bousher, V. Brendler, P. Hedemann Jensen, and S. Nordlinder. (2001). Evaluation and ranking of restoration strategies for radioactively contaminated sites. *Journal of Environmental Radioactivity*, 56, 33–50.

Concluding Remarks

As demonstrated in the previous 10 chapters, much is at present known of the fate and likely consequences in inhabited areas of radioactive contamination released to the atmosphere in connection with nuclear or radiological emergencies. Many important mechanisms and processes have been identified and described, and for instance some of the findings related to atmospheric contaminant dispersion, deposition and migration in the environment are of a generic nature and essentially apply to any type of airborne pollution. Also some of the instruments suggested to balance criteria and facilitate consideration of, for example, social and ethical perspectives in the development of robust and defendable countermeasure strategies could have very wide-ranging applicability beyond the originally intended scope.

A requirement in decision support following an emergency is the provision of timely and reasonably accurate consequence prognoses. Naturally, an important aspect in this context is an account of the doses that would be received over time in different areas and under different circumstances, with and without implementation of countermeasures. As illustrated in Chapter 6, current state-of-the-art consequence assessment systems have built-in models, in which measurement data and/or probabilistic methodologies to define source term and initial contamination parameters are used together with methodologies reflecting physical and chemical processes (Chapters 2–5), and libraries of dose conversion factors and related assumptions, for dose estimation. Traditionally, inhalation doses received during the passage of the contaminated plume and external doses from gamma-emitting radionuclides are perceived as important, but recent studies have shown that, for instance in connection with some conceivable malicious dispersion scenarios, a wide range of dose pathways may be of about equal importance.

The increasing diversity in the nature of radiological and nuclear threats to inhabited areas sets new demands on emergency preparedness, and it is clear that consequence assessment methodologies based on the 'classical' assumptions of high-altitude long-range dispersion of small particulate contaminants, which will be strongly fixed on surfaces after deposition, are in many thinkable situations invalid. Different processes leading to the atmospheric release of contaminants (different types of explosions, fire, nebulisation, etc.) can create particles or gases of very different physicochemical forms. The physicochemical form (e.g. particle size, solubility) will in turn be decisive in governing dispersion, deposition and post-deposition migration, and thus dose.

It has for instance been estimated that external doses received at long distance from a major release from a PWR reactor accident would over the first few months be dominated by contributions from ^{103}Ru (Kelly, 1987). However, as mentioned in Chapter 5, the volatility of ruthenium, and thus its aerosol characteristics, are highly sensitive to the accident release process. Large particles would lead to high deposits close to the accident site, whereas small particles would disperse over large areas and give rise to lower individual doses to a larger population. A considerable improvement in the reliability of dose modelling could thus be made by better accommodating the release process (e.g. by accident category types) in the description of the produced contaminants. As mentioned in Chapters 6 and 8, smaller releases with less energy and other radionuclides (e.g. an explosion of a 'dirty bomb') are expected to produce contaminants with completely different characteristics from those traditionally assumed in operational consequence assessment systems. Revised models and parameterisation are thus required to adequately predict dose contributions from such releases, and enable optimisation of intervention. This and a number of other shortcomings in current dose modelling are described in Section 8 of Chapter 6. These also reflect gaps in knowledge in relation to deposition, clearance and migration, as treated in Chapters 2–5.

It should also be noted that a malicious attack may comprise several different types of contaminating agents, as for instance indicated by the finding of both ^{90}Sr and nerve gas in a taxi in Georgia in 2002 (see Chapter 1). There is therefore a need to enable consideration of combined exposures, and evaluate consequences and countermeasures as a whole.

Pre-emergency planning (e.g. customisation of generic preparedness guidance material for specific areas, integration of stakeholder groups in establishing an operational preparedness, implementation and training in use of decision-support systems and acquisition of equipment and materials to enable timely countermeasure implementation) is of the utmost importance. The importance of preventive measures to reduce the likelihood of terror attacks and accidents should also be stressed.

As described in Chapter 9, non-radiological aspects may be of great importance, but the consideration of these in identification of optimal solutions for different contamination situations involving inhabited areas is far from harmonised between different (even neighbouring) countries, which can lead to both conflicts in perception and trust, and development of strategies that may not be based on a sustainable and defendable platform. For instance, the process of integration of stakeholders in decision-making has reached a mature level in some European countries, whereas it has not been initiated in others. Also in the use of instruments to guide optimisation of countermeasure strategies (e.g. multi-attribute analysis tools) there remain unresolved problems, as discussed in Chapter 10.

The role of measurement and monitoring is discussed primarily in Chapter 7, but also in other relevant places, for instance in relation to implementation of countermeasure strategies (Chapter 10). Of particular importance in this context is to enable optimisation of monitoring strategies, so that the right parameters are assessed in the right places and at the right time. This requires careful consideration of, for instance, the type and positioning of stationary monitoring equipment, with an objective of providing improved source identification through backtracking and data assimilation. In the European context, an effort is currently being made to address this particular issue. Large areas could become contaminated and very large monitoring efforts required over short times. Maintenance of large measurement capacities (skilled personnel – also to interpret results, equipment and transportation) and collaboration and harmonisation across borders is therefore essential. This is one important reason to keep local radiological centres alive; as my late colleague, Henning Dahlgaard, put it, 'a nuclear preparedness plan without working scientific projects is like an airforce without trained fighter pilots'. Another very good reason is that it is clear from recent developments that challenges in the radiological field are in general far from static, but necessitate continuous research and development programmes.

REFERENCE

Kelly, G. N. (1987). The importance of the urban environment for accident consequences. *Radiation Protection Dosimetry*, 21(1/3), 13–20.

Kasper G. Andersson
Editor

Author Index

Subject Index